P. Stavroulakis (Ed.)

Neuro-Fuzzy and Fuzzy-Neural Applications in Telecommunications

Springer
Berlin
Heidelberg
New York
Hong Kong
London
Milan
Paris
Tokyo

Peter Stavroulakis (Ed.)

Neuro-Fuzzy and Fuzzy-Neural Applications in Telecommunications

With 225 Figures

Springer

Professor Peter Stavroulakis
Technical University of Crete
Aghiou Markou
731 32 Chania, Crete
Greece

ISBN 3-540-40759-6 Springer-Verlag Berlin Heidelberg New York

Library of Congress Cataloging-in-Publication-Data
Stavroulakis, Peter.
Neuro-Fuzzy and Fuzzy-Neural Applications in Telecommunications / Peter Stavroulakis.
p. cm. – (Signals and communications technology) (Engineering online library)
Includes bibliographical references and index.
ISBN 3-540-40759-6 (alk. paper)
1. Telecommunication. 2. Fuzzy systems. 3. Neural networks (Computer science) I. Title. II. Series. III. Series: Engineering online library

TK5101.S685 2004
621.382–dc22

Springer-Verlag is a part of Springer Science+Business Media
springeronline.com

Typesetting: Fotosatz-Service Köhler GmbH, Würzburg
Coverdesign: design & production, Heidelberg

Printed on acid-free paper 62/3020/M 5 4 3 2 1 0

This book is dedicated to my wife Nina and my four sons, Peter, Steven, Bill and Stelios, whose patience lasted so long and was a repetition of similar experiences from other books I have written recently, so they lately started asking me if I had changed my profession from a teacher to a writer.

Acknowledgement

I would like to thank my assistant Mr H. Kosmidis who worked long hours helping put the material in a publishing form, working especially with the figures. Special thanks are due to my personal assistant Miss A. Karatsivi, whose help has been invaluable, many times taking over the flow of the work required to put this book together.

I am also fortunate to have assistants like Aris Papadakis and Kostas Dalamagidis who helped me collect and edit major portions of the appendix. Special thanks are due to my colleague and friend Professor Spyros Tzafestas who allowed to use figures and some material from his book (in Greek) titled Computational Intelligence. Last but not least, I would like to thank all contributors whose diligent work made this book possible.

Preface

Neurofuzzy and fuzzyneural techniques as tools of studying and analyzing complex problems are relatively new even though neural networks and fuzzy logic systems have been applied as computational intelligence structural elements for the last 40 years. Computational intelligence as an independent scientific field has grown over the years because of the development of these structural elements.

Neural networks have been revived since 1982 after the seminal work of J.J. Hopfield and fuzzy sets have found a variety of applications since the publication of the work of Lotfi Zadeh back in 1965. Artificial neural networks (ANN) have a large number of highly interconnected processing elements that usually operate in parallel and are configured in regular architectures. The collective behavior of an ANN, like a human brain, demonstrates the ability to learn, recall, and generalize from training patterns or data. The performance of neural networks depends on the computational function of the neurons in the network, the structure and topology of the network, and the learning rule or the update rule of the connecting weights. This concept of trainable neural networks further strengthens the idea of utilizing the learning ability of neural networks to learn the fuzzy control rules, the membership functions and other parameters of a fuzzy logic control or decision systems, as we will explain later on, and this becomes the advantage of using a neural based fuzzy logic system in our analysis.

On the other hand, fuzzy systems are structured numerical estimators. They start from highly formalized insights about the psychology of categorization and the structure of categories in the real world and they articulate fuzzy IF-THEN rules as a kind of expert knowledge. As a general principle, fuzzy logic is based on the incompatibility principle which suggests that complexity and ambiguity are correlated. As we learn more and more about a system, its complexity decreases and our understanding increases. The major tasks encountered in using fuzzy systems involve determination of fuzzy logic rules and the membership functions. These fuzzy systems which are based on these two basic elements are also called fuzzy inference systems. The fuzzy logic rules and membership functions of the fuzzy system can be used to find and interpret the structure and the weights of neural networks. Fuzzy neural networks have higher training speed and are more and more robust than conventional neural systems.

In general, integrating fuzzy systems with ANNs and ANNs with fuzzy systems, we maximize the learning and adaptive capabilities of the combined system which are not available in neither of the systems from which the integrated system came from. Which of the two tools (neurofuzzy of fuzzyneural) will be used in a particular situation depends on the particular application. If for example some expert information is available at the outset and is easy to develop some rules, we can use a neurofuzzy system for which the neural network has been used to further refine the fuzzy rules.

In this book, those particular tools have been used to solve problems in Telecommunications which cover a wide variety of real world cases. Chapter 2 and Appendices A, B, C deal with the background material on neural networks and fuzzy systems as well as with their integration. Chapter 3 deals with the application of neurofuzzy techniques in speech coding and recognition. Image/video compression using neurofuzzy techniques are covered in Chapter 4. In Chapter 5 a neurofuzzy system is used to study a source location and tracking problem in wireless communications. Fuzzyneural techniques are applied to handoffs problems in cellular, in Chapter 6 and finally Chapter 7 covers traffic control of ATM networks using a powerful neurofuzzy system.

Peter Stravroulakis

Table of Contents

List of Contributors

Benjapolakul, Watit, Prof.
Homnan, Bongkarn, Prof.
Niruntasukrat, Aimaschana, Prof.
Department of Electrical Engineering, Faculty of Engineering
Chulalongkorn University
Bangkok 10330
Thailand
Email: watit@ee.eng.chula.ac.th
Watit.B@chula.ac.th
Tel: 66 22186902
Fax: 66 22186912

Beritelli, Francesco, Prof.
Serrano, Salvatore, Prof.
Department of Engineering Information and Telecommunications
University of Catania
V. le A. Doria 6, 95125
Italy
Email: beritelli@diit.unict.it
Tel: 39 0957382367
Fax: 39 0957382397

Lee, Shie-Jue, Prof.
Lee, Wan-Jui, Prof.
Ouyang, Chen-Sen, Prof.
Department of Electrical Engineering
National Sun Yat-Sen University
No.70 Lien-Hai Rd, Kaohsiung, ROC
Taiwan
Email: leesj@ee.nsysu.edu.tw
wrlee@water.ee.nsysu.edu.tw
ouyang@water.ee.nsysu.edu.tw

Perez-Neira, Ana, Prof.
Bas, Joan, Prof.
Lagunas, Miguel, Prof.
Department of Signal Theory and Communications
Campus Nord UPC. Modulo D-5
C/Jordi Girona 1-3, 08034 Barcelona
Spain
Email: {anuska,jbas}@gps.tsc.upc.es
Tel: 34 934016436
Fax: 34 934016447

Russo, Marco, Prof.
Department of Physics
University of Messina
Italy
Email: m.russo@ct.infn.it

Tripathi, Nishith D., Prof.
Reed, Jeffrey H., Prof.
VanLandingham Hugh F., Prof.
Bradley Department of Electrical Engineering
Virginia Polytechnic Institute and State University
Blacksburg, VA
USA
Email: nishitht@yahoo.com

1 Introduction

When King Philip called Aristotle from Plato's Academy in Athens to his Palace in Pella (Northern Greece) and gave him almost unlimited funds in order to tutor his son Alexander the Great and provide him with the best education and instruction possible at that time, he did not know that he had initiated a new research project of unknown global consequences.

Aristotle, of course, accepted the project and overwhelmed himself with such an honor and serious responsibility. The central theme of the instruction given to Alexander, was: **"It is the mark of an instructed mind to rest satisfied with that degree of precision which the nature of the subject admits, and not to seek exactness, where only an approximation of the truth is possible".** Aristotle, therefore, initiated Alexander to fuzzy logic, but in some sense to adaptive learning as well. **"Not seek exactness, where it does not exist"**. In other words, adapt to the situation at hand. Later on, when Alexander succeeded his father and decided to spread the accumulated Hellenic knowledge, culture and civilization to the rest of the world, in one of his expeditions he was confronted with the most difficult problem of unknotting the Gordian knot. He used a Neuro-fuzzy concept to solve the problem. He had to adapt to the situation (neural network concept) in which in front of his Generals, Soldiers, and Enemies, he could not afford a failure and automatically used the if-then rule thinking (fuzzy logic), as follows: **If I try unknotting it and don't succeed, then I will be embarrassed and doubted beyond repair, hence cut the knot.** Since then, people have been using neurofuzzy systems logic, without knowing it.

Neural networks and fuzzy systems, are the structural elements of Computational Intelligence (CI). Actually, CI consists of three components, where the third one is genetic algorithms. In this book, we will concentrate on the first two and their combination, as they apply to the design and implementation of communication systems and to the analysis and study of those aspects of the systems under study, which in general we have no mathematical model.

Neural networks started to become a formal field of study following the work of McCulloch and Pitts [1] in 1943. Researchers like Hebb in 1949 [2], Minsky in 1967 [4] and Rosenblatt in 1958 [3], among others, founded this field on a strong mathematical basis. In 1960, Widrow and Hoff [5] developed the learning algorithm based on the Least Mean Square via the Adaptive Linear Element (ADALINE). The high interest of scientists in this field was suddenly reduced, due to a paper by Minski and Papert [6], who doubted the usefulness of perception as a solution to the problem of pattern recognition. The interest returned in 1982, after the publication of the work of Hopfield. Since then, neural networks have been used to solve many complex problems in diverse fields. Their distributed nature and learning characteristics make them essential tools for analysis in many situations, where a mathematical model for the problem under study, is either not possible or very difficult and complex. An analysis of their essential characteristics is given in Appendix A, as well as their structural relation to the fuzzy logic based systems, with which they will be combined to create a new tool of analysis which is the subject matter of this book. For more details, the reader is referred to [7–8].

As far as fuzzy logic based systems are concerned, their mathematical foundation was established by Loft Zadeh [9–10], who came up with the principle of incompatibility which is in some sense the same concept as that developed by Aristotle, more than 2000 years ago. Since then, people have been using both neural networks and fuzzy logic based systems separately, or in combination to solve many engineering problems [11–13]. In Appendix B, we discuss the most important characteristics of fuzzy logic based systems, which can be combined with relevant characteristics of neural networks and create a new, more powerful tool which is the subject matter of this book, as it relates to Telecommunications. Chapter 2 covers the mathematical foundation for the creation of neurofuzzy or fuzzyneural schemes and is followed by chapters that cover among other topics, neurofuzzy applications in speech coding, image compression using neurofuzzy techniques, a neurofuzzy approach to source location and tracking, neurofuzzy applications in Handoff and a neurofuzzy system application for access control in asynchronous transfer mode networks.

References

[1] McCulloch W S and Pitts W (1943) A locical caculers of ideas immonent in nervous activity.

[2] Hebb D O (1949) The Organization of Behavior: A Neuropsychological Theory. New York: John Wiley

[3] Rosenblatt F (1958) The perception: A probalistic model for information storage and organization is the brain. Phychol Rev. 65: 386–408.

[4] Minsky M L (1967) Computation: Finite and Infinite Machines. Englewood Cliffs, N.J. Prentice Hall.

[5] Widzow B, Hoff M E, Jr. (1960) Adaptive switching circuits. Proc. IRE West Elect. Show Conv. Ree. Part 4, 96–104, New York.

[6] Minsky M L and Papert S A (1988) Perceptrons, Cambridge MA: MIT Press.

[7] Cichocki A, Unbehanen R (1993) Neural Networks for Optimization and Signal Processing. John Wiley, New York.

[8] Haykin (1994) Neural Networks: A Comprehensive Foundation. Macmillan, New York.

[9] Zadeh L A (1965) Fuzzy Sets, Inf. Control.

[10] Zadeh L A (1973) Outline of a New Approach to the Analysis of Complex Systems and Decision Processes. IEEE Trans. Syst. Man Cybern., SMC.

[11] Lin C T, Lee C S G (1996) Neural Fuzzy Systems. John Wiley, New York.

[12] Ross T J (1995) Fuzzy Logic with Engineering Applications. McGraw-Hill, New York.

[13] Kosko B (1997) Fuzzy Engineering. Prentice Hall, Upper Saddle River, N.J.

2 Integration of Neural and Fuzzy

Peter Stavroulakis

2.1 Introduction

In this chapter, the integration of neural networks and fuzzy systems will be discussed. A substantial portion of this material comes from reference [1]. The combination of the techniques of fuzzy logic systems and neural networks suggests the novel idea of transforming the burden of designing fuzzy logic control and decision systems to the training and learning of connectionist neural networks. This neuro-fuzzy and/or fuzzy-neural synergistic integration reaps the benefits of both neural networks and fuzzy logic systems. That is, the neural networks provide connectionist structure (fault tolerance and distributed representation properties) and learning abilities to the fuzzy logic systems, and the fuzzy logic systems *provide* the neural networks with a structural framework with high-level fuzzy IF-THEN rule thinking and reasoning. These benefits can be witnessed in three major integrated systems: neural fuzzy systems, fuzzy neural networks, and fuzzy neural hybrid systems. These three integrated systems, along with their applications, will be discussed and explored in the next six chapters, as well as in the Appendices.

With the advance in neural network models and their applications to machine learning, a considerable interest has been on unifying neural networks and fuzzy set theory [2, 3]. In this section, we will briefly discuss several studies in which neural networks have been utilized to come up with membership functions from a given set of data.

Takagi & Hayashi [4] discuss a Neural Network that generates nonlinear, multi-dimensional membership functions, which is a membership function generating module of a larger system that utilized fuzzy logic. They claim that the advantage of using nonlinear multi-dimensional membership functions is in their effect of reducing the number of fuzzy rules in the rule base.

Yamakawa & Furukawa [5] present an algorithm for learning membership functions, using a model of the *fuzzy neuron.* Their method uses example-based learning and optimization of cross-detecting lines. They assign *trapezoidal* membership functions and automatically come up with its parameters. The context is handwriting recognition. They also report some computational results for their algorithm.

On the experimental side, Erickson, Lorenzo & Woodbury [6] claim that fuzzy membership functions and fuzzy set theory *explain better* the classification of taste responses in brain stem. They analyze previously published data and allow each neuron to belong to *several* classifications *to a degree.* This degree is measured by the neuron's response to the stimuli. They show that their model based on fuzzy set theory explains the data better than other statistical models.

Furukawa & Yamakawa [7] describe two algorithms that yield membership functions for a fuzzy neuron and their application to the recognition of hand writing. The crossing points of two (trapezoidal) membership functions are *optimized* for the task at hand.

Fuzzy systems and neural networks are both numerical model-free estimators and dynamical systems. They share the ability of improving the intelligence of systems working in uncertain, imprecise, and noisy environments. Fuzzy systems and neural networks estimate sampled functions and behave as associative memories. Both have an advantage over traditional statistical estimation and adaptive control approaches to function estimation. They estimate a function without requiring a mathematical description of how the output functionally depends on the input; that is, they learn from numerical examples. Both fuzzy and neural approaches are numerical in nature, can be processed using mathematical tools, can be partially described with theorems, and admit an algorithmic characterization that favors silicon and optical implementation. These properties distinguish fuzzy and neural approaches from the symbolic processing approaches of artificial intelligence (AI). To a certain extent, both systems and their techniques have been successfully applied to a variety of real-world systems and devices.

Fuzzy logic and neural networks are complementary technologies. Neural networks extract information from systems to be learned or controlled (refer Appendix A), while fuzzy logic techniques most often use verbal and linguistic information from experts (refer Appendix B). A promising approach to obtaining the benefits of both fuzzy systems and neural networks and solving their respective problems, is to combine them into an integrated system. For example, one can learn rules in a hybrid fashion and then calibrate them for a better whole-system performance. The common features and characteristics of fuzzy systems and neural networks warrant their integration.

The integrated system will possess the advantages of both neural networks (e.g., learning abilities, optimization abilities, and connectionist structures) and fuzzy systems (e.g., humanlike IF-THEN rules thinking and ease of incorporating expert knowledge). In this way, we can bring the low-level learning and computational power of neural networks into fuzzy systems and also high-level, humanlike IF-THEN rule thinking and reasoning of fuzzy systems into neural networks. Thus, on the neural side, more and more transparency is pursued and obtained either by prestructuring a neural network to improve its performances, or by a possible interpretation of the weight matrix following the learning stage. On the fuzzy side, the development of methods allowing automatic tuning of the parameters that characterize the fuzzy system can largely draw inspiration from similar methods used in the connectionist community. Thus, neural networks can improve their transparency, bringing them closer to fuzzy systems, while fuzzy systems can self-adapt, bringing them closer to neural networks.

Integrated systems can *learn* and *adapt*. They learn new associations, new patterns and new functional dependencies. They sample the flux of experience and encode new information. They compress or quantize the sampled flux into a small but statistically representative set of prototypes or exemplars. Broadly speaking, we may characterize the efforts at merging these two technologies, in three categories:

1. **Neural fuzzy systems**, the use of neural networks as tools in fuzzy models.
2. **Fuzzy neural networks**, fuzzification of conventional neural network models.
3. **Fuzzy-neural hybrid systems**, incorporating fuzzy technologies and neural networks into hybrid systems, which include Genetic Algorithms.

In the first approach, *neural fuzzy systems* (Appendix C.6) aim at providing fuzzy systems with the kind of automatic tuning methods typical of neural networks, but without altering their functionality (e.g., fuzzification, defuzzification, inference engine, and fuzzy logic base). In neural fuzzy systems, neural networks are used in augmenting numerical processing of fuzzy sets, such as membership function elicitation and realization of mappings between fuzzy sets that is utilized as fuzzy rules. Since neural fuzzy systems are inherently fuzzy logic systems, they are mostly used in control applications.

In the second approach, *fuzzy neural networks* (Appendix C.1–C.4) retain the basic properties and architectures of neural networks and simply 'fuzzify' some of their elements. In fuzzy neural networks, a crisp neuron can become fuzzy and the response of the neuron to its lower-layer activation signal can be of a fuzzy relation type, rather than a sigmoid type. One can find examples of this approach where domain knowledge becomes formalized in terms of fuzzy sets and later can be applied to enhance the learning algorithms of the neural networks or augment their interpretation capabilities. Since the neural architecture is conserved, what still varies is some kind of synaptic weights connecting low-level to high-level neurons. Since fuzzy neural networks are inherently neural networks, they are mostly used in pattern recognition applications.

Finally, in the third approach (Appendix C.5), both fuzzy techniques and neural networks play a key role in hybrid systems. They do their own jobs in serving different functions in the system. Making use of their individual strengths, they incorporate and complement each other in achieving a common goal. The architectures of fuzzy-neural hybrid systems are quite application-oriented, and they are suitable for both control and pattern-recognition applications. All of these techniques will be explained, shown and used in examples in the sections that follow, before we start applying neurofuzzy and fuzzyneural techniques in real world Telecommunication problems, as in Chapters 3–7.

In this chapter, we study basic hybrid systems of every category. Specifically we start with the neuro-fuzzy systems in section 2.2.1, where we present three models (ANFIS, GARIC and fuzzy ART-ARTMAP) and we show the universal approach of non linear functions of the ANFIS system. Examples also show the efficiency of these hybrid systems.

2.2 Hybrid Artificial Intelligent Systems

Artificial intelligence (AI) is the computer area that includes calculations and reasoning under certain circumstances of inaccuracy, uncertainty and partial truth, and achieves stoutness and low cost solutions [8–12]. Each of the three AI components: neural networks (NN), fuzzy logic (FL), and genetic algorithms (GA), has certain properties and advantages:

1) NN allow a system to learn
2) FL allows experiential knowledge to be used in the system
3) GA allows the system to improve itself.

By combining these components, we can design and build hybrid AI systems which are able to efficiently solve complex problems. Hybrid AI systems combine the properties of each one and overcome their restrictions or disadvantages. Hybrid techniques can build intelligent systems that have many applications. The four possible combinations of NN, FL and GA for the development of hybrid techniques and AI systems, are shown in Fig. 2.2.1.

The symbols NFS, NGS, etc in Fig 2.2.1 have the following meaning: NFS neuro-fuzzy systems, NGS neuro-genetic systems, FGS: fuzzy genetic systems, NFGS: neuro-fuzzy-genetic systems.

NN are used in NFS in order to learn the membership functions and/or adjust the structure of the fuzzy systems (rules, etc). GA are used in FGS in order to find the best structure and subsequently for the adjustment of the parameters (normalization coefficients, structure and number of fuzzy sets, number of fuzzy rules and their linguistic description). On the contrary, fuzzy logic may be used to improve the behavior of GA, for example by using fuzzy operands in the development of genetic operations or by using other fuzzy criteria in the genetic procedures. Genetic algorithms are used in NGS in order to automate the neural network design (choice of weights) or the genetic selection of their topology or by making the best selection of the learning parameters (i.e. learning rate or the momentum). Finally, we can design neuro-fuzzy-genetic systems

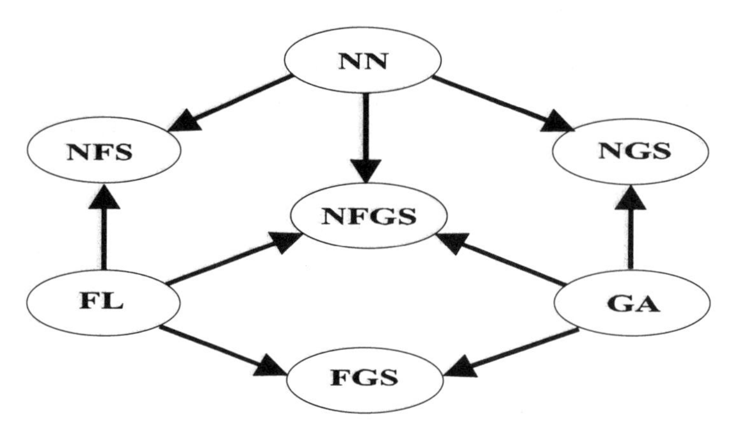

Fig. 2.2.1. Hybrid artificial intelligent systems

(NFGS) by combining NN, FL and GA, which integrate the advantages and properties of all of them, in order to maximize their efficiency. In this book, GA will not be studied.

2.2.1 Neuro-Fuzzy Systems

The pure fuzzy logic systems have the following two disadvantages:

1) They do not provide a certain method of specifying the membership functions.
2) They do not provide a learning or adaptation component.

These two disadvantages are overcome if we use neural networks for guiding the fuzzy logic. Realistically, NN can be trained to automatically choose the membership functions (that is the fuzzy sets) and to choose the number and/or the shape of the fuzzy rules. For this purpose, numerous techniques have been developed with relative fluctuations in generality, simplicity and applicability.

Here, we will study the following techniques:

Adaptive Neuro Fuzzy Inference System (ANFIS)
Generalized Adaptive Reinforced Intelligent Control (GARIC)
Adaptive Resonance Theory (Fuzzy ART- ARTMAP)

2.2.1.1 Adaptive Neuro-Fuzzy Inference System (ANFIS)

The ANFIS is a platform for adaptive neuro-fuzzy logic. This platform depicts the fuzzy system operation in an NN, which has five layers and is trained with a back propagation algorithm. The fuzzy rules can be of either of a Takagi-Sugeno or Mamdani type, and logic can use either the multiplication or minimum rule. The membership functions can be of any type (conical, Gaussian, trapezoidal or triangular type).

A. Neuro-fuzzy ANFIS System for the Takagi-Sugeno model
Here, we will describe the adaptive neuro-fuzzy system ANFIS, considering Takagi-Sugeno type rules:

$$\begin{aligned}&\text{IF } x_1 \text{ is } A_1^j \text{ AND} \ldots \text{AND} x_N \text{ is } A_N^j\\&\text{THEN } y_j = f_j = c_{j1}x_1 + c_{j2}x_2 + \ldots + c_{jN}x_N + r_j\end{aligned} \tag{2.2.1}$$

for $j = 1, 2, \ldots, M$ (M is the number of rules), where A_i^j are the fuzzy sets and c_{ji}, r_j are constant coefficients ($i = 1, 2, \ldots, N$ $j = 1, 2, \ldots, M$).

The membership function of A_i^j is considered to be conical, so:

$$\mu_{A_i^j}(x_i) = \frac{1}{\left[1 + \left(\frac{x_i - m_i^j}{\alpha_i^j}\right)^2\right]^{b_i^j}} \tag{2.2.2}$$

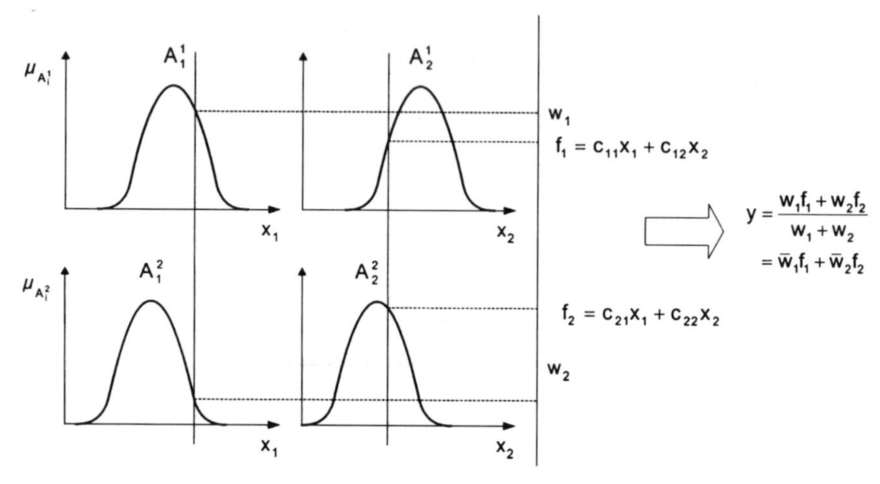

Fig. 2.2.2. Takagi-Sugeno logic in the case of two rules with two input variables

where (α_i^j, b_i^j, m_i^j, $i = 1, 2, \ldots N$; $j = 1, 2, \ldots, M$ are the parameters that determine the fuzzy sets A_i^j.

Figure 2.2.2 shows the logic mechanism of the Takagi-Sugeno type fuzzy model in the case of two rules with two input variables x_1 and x_2.

In the case where the logic multiplication rule is used in Fig. 2.2.2, we have:

$$w_1 = \mu_{A_1^1}(x_1)\mu_{A_2^1}(x_2) \text{ and } w_2 = \mu_{A_1^2}(x_1)\mu_{A_2^2}(x_2) \tag{2.2.3}$$

$$\begin{aligned} f_1 &= c_{11}x_1 + c_{12}x_2 \text{ we consider } r_1 = 0 \\ f_1 &= c_{21}x_1 + c_{22}x_2 \text{ we consider } r_2 = 0 \end{aligned} \tag{2.2.4}$$

and the result is:

$$f = \frac{w_1 f_1 + w_2 f_2}{w_1 + w_2} = \overline{w_1} f_1 + \overline{w_2} f_2 \tag{2.2.5}$$

where

$$\overline{w_j} = \frac{w_j}{w_1 + w_2} \quad (j = 1, 2) \tag{2.2.6}$$

In the generalized form of M rules with N variables we have ($j = 1, 2, \ldots, M$):

$$w_j = \prod_{i=1}^{N} \mu_{A_i^j}(x_i) \tag{2.2.7}$$

$$\overline{w_j} = w_j / \sum_{k=1}^{M} w_k \tag{2.2.8}$$

$$\overline{w_j f_j} = \overline{w_j}\, [c_{j1}x_1 + \ldots + c_{jN}x_N + r_j] \tag{2.2.9}$$

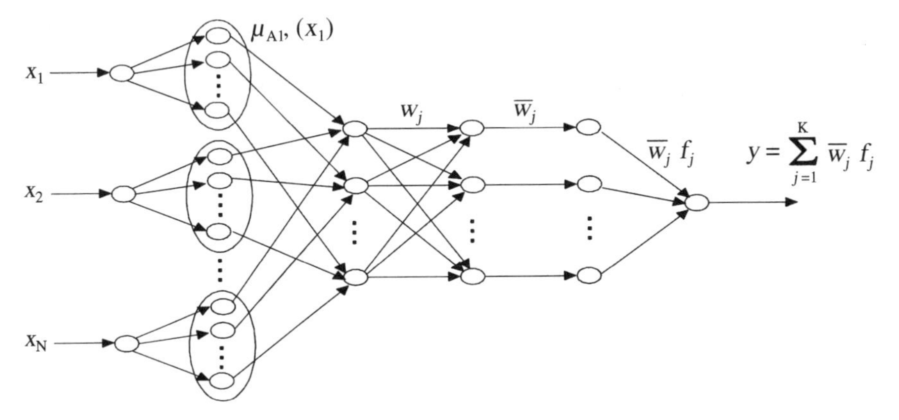

Fig. 2.2.3. Structure of the ANFIS System for the Takagi-Sugeno model

and

$$y=\sum_{j=1}^{M}\overline{w_j}f_j=\sum_{j=1}^{M}w_jf_j\,/\sum_{j=1}^{M}w_j \qquad (2.2.10)$$

The structure of the ANFIS neural network that realizes Eqs. (2.2.7)–(2.2.10), is given in Fig. 2.2.3.

The layers of the ANFIS NN, perform the following functions:

Layer 1:
Every node of layer 1 corresponds to a different fuzzy set input A_i^j. So if A_i^j are of the conical form (2.2.2), then every node corresponds to the three parameters $(\alpha_i^j, b_i^j, m_i^j)$.

Layer 2:
This layer has as many nodes as is the number of the system's rules (that is M nodes). Every j node multiplies the signals $\mu_{A_i^j}(x_i)$ that are received from layer 1, and produces the size w_j, according to (2.2.7), which shows the triggering power (adaptiveness) of rule j.

Layer 3:
Node j $(j=1,2,\ldots,M)$ of layer 3 calculates the normalized triggering power of $\overline{w_j}$ of rule j, according to (2.2.8).

Layer 4:
Node j of layer 4 calculates the product $\overline{w_j}f_j$ according to (2.2.9), where f_i is the inference of rule j $(j=1,2,\ldots,M)$.

Layer 5:
This layer (defuzzyfication layer) calculates the total output y of the system according to (2.2.10), that is the sum of all $\overline{w_j}f_j$ received from layer 4.

During the training process, the NN updates its parameters (parameters $(\alpha_i^j, b_i^j, m_i^j)$ that define the shape of the membership function of A_i^j and the parameters c_{ji} and r_j of the inference) with the back propagation algorithm.

Note: If we combine the third and fourth layer in one, we get a NN with only four layers. Also, the normalization of w_j can be done in the last layer.

B. ANFIS system for the Mamdani model

The ANFIS system which is based on the Takagi-Sugeno model is the simplest, however in many cases, the use of the Mamdani model (fuzzy rules) is necessary:

$$\begin{aligned} R_j\text{: IF } x_1 \text{ is } A_1^j \text{ AND ... AND} x_N \text{ is } A_N^j \\ \text{THEN} y_j \text{ is } B^j \text{ for} j=1,2,...,M \end{aligned} \tag{2.2.11}$$

If $\overline{y_j}$ is the center of gravity (COG) of the fuzzy set B^j, then the total output y of the fuzzy system (set of rules) (2.2.11), is given by the following equation:

$$y=f(x)=\frac{\sum_{j=1}^{M} w_j \overline{y_j}}{\sum_{j=1}^{M} w_j}=\frac{g}{h} \tag{2.2.12}$$

with $X = [x_1, x_2, \ldots, x_N]^T$, where

$$w_j=\prod_{i=1}^{N} \mu_{A_i^j}(x_i) \tag{2.2.13}$$

if the logic multiplication rule is used, and

$$w_j=\min\{\mu_{A_1^j}(x_1),\mu_{A_2^j}(x_2),\ldots,\mu_{A_N^j}(x_N)\} \tag{2.2.14}$$

if the minimum rule is used.

The form of the membership function of A_i^j can be conical (2.2.2) or Gaussian:

$$\mu_{A_i^j}(x_i)=\alpha_i^j \exp\left[-\left(\frac{x_i-m_i^j}{\sigma_i^j}\right)^2\right] \tag{2.2.15}$$

where $(\alpha_i^j, m_i^j, \sigma_i^j)$ are the parameters that define the function shape.

Equation (2.2.12) together with (2.2.13) or with (2.2.14), may be realized just like Eq. (2.2.10) with (2.2.7)–(2.2.9), using a multilayer forward feedback type NN of five layers, as shown in Fig. 2.2.4.

If we consider that the membership functions have the Gaussian form with $\alpha_i^j = 1$ (for all i and j), then the NN has to adjust the parameters $\overline{y_j}$, m_i^j and σ_i^j so that the square error is minimised:

$$e(t)=\frac{1}{2}[f(\mathbf{x}^{\mathbf{p}})-y_d]^2 \tag{2.2.16}$$

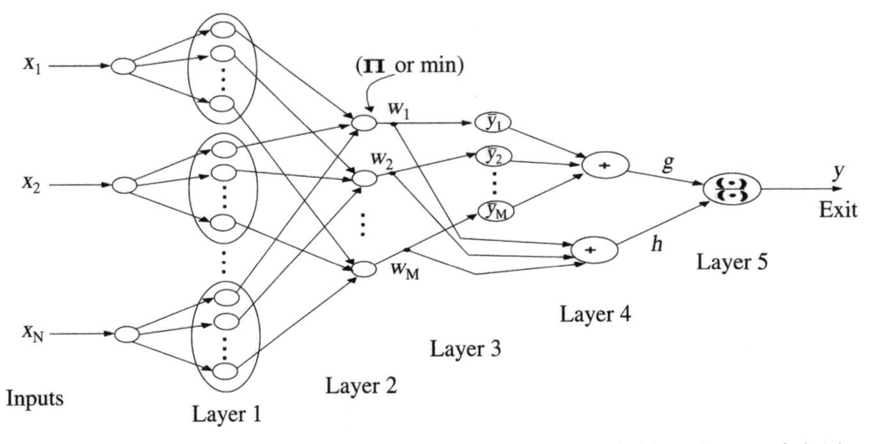

Fig. 2.2.4. ANFIS system for the Mamdani model, using the multiplication rule (II) or minimum rule (min)

where p is the input pattern indicator $\boldsymbol{x}^p$ and y_d is the desired output. The training (updating) of $\overline{y_j}$ is performed using the following rule:

$$\overline{y_j}(t+1)=\overline{y_j}(t)-\gamma\frac{\partial e(t)}{\partial \overline{y_j}} \tag{2.2.17}$$

for $j = 1, 2, \ldots, M$ and $t = 0, 1, 2, \ldots$.

We can see that f in (2.2.12) (and consequently e as well) depends on $\overline{y_j}$ only through the numerator g, so we have:

$$g=\sum_{j=1}^{M}w_j\,\overline{y_j},h=\sum_{j=1}^{M}w_j \tag{2.2.18}$$

and in the case of the multiplication rule (II):

$$w_j=\prod_{i=1}^{N}\exp\left[-\left(\frac{x_i^p-m_i^j}{\sigma_i^j}\right)^2\right] \tag{2.2.19}$$

So we have:

$$\frac{\partial e(t)}{\partial \overline{y_j}}=(f-y_d)\frac{\partial f}{\partial g}\frac{\partial g}{\partial \overline{y_j}}=(f-y_d)\left(\frac{1}{h}\right)w_j$$

so (2.2.17) becomes:

$$\overline{y_j}(t+1)=\overline{y_j}(t)-\gamma\left(\frac{f-y_d}{h}\right)w_j \tag{2.2.20}$$

Updating of m_i^j is done by using the learning rule:

$$m_i^j(t+1) = m_i^j(t) - \gamma \frac{\partial e(t)}{\partial m_i^j} \tag{2.2.21}$$

for $i = 1, 2, \ldots, M$ and $j = 1, 2, \ldots, M$.

In this case we have:

$$\frac{\partial e(t)}{\partial m_i^j} = (f - y_d) \frac{\partial f}{\partial w_j} \frac{\partial w_j}{\partial m_i^j} \tag{2.2.22}$$

where

$$\frac{\partial f}{\partial w_j} = \frac{\frac{\partial g}{\partial w_j} h - g \frac{\partial h}{\partial w_j}}{h^2} = \frac{\overline{y}_j h - g}{h^2} = \frac{\overline{y}_j - \frac{g}{h}}{h} = \frac{\overline{y}_j - f}{h} \tag{2.2.23}$$

and

$$\frac{\partial w_j}{\partial m_i^j} = \frac{w_j}{\exp\left[-\left(\frac{x_i^p - m_i^j}{\sigma_i^j}\right)^2\right]} \frac{\partial}{\partial m_i^j} \left\{ \exp\left[-\left(\frac{x_i^p - m_i^j}{\sigma_i^j}\right)^2\right] \right\} = \frac{2 w_j (x_i^p - m_i^j(t))}{(\sigma_i^j)^2} \tag{2.2.24}$$

Consequently, (2.2.21) based on (2.2.22)–(2.2.24) becomes:

$$m_i^j(t+1) = m_i^j(t) - 2\gamma \frac{f - y_d}{h} (\overline{y}_j - f) w_j \frac{(x_i^p - m_i^j(t))}{(\sigma_i^j)^2} \tag{2.2.25}$$

Likewise, we can determine the learning rule for σ_i^j, which is:

$$\sigma_i^j(t+1) = \sigma_i^j(t) - \gamma \frac{\partial e(t)}{\partial \sigma_i^j} = \sigma_i^j(t) - \gamma (f - y_d) \frac{\partial f}{\partial w_j} \frac{\partial w_j}{\partial \sigma_i^j} \tag{2.2.26}$$

where,

$$\frac{\partial f}{\partial w_j} = \frac{\overline{y}_j - f}{h} \tag{2.2.27}$$

and

$$\frac{\partial w_j}{\partial \sigma_i^j} = 2 w_j \frac{(x_i^p - m_i^j(t))^2}{(\sigma_i^j(t))^3} \tag{2.2.28}$$

Introducing (2.2.27) and (2.2.28) in (2.2.26), the following learning rule is formed:

$$\sigma_i^j(t+1) = \sigma_i^j(t) - 2\gamma (f - y_d) w_j \frac{(\overline{y}_j - f)}{h} \frac{(x_i^p - m_i^j(t))^2}{(\sigma_i^j(t))^3} \tag{2.2.29}$$

From all the above we conclude that the training algorithm of the fuzzy system has two phases. In the first phase (forward) for an input pattern $\boldsymbol{x}^p$ we go forward through the layers of the network to find w_j, g, h and f (Eqs. (2.2.12), (2.2.18) and (2.2.19)). In the second phase (back propagation) we update the network parameters y_j, m_i^j and σ_i^j ($i = 1, 2, \ldots, N$ and $j = 1, 2, \ldots, M$) according to the learning rules (2.2.19), (2.2.25) and (2.2.29).

2.2.1.2 Generalized Adaptive Reinforced Intelligent Control (GARIC)

The hybrid system GARIC results from the combination of neural networks, fuzzy logic and adaptive learning. Basically, it is a new method of learning and updating of the fuzzy system's parameters with the use of adaptive signals. This system expands the pure neural system of adaptive learning AHC (Adaptive Heuristic Critic) in addition to including expert knowledge expressed with fuzzy rules.

The architecture of the system has the form shown in Fig. 2.2.5 and includes three parts:

The neural Action Selection Network (ASN)
The neural Action Evaluation Network (AEN)
The Action Stochastic Modifier (ASM)

The ASN depicts a state vector in a proposed action F through fuzzy reasoning, while AEN shows a state vector and a failure signal in a scalar score v, which shows how good the state is. This score is used in the production of an inner adaptive signal r'.

Action Stochastic Modifier
The Action Stochastic Modifier (ASM) accepts the proposed action F and the adaptive inner signal r' and creates, stochastically, the final action signal F' which is applied on the controlled system. F' is a Gaussian random variable with average value F and standard variation $\sigma(r'(t-1))$, where $\sigma(.)$ is a non-

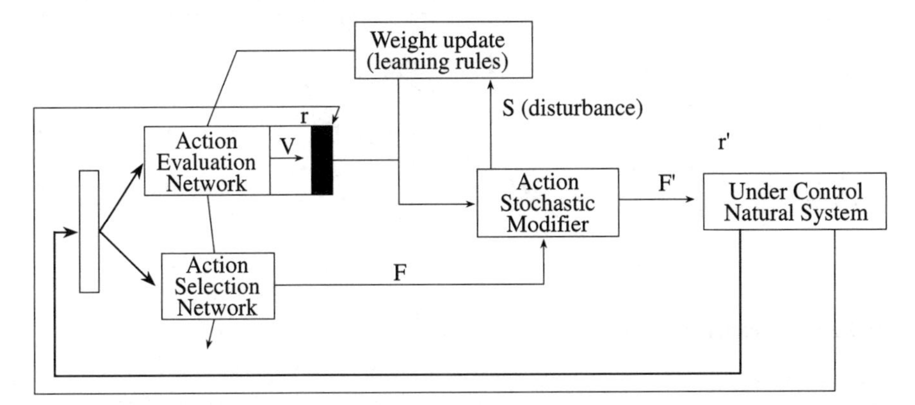

Fig. 2.2.5. Architecture of the GARIC system with Adaptive Heuristic Critic (AHC) [r' = inner adaptive signal]

negative monotonous descending function (e.g. exp(−r′)). The stochastic disturbance

$$s(t) = \frac{F' - F(t)}{\sigma(r'(t-1))} \tag{2.2.30}$$

results in a better state space exploration and in better generalization. Obviously, (2.2.30) means that $s(t)$ is just a normalized variation from the proposed action F. The $s(t)$ is used as a factor in the ACN learning.

Action Selection Network

The ASN form (for two inputs x_1, x_2) is shown in Fig. 2.2.6 and includes five layers, each one of which performs one step of the fuzzy logic, as in the ANFIS network.

The membership functions of the fuzzy sets (language variables) input, may be conical, Gaussian or triangular. Here we will use the triangular form:

$$\mu_{C,S_L,S_R}(x) = \begin{cases} 1 - \frac{|x-c|}{s_R}, x \in [c, c+s_R] \\ 1 - \frac{|x-c|}{s_L}, x \in [c-s_L, c] \\ 0, \quad \text{in other cases} \end{cases} \tag{2.2.31}$$

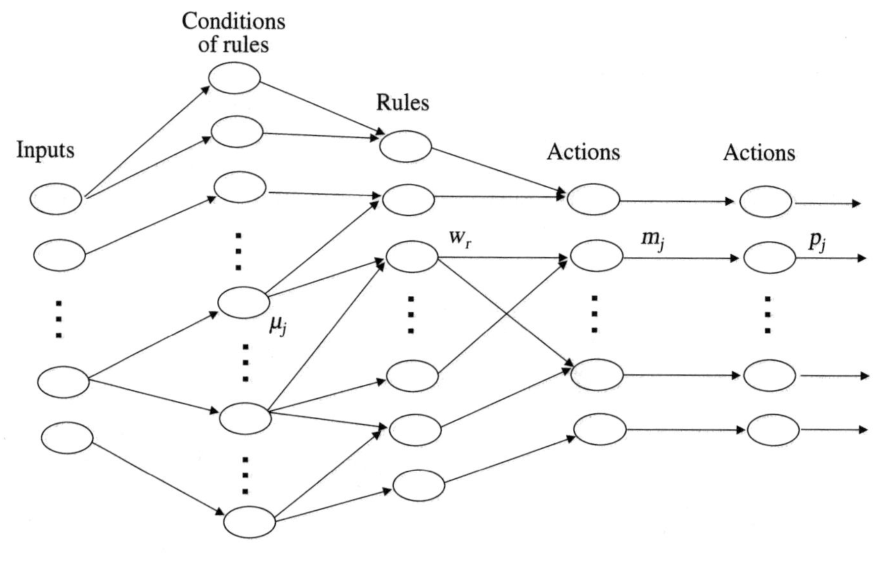

Fig. 2.2.6. Architecture of NN ASN

The function of ASN's layers, is as follows:

Layer1

The first layer is the input layer in which we use the components $x_1, x_2, \ldots, x_n$ of the state vector $x = [\, x_1, x_2, \ldots, x_n]^T$ of the system, which take on true values. The variables $x_1, x_2, \ldots, x_n$ may be considered as the language variables that we are interested in. Layer 1 does not include any calculations.

Layer2

Every node of layer 2 corresponds to one possible language value of an input variable, and calculates the equivalent triangular membership function (2.2.31), where c is the center and s_L, s_R is the left and right opening, respectively.

Layer3

This layer realizes a connection (AND) of all the conditions of the left side of every rule which occupies one node of the layer. The inputs of every node of layer 3 come from all the nodes of layer 2 which participate in the left side (IF) of the equivalent rule. Every j node of layer 3 calculates the value of the triggering force w_j of the j rule with the following equation:

$$w_j = \frac{\sum_i \mu_i^j e^{-\kappa\mu_i^j}}{\sum_i e^{-\kappa\mu_i^j}} \tag{2.2.32}$$

where i extends to every node of layer 2 that is connected to node j. The operator in (2.2.32) is an intersection operator (AND) of fuzzy sets, but not as restrictive as the 'min' operator. That is why it is called 'Softmin'. It is also differentiable, which is needed in neural learning (back propagation).

In the operator of (2.2.32), μ_i^j is the adaptation rank between one condition of the IF part of rule j and the equivalent input variable x_j. Parameter k adjusts the hardness of 'softmin' and when $\kappa \to \infty$, the operator approaches the 'min' operator. Let us consider that $i = 1, 2$. Then:

$$w_j = \frac{\mu_1 e^{-\kappa\mu_1} + \mu_2 e^{-\kappa\mu_2}}{e^{-\kappa\mu_1} + e^{-\kappa\mu_2}} = \begin{cases} \dfrac{\mu_1 e^{-\kappa(\mu_1-\mu_2)} + \mu_2}{e^{-\kappa(\mu_1-\mu_2)}+1} \to & \mu_2 \text{ if } \mu_1 > \mu_2 \\ \dfrac{\mu_2 e^{-\kappa(\mu_2-\mu_1)} + \mu_1}{e^{-\kappa(\mu_2-\mu_1)}+1} & \mu_1 \text{ if } \mu_2 > \mu_1 \end{cases} \tag{2.2.33}$$

Layer 4

The nodes of layer 4 correspond to the possible output actions (conclusions) with inputs deriving from all the rules that suggest every particular action. The output of every node i of this layer is calculated with the following equation:

$$m_i = \frac{2}{1+e^{-\sum_j (w_j - 0.5)}} - 1, -1 \le m_i \le 1 \tag{2.2.34}$$

m_i gives the weight of the (output) action i, based on the contribution w_j of all the rules that result in this action.

Layer 5

This layer has as many nodes as there are actions (that is equal to the nodes of layer 4). Every node accepts the weight action of m_i from layer 4 and generates the probability p_i that the equivalent action will be chosen, according to the Boltzman distribution:

$$p_i = \frac{e^{m_i/T}}{\sum_i e^{m_i/T}} \tag{2.2.35}$$

where T is the "temperature" of the system. The final control action "α" that must be applied on the system, can be found by random choice based on the probability vectors. As in the algorithm of simulated annealing, temperature T is chosen to be high at the beginning, so that all the output actions have a similar probability of selection (when $T \to \infty$, then $e^{m_i/T} \to \infty$ regardless of i). This means that at the beginning, the stochastic nature of the network is powerful and it examines every possible action. As the learning continues, the temperature T falls gradually, weakening the stochastic nature and decreasing the probabilities of choosing the action with the larger potential. The ASN learns the parameters c, s_R, s_L of the input membership functions. So the internal adaptive signal r' is transferred backwards and the equivalent parameters are updated. These parameters are used in the fuzzy rules which suggest the output action 'α'. If we consider only the chosen output action 'α' and we apply the back propagation learning rule for the θ parameter (where θ can be the c, s_R, or the s_L) we have:

$$\Delta\theta = \gamma r' \frac{\partial m_\alpha}{\partial \theta} = \gamma r' \sum_j \frac{\partial m_\alpha}{\partial w_j} \frac{\partial w_j}{\partial \theta} = \gamma r' \sum_j \frac{\partial m_\alpha}{\partial w_j} \cdot \frac{\partial w_j}{\partial \mu} \cdot \frac{\partial \mu}{\partial \theta} \tag{2.2.36}$$

where γ is the learning rate of the ASN. Obviously, all the partial derivatives included in (2.2.36) can be calculated in every node by using the equivalent equations, as we move forward. So from (2.2.34) we have:

$$\frac{\partial m_\alpha}{\partial w_j} = \frac{1}{2}(1+m_\alpha)(1-m_\alpha) \tag{2.2.37}$$

where j refers to every rule that suggests the output (action) 'α'. We can observe that the partial derivative of m_α as per w_j is not depended on w_j. Likewise from (2.2.32) we find:

$$\frac{\partial w_j}{\partial \mu_q} = \frac{e^{-\kappa\mu_q}[1+\kappa(w_j - \mu_q)]}{\sum_i e^{-\kappa\mu_i}} \tag{2.2.38}$$

Finally, $\partial\mu/\partial\theta$ for $\theta = c$, s_L, s_R, are shown in Table 2.2.1.

Table (2.2.1). Derivatives of the membership function μ with respect to its parameters

x	$\frac{\partial\mu}{\partial c}$	$\frac{\partial\mu}{\partial s_L}$	$\frac{\partial\mu}{\partial s_R}$
$[c, c+s_R]$	$1/s_R$	0	$(x-c)/s_R^2$
$[c-s_L, c]$	$-1/s_L$	$(x-c)/s_L^2$	0
otherwise	0	0	0

By inserting (2.2.37), (2.2.38) and the derivatives of Table 2.2.1 in (2.2.36), we calculate in every step the necessary updates (corrections) of the free parameters of the ASN.

Action Evaluation Network (AEN)

The AEN, which is the Adaptive Critic Element (ACE) of the system, foresees the reinforcing signals which correspond to the various input states. The only information that the AEN receives, is the state of the system under control, that is the values of state variables, and if there has been a failure. AEN is a typical 2-layer NN with n_h hidden nodes and $n+1$ input nodes $x_0, x_1, \ldots, x_n$, where x_0 is the threshold input. Every hidden node accepts $n+1$ inputs and $n+1$ weights, while every output node accepts $n+n_h+1$ inputs and has $n+n_h+1$ weights. The structure of the neural evaluation network has the form shown in Fig. 2.2.7.

The output v of the AEN is suitably degraded and is combined with the exterior reinforcing signal (failure signal) r to give an inner reinforcing signal r'. Specifically, the inner reinforcing signal r' is calculated as follows:

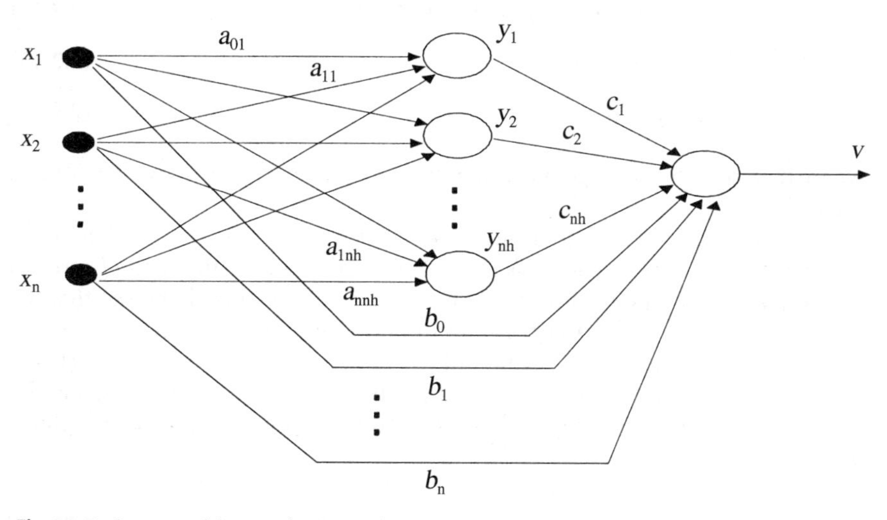

Fig. 2.2.7. Structure of the neural action evaluation network

$$r'(t+1)=\begin{cases}0 & \text{initial state}\\ r(t+1)-v(t,t) & \text{failure state}\\ r(t+1)-\alpha v(t,t+1)-v(t,t) & otherwise\end{cases} \qquad (2.2.39)$$

where $\alpha\,(0\leq\alpha\leq 1)$ is the degradation coefficient.

Equation (2.2.40) shows that the value v which is produced one time step later, is given less weight compared to the current value of v. The outputs of the hidden nodes are:

$$y_i(t,t+1)=f_h\left(\sum_{j=1}^{n}a_{ij}(t)x_j(t+1)\right) \qquad (2.2.40)$$

where $f_h(s) = 1/(1 + e^{-s})$ and t, $t + 1$ are successive time steps. The AEN output node accepts inputs both from the hidden nodes and directly from the input nodes $(x_1, x_2, \ldots, x_n)$.

So:

$$v(t,t+1)=\sum_{i=1}^{n}b_i(t)x_i(t+1)+\sum_{i=1}^{n_h}c_i(t)y_i(t,t+1) \qquad (2.2.41)$$

The double argument of time (t, $t + 1$) is used to avoid instability in the weight updating.

Learning occurs with the algorithm of Adaptive Heuristic Critic (AHC) for the output node, and with the algorithm of back propagation for the hidden nodes.

Specifically, the learning rules (updating) of weights a_{ij}, b_I and c_I are the following:

$$a_{ij}(t+1)=a_{ij}(t)+\beta_a r'(t+1)y_i(t,t)(1-y_i(t,t))\operatorname{sgn}[c_i(t)]x_j(t) \qquad (2.2.42)$$

$$b_i(t+1)=b_i(t)+\beta_b r'(t+1)x_i(t) \qquad (2.2.43)$$

$$c_i(t+1)=c_i(t)+\beta_c r'(t+1)y_i(t,t) \qquad (2.2.44)$$

where $\beta_\alpha > 0, \beta_b > 0, \beta_c = \beta_b > 0$ and indicators i, j vary, as shown in (2.2.40) and (2.2.41).

We note that because there is no direct measurement (that means there is no knowledge of the correct action), $r'(t + 1)$ is used as a measurement of error in the output weight update. If $r'(t + 1)$ is positive, the weights change so that the output v is increased for a positive input and vice versa. Here we use the sign (sgn(.)) of the weights $c_i(t)$ of the hidden nodes and not their values. This is done because it has been proven both by experience and experiments, that the learning algorithm is stronger when the sign instead of value of the weight, is used.

2.2.1.3 Fuzzy ART and ARTMAP

Fuzzy ART is the fuzzy extension of the basic adaptive coordination neural network, for the case of fuzzy input vector. Fuzzy ART is a non-supervised learning algorithm. Fuzzy ARTMAP is an extension of fuzzy ART, which operates with supervised learning. Generally, the ART networks (ART1, ART2 and ART3) include properties of production systems (based-on-rules). Fuzzy ARTMAP gives an accurate mathematical realization of the ART concept, which is computationally very powerful and overcomes in efficiency, many neural, genetic or techniques based on knowledge.

A. Fuzzy ART system

The fuzzy ART system includes a field (set) of nodes called F_0, which represent the current input vector, and a field F_1 which receives inputs from both F_0 and F_2, which represents the active code or category. The action vector of F_0 is symbolized by:

$\boldsymbol{I} = [I_1, \ldots, I_M]$, $I_i \in [0,1]$ for $i = 1, 2, \ldots, M$. The action vector of F_1 is symbolized by $\boldsymbol{x} = [x_1, x_2, \ldots, x_M]$ and the action vector of F_2 is symbolized by $\boldsymbol{y} = [y_1, y_2, \ldots, y_N]$. The number of nodes in every field is arbitrary.

To every node j of F_2 ($j = 1, 2, \ldots, N$) corresponds an adaptive weight vector $v_j = [v_{j1}, \ldots, v_{jM}]$, whose initial value is chosen as:

$$\boldsymbol{v}_j(0) = [v_{j1}(0), v_{j2}(0), \ldots, v_{jM}(0)] = [1, 1, \ldots, 1] \tag{2.2.45}$$

Initially every category is called 'non-binded'. After the selection of a category for codification (sorting), that category is 'put aside'. As we prove next, every weight v_{ji} does not increase with time and consequently it converges towards a certain limit. The weight v_{ji} of fuzzy ART encloses both the 'upper-to-lower' and 'lower-to-upper' weights of ART1. The parameters of the system are the learning rate $\gamma \in [0, 1]$, the safety threshold $\rho \in [0, 1]$ and the choice parameter $a > 0$.

For every input I and node j of F_2 we define a selection function S_j as:

$$S_j(\boldsymbol{I}) = \frac{|\boldsymbol{I} \wedge \boldsymbol{v}_j|}{a + |\boldsymbol{v}_j|} \tag{2.2.46}$$

where the operator '$\wedge$' is the AND operator of fuzzy sets:

$(p_i \wedge q_i) = \min(p_i, q_i)$ and the metric $|\cdot|$ is defined as:

$$|\boldsymbol{p}| = \sum_{i=1}^{M} |p_i|$$

We say that the system makes a category selection when at most one node of F_2 can be triggered at a certain time. The chosen category receives the index J:

$$S_J(\boldsymbol{I}) = \max\{S_J(\boldsymbol{I}), j = 1, 2, \ldots, N\}$$

If more then one S_J are maximum, the category j with the smallest index, is chosen. Specifically, the nodes are chosen in order $j = 1, 2, 3, \ldots$. When category J is chosen, then $y_J = 1$ and $y_J = 0$ for $j \neq J$. In a selection system the action vector $\boldsymbol{x}$ of F_1 is defined as:

$$\boldsymbol{x} = \begin{cases} \boldsymbol{I}, & \text{if } F_2 \text{ is inactive} \\ \boldsymbol{I} \wedge \boldsymbol{v}_j, & \text{if the } J \text{ node of } F_2 \text{ has been chosen} \end{cases} \tag{2.2.47}$$

If the winner node J satisfies the threshold criterion

$$\frac{|\boldsymbol{I} \wedge \boldsymbol{v}_j|}{|\boldsymbol{I}|} \geq \rho \tag{2.2.48}$$

then we have resonance. That is, when category J is chosen according to (2.2.46) we have resonance when:

$$|\boldsymbol{x}| = |\boldsymbol{I} \wedge \boldsymbol{v}_j| \geq \rho |\boldsymbol{I}| \tag{2.2.49}$$

and the sorting is completed. If we don't have resonance, that is if $\frac{|\boldsymbol{I} \wedge \boldsymbol{v}_j|}{|\boldsymbol{I}|} < \rho$ which means:

$$|\boldsymbol{x}| = |\boldsymbol{I} \wedge \boldsymbol{v}_j| < \rho |\boldsymbol{I}| \tag{2.2.50}$$

then, the best adapting sample has not yet been found and the system proceeds to find a new category that satisfies the criterion (2.2.47), by making S_j zero, that is setting $S_J(I) = 0$.

When the search is over, the weight $\boldsymbol{v}_j$ is updated by the following rule:

$$\boldsymbol{v}_J^{\text{new}} = \gamma(\boldsymbol{I} \wedge \boldsymbol{v}_J^{\text{old}}) + (1-\gamma)\boldsymbol{v}_J^{\text{old}} \tag{2.2.51}$$

where ρ is the learning rate. The algorithm (2.2.50) can be depicted and realized with a neural network. Fast learning corresponds to $\gamma = 1$.

For an efficient sorting of input patterns that is disturbed by noise, it is useful to set $\gamma = 1$ when node J is not selected and to choose $\gamma < 1$ after it is selected. Then we have:

$$\boldsymbol{v}_J^{\text{new}} = \boldsymbol{I} \tag{2.2.52}$$

when the node J becomes active for the first time.

The system uses complement code, which is a normalization rule that maintains the information width. The complement code represents both the ON response and the OFF response in an input vector $\boldsymbol{\alpha}$ (Fig. 2.2.8)

We assume that $\boldsymbol{\alpha}$ on its own depicts the ON response, so its complement $\alpha^c(a_i^c = 1 - a_i)$, depicts the OFF response. The input $\boldsymbol{I}$ of the F_1 field expressed in complement code, is:

$$\boldsymbol{I} = [\boldsymbol{a}, \boldsymbol{a}^c] = [a_1, \ldots, a_M, a_1^c, \ldots a_M^c] \tag{2.2.53}$$

so we have:

$$|\boldsymbol{I}| = |[\boldsymbol{a}, \boldsymbol{a}^c]| = \sum_{i=1}^{M} a_i + M - \sum_{i=1}^{M} a_i = M$$

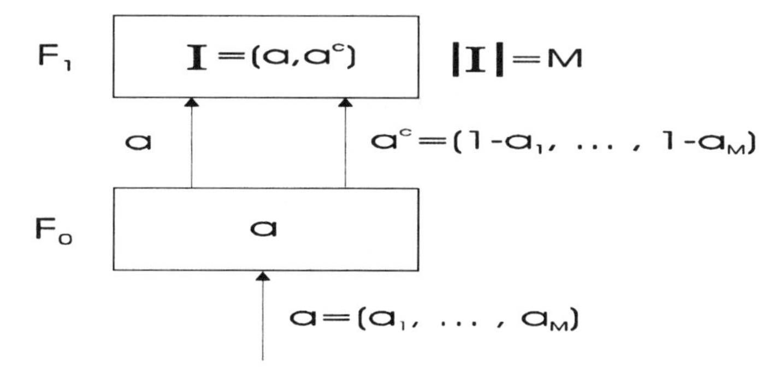

Fig. 2.2.8. Complement code. It uses both the input vector $\boldsymbol{\alpha}$ (ON) and its complement $\boldsymbol{\alpha}^c$ (OFF), to normalize the input patterns

This means that the vectors expressed in complement code, are automatically normalized. When this codification is used, the initial weight conditions (2.2.45) become:

$$v_{j1}(0) = \ldots = v_{j,2M}(0) = 1 \tag{2.2.54}$$

We note that the system ART1 can be trained to sort binary patterns only, while fuzzy ART can sort both binary and analog patterns. Comparing (2.2.46) with (2.2.48), we see that this capability is achieved by substituting the binary intersection II with the fuzzy intersection "Λ" = "min".

It is, of course known, that the min operator (fuzzy AND) is diverted to the binary operation AND, when the input patterns are binary. It is also noted that as the parameter α approaches 0, the selection function S_j represents the rate by which the weight v_j is a fuzzy set. A summary of comparison between ART1 and ART, is shown in Fig. 2.2.9.

ART1 (binary)	**ART (analog)**
Category selection function	
$S_j(\mathbf{x}) = \dfrac{\lvert \mathbf{x} \cap \mathbf{v}_j \rvert}{a + \lvert \mathbf{v}_j \rvert}$	$S_j(\mathbf{x}) = \dfrac{\lvert \mathbf{x} \wedge \mathbf{v}_j \rvert}{a + \lvert \mathbf{v}_j \rvert}$
Adaption/threshold criterion	
$S_j(\mathbf{x}) = \dfrac{\lvert \mathbf{x} \cap \mathbf{v}_j \rvert}{a + \lvert \mathbf{v}_j \rvert} \geq \rho$	$S_j(\mathbf{x}) = \dfrac{\lvert \mathbf{x} \wedge \mathbf{v}_j \rvert}{a + \lvert \mathbf{v}_j \rvert} \geq \rho$
Fast learning (γ=1)	
$\mathbf{v}_J^{new} = \mathbf{x} \cap \mathbf{v}_J^{old}$	$\mathbf{v}_J^{new} = \mathbf{x} \wedge \mathbf{v}_J^{old}$

Fig. 2.2.9. Comparison of algorithms ART1 and fuzzy ART

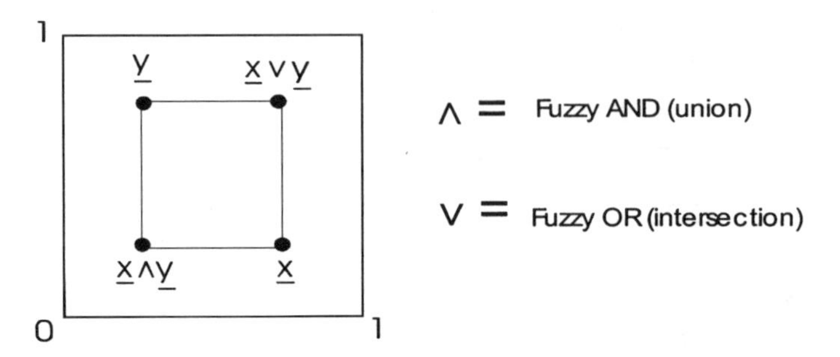

Fig. 2.2.10. The functions of fuzzy AND and fuzzy OR, produce hyper-rectangles of categories

Due to the normalization of the input patterns, the categories that are formed from fuzzy ART are hyper-rectangular, as shown in Fig. 2.2.10, in the case of two dimensions (2-D), where the values "min" and "max" determine the acceptable region of changes in every dimension.

$$+$$

$$[\boldsymbol{x}=(x_1,x_2), y=(y_1,y_2), (\boldsymbol{x}\wedge \boldsymbol{y})_1=\min(x_1,y_1), (\boldsymbol{x}\wedge \boldsymbol{y})_2=\min(x_2,y_2), (\boldsymbol{x}\vee \boldsymbol{y})_1= \\ =\max(x_1,y_1), (\boldsymbol{x}\vee \boldsymbol{y})_2=\max(x_2,y_2)]$$

A. Fuzzy ARTMAP system

The fuzzy model ARTMAP is made up of two fuzzy components ART (ARTa and ARTb), as shown in Fig. 2.2.11.

During supervised learning, ARTa receives a wave of input patterns (vectors) $\{\boldsymbol{a}^{(p)}\}$, while ARTb receives a wave of input patterns $\{\boldsymbol{b}^{(p)}\}$, where $\boldsymbol{b}^{(p)}$ is the correct expectation, given $\boldsymbol{a}^{(p)}$. ARTa and ARTb are interconnected by a network of relational learning and an internal regulator which ensures autonomous operation in real time. The regulator (controller) creates the minimum number of categories of ARTa, that is the minimal number of hidden nodes that are needed to achieve the desired precision. This is achieved with the use of a *minimax* learning algorithm, which minimizes the expectation error and maximizes the generalization of the category, simultaneously. Thus a "trial and error" process leads to the prediction of the size of the category, using only local operations. It works by increasing the threshold parameter ρ_α of the ARTa, with the minimal value needed in order to correct a prediction error in ARTb. The parameter ρ_α regulates the minimum certainty that ARTa should have in a category or assumption that is activated by input $\boldsymbol{a}^{(p)}$ in order to accept it (select it) and quit searching to find a better one through the automatic control process. As in ART1, small values of ρ_α lead to the creation of more general categories, that also lead to greater generalization and higher coding compression. A prediction failure in ARTb, increases the minimum value of ρ_α with the smaller quantity that is needed to put in operation the control process in ARTa through a "match tracking mechanism" (Fig. 2.2.10). The adaptation tracking minimizes the minimal desired generalization, so that the error of prediction is corrected. Following

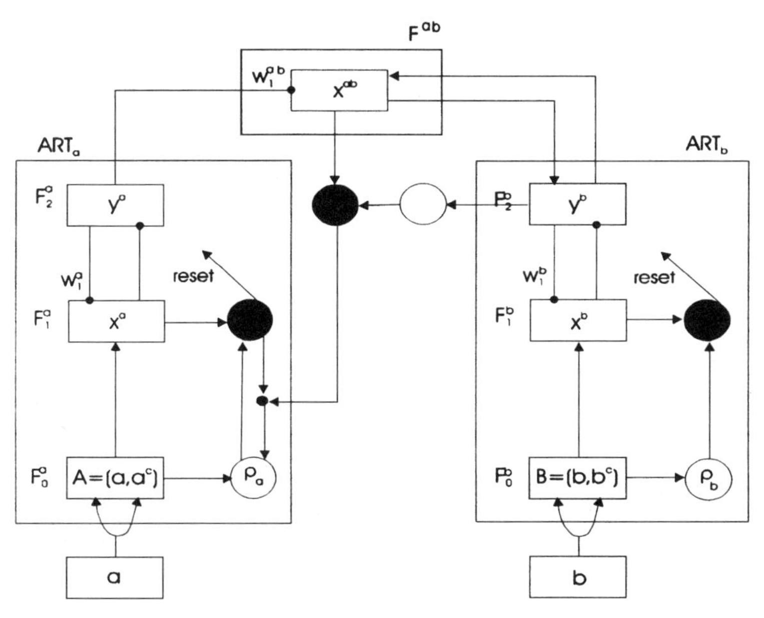

Fig. 2.2.11. Architecture of fuzzy ARTMAP system

this process new category is selected in ARTa, which focuses on a new set of properties of inputs $\boldsymbol{a}^{(p)}$ that can better predict $\boldsymbol{b}^{(p)}$. The combination of "match tracking" and "fast learning", allows the ARTMAP system to learn a different prediction for a rare event instead of being satisfied with a "cloud" of similar frequent events within which it is implanted.

As shown in Fig. 2.2.11 the ARTa and $\mathrm{ART_b}$ components are connected to the F^{ab} component which is called "field mapping" and it is used in shaping correlations of predictions between the categories and materializing the "match tracking algorithm", while the threshold parameter of ARTa increases, when a "predicted non-adaptation" occurs in ARTa. The interaction that is introduced by component F^{ab} is realized as follows.

Fuzzy ARTMAP Algorithm

ARTa and $\mathrm{ART_b}$: The inputs in ARTa and $\mathrm{ART_b}$ are depicted with complementary a coding.

For ARTa we have $\boldsymbol{I} = [\boldsymbol{a}, \boldsymbol{a}^c]$, while for $\mathrm{ART_b}$ it is $\boldsymbol{I} = \boldsymbol{B} = [\boldsymbol{b}, \boldsymbol{b}^c]$. The variables and the weights of ARTa, $\mathrm{ART_b}$ and F^{ab} are symbolized by superscripts a, b and ab respectively:

For ARTa:
$$\boldsymbol{x}^a = [x_1^a, \ldots, x_{2Ma}^a]$$
$$\boldsymbol{y}^a = [y_1^a, \ldots, y_{Na}^a]$$
$$\boldsymbol{v}_j^a = [v_{j1}^a, \ldots, v_{j2Ma}^a]$$

For ART$_b$: $\boldsymbol{x}^b = [x_1^b, \ldots, x_{Mb}^b]$

$\boldsymbol{y}^b = [y_1^b, \ldots, y_{Nb}^b]$

$\boldsymbol{v}_K^b = [v_{K1}^b, \ldots, v_{K2Mb}^b]$

For F^{ab}: $\boldsymbol{v}_j^{ab} = [v_{j1}^{ab}, \ldots, v_{jNb}^{ab}]$, where $\boldsymbol{v}_j^{ab}$ is the weight vector from node j of F_2^a to F^{ab}.

Initial conditions
In all vectors $\boldsymbol{x}^a, \boldsymbol{y}^a, \boldsymbol{x}^b, \boldsymbol{y}^b$ and $\boldsymbol{x}^{ab}$ an initial value of zero is given.

Activation of Field mapping
F^{ab} is activated whenever one of the categories ARTa or ART$_b$ is allowed. If node J of F_2^a is selected then the weights $\boldsymbol{v}_j^{ab}$ activate F^{ab}. If node K of F_2^a is activated, then node K of F^{ab} is activated via a connection between F_2^a and F^{ab}. If ARTa and ART$_b$ (both) are activated, then F^{ab} becomes active only if ARTa predicts the same category as ART$_b$ through the weights $\boldsymbol{v}_j^{ab}$. The output vector $\boldsymbol{x}^{ab}$ of F^{ab} is given by:

$$\boldsymbol{x}^{ab} = \begin{cases} \boldsymbol{y}^b \wedge \boldsymbol{v}_J^{ab} & \text{if node } J \text{ of } F_2^a \text{ is active and } F_2^b \text{ is active} \\ \boldsymbol{v}_J^{ab} & \text{if node } J \text{ of } F_2^a \text{ is active and } F_2^b \text{ is not active} \\ \boldsymbol{y}^b & \text{if } F_2^a \text{ is not active and } F_2^b \text{ is active} \\ 0 & \text{if } F_2^a \text{ and } F_2^b \text{ are not active} \end{cases} \quad (2.2.55)$$

From (2.2.55) we can see that when $\boldsymbol{v}_j^{ab}$ is not confirmed by $\boldsymbol{y}^{b,}$ we have $\boldsymbol{x}^{ab} = 0$. Such a non-adaptive event activates a new search for a better category in ARTa, as follows.

Match tracking
In the beginning of every input pattern presentation, the threshold parameter ρ_α of ARTa is equal to a base value $\bar{\rho}_\alpha$. The threshold parameter of field mapping F^{ab} is ρ_{ab}. If

$$|\boldsymbol{x}^{\alpha b}| < \rho_{ab} |\boldsymbol{y}^b| \quad (2.2.56)$$

then ρ_{ab} increases until it slightly exceeds $|A \wedge \boldsymbol{v}_J^a| \cdot |A|^{-1}$, where A is the input to F_1^a in the form of complementary code. Then:

$$|\boldsymbol{x}^\alpha| = |A \wedge \boldsymbol{v}_j^a| < \rho_a |A| \quad (2.2.57)$$

where J is the index of the active node F_2^a as in (2.50). When this happens the ARTa survey results either in the activation of another node J of F_2^a with

$$|\boldsymbol{x}^\alpha| = |A \wedge \boldsymbol{v}_j^a| \geq \rho_a |A| \text{ and } |\boldsymbol{x}^{\alpha b}| = |\boldsymbol{y}^b \wedge \boldsymbol{v}_j^{ab}| \geq \rho_{ab} |\boldsymbol{y}^b| \quad (2.2.58)$$

or if such a node does not exist, it results in the closing of F_2^a for the rest of the input pattern presentations.

Learning of field mapping
The weights v_{jk}^{ab} of the connections $F_2^a \rightarrow F^{ab}$ are initially chosen as:

$$v_{jk}^{ab}(0) = 1 \tag{2.2.59}$$

During "resonance", with the category J of ARTa being active, $\boldsymbol{v}_j^{ab}$ approaches the field mapping vector $\boldsymbol{x}^{ab}$. With fast learning, when node J learns the way to find the category K of ART_b, this correlation is permanent, that is we have always $v_{jk}^{ab} = 1$.

2.2.1.4 The System ANFIS as a Universal Approximator

In this section, we study the interesting property of the ANFIS system by which we are able to approximate with as much accuracy as we want, any arbitrary function. Specifically, we will prove that if the number of rules is not restricted, a Takagi-Sugeno model of zero rank has infinite approximation power of an arbitrary non-linear function in a compact set. The proof will be given with the help of the Stone-Weirstarss theorem, which is:

Stone-Weirstarss theorem
Let there be a compact field D of the N-dimensional space and F a set of continuous real functions (*) f on D, which satisfies the following criteria:

1. ***Unitary function:*** The constant $f(x) = 1$ belongs to F.
2. ***Separability:*** For any two points $x_1 \neq x_2$ in D, there is a $f \in F$ such that $f(x_1) \neq f(x_2)$.
   * We call real functions, the functions that take on real values.
3. ***Algebraic closure:*** If f and g are any two functions in F, then the functions fg and $\alpha f + bg$ belong to F, for any real numbers a and b.

Then the set of functions F is dense in the set $C(D)$ of the continuous real functions on D. That is to say, for each $\varepsilon > 0$ and each function $g \in C(D)$ there is a function f in F such that:

$$\begin{aligned} &|g(x) - f(x)| < \varepsilon \\ &\text{for all } x \in D \end{aligned} \tag{2.2.60}$$

We will now check if the three criteria above, are satisfied from the model Takagi-Sugeno. At first, in fuzzy reasoning systems, the field D in which we work is almost always compact, because it is well known in the analysis of real functions that each closed and bounded set in R^N is compact.

Unitary function
Our fuzzy system must be able to calculate the unitary function $f(x) = 1$. An obvious way of calculating this function is by setting the result of each rule equal to 1. A fuzzy system with only one rule is enough to satisfy this criterion.

Separability
Our fuzzy system must be able to calculate functions which take different values at different points. This can be achieved by any fuzzy reasoning system, with a suitable selection of its parameters.

Algebraic closure (Addition)

Our fuzzy systems must not alter by addition and multiplication. Let's assume we have two fuzzy Takagi-Sugeno systems S and $\hat{S}$ with two rules each. Their outputs y and $\hat{y}$ are:

$$S: y = \frac{w_1 f_1 + w_2 f_2}{w_1 + w_2}$$

$$\hat{S}: \hat{y} = \frac{\hat{w}_1 \hat{f}_1 + \hat{w}_2 \hat{f}_2}{\hat{w}_1 + \hat{w}_2} \tag{2.2.61}$$

Then the sum $ay + b\hat{y}$ is equal to:

$$ay + b\hat{y} = a\frac{w_1 f_1 + w_2 f_2}{w_1 + w_2} + b\frac{\hat{w}_1 \hat{f}_1 + \hat{w}_2 \hat{f}_2}{\hat{w}_1 + \hat{w}_2} \tag{2.2.62}$$

$$= \frac{w_1\hat{w}_1(af_1 + b\hat{f}_1) + w_1\hat{w}_2(af_1 + b\hat{f}_2) + w_2\hat{w}_1(af_2 + b\hat{f}_1) + w_2\hat{w}_2(af_2 + b\hat{f}_2)}{w_1\hat{w}_1 + w_1\hat{w}_2 + w_2\hat{w}_1 + w_2\hat{w}_2}$$

Consequently, we can put together a Takagi-Sugeno system based on four rules, which calculates the $ay + b\hat{y}$. The triggering force and the output of each rule are defined as $w_i\hat{w}_j$ and $af_i + b\hat{f}_j$ $(i, j = 1, 2)$, respectively.

Algebraic Closure (Multiplication)

The multiplication of y and $\hat{y}$ is equal to:

$$y\hat{y} = \frac{w_1\hat{w}_1 f_1\hat{f}_1 + w_1\hat{w}_2 f_1\hat{f}_2 + w_2\hat{w}_1 f_2\hat{f}_1 + w_2\hat{w}_2 f_2\hat{f}_2}{w_1\hat{w}_1 + w_1\hat{w}_2 + w_2\hat{w}_1 + w_2\hat{w}_2} \tag{2.2.63}$$

We therefore see, that we can put together a Takagi-Sugeno fuzzy system of four rules which can calculate the $y\hat{y}$, where $w_i\hat{w}_j$ and $f_i\hat{f}_j$ $(i, j = 1, 2)$ is the triggering signal and the output of every rule, respectively.

From the above analysis, we can see that the ANFIS/Takagi-Sugeno systems that calculate the $ay + b\hat{y}$ and $y\hat{y}$ belong to the same group as the S and $\hat{S}$ systems, only when the membership functions that are used are "invariable through multiplication". A category of such membership functions, is the Gaussian:

$$\mu_{A_i}(x) = a_i \exp\left[-\left(\frac{x - m_i}{\sigma_i}\right)^2\right]$$

Another category of membership functions that is inveriable through multiplication is the membership functions of cut sets, which takes values 0 or 1.

Due to the fact that with the above choice of membership functions the Takagi-Sugeno model of zero degree satisfies the four points of the Stone-Weierstrass theorem, the theorem is valid and we have:

"For each point $\varepsilon > 0$ and any function g that takes on real values, there is a Takagi-Sugeno model S of zero degree with Gaussian membership functions such that:

$$|g(x) - S(x)| < \varepsilon \tag{2.2.64}$$

for all $\mathbf{x}$ *in the related compact set D".*

The above result regarding function universal approximation also applies to the ANFIS with rules of Mamdani type and Gaussian membership functions (section 2.2.1B). Thus we have the following theorem:

"For each point $\varepsilon > 0$ and any function g that takes on real values in a compact set $D \subset R^N$, there is a Mamdani fuzzy model with Gaussian membership functions (refer (2.2.12) and (2.2.13)):

$$f(\mathbf{x}) = \frac{\sum_{j=1}^{M} \bar{y}_j \left\{ \prod_{i=1}^{N} a_i^j \exp\left[-\left(\frac{x - m_i^j}{\sigma_i^j} \right)^2 \right] \right\}}{\sum_{j=1}^{M} \left\{ \prod_{i=1}^{N} a_i^j \exp\left[-\left(\frac{x - m_i^j}{\sigma_i^j} \right)^2 \right] \right\}} \tag{2.2.65}$$

such that:

$$| g(\mathbf{x}) - f(\mathbf{x}) | < \varepsilon \tag{2.2.66}$$

for all $\mathbf{x}$ in the compact set D".

2.2.2 Examples [12–18]

Example 2.2.2.1 (Modeling of a 2-dimensional non linear sine function)
Let us study the application of an ANFIS/Takagi-Sugeno system for the modeling (approach) of two input function of x, y:

$$z = \frac{\sin(x)\sin(y)}{xy} \tag{2.2.67}$$

with uniformly distributed input data in the 2-D region $[-10, 10] \times [-10, 10]$.

Solution
The application was done in the computer (MATLAB) [1] with 121 pairs of input data. The ANFIS system that was used, has 16 rules with four membership functions for each input variable. The model has 72 parameters in total to choose from (24 parameters of the left hand side of the rules and 48 of the linear right hand side of the rules).

Figure 2.2.12 compares the RMSE error (root mean square error) that results from the ANFIS system to the error that is given by a multilayer Perceptron network (2-18-1) with the back propagation algorithm. Each curve of the figure is the average value of ten curves that resulted from 10 runs of each algorithm. For the Perceptron, these ten runs began from different sets of random initial weights. For the ANFIS, the runs correspond to 10 κ values in the region [0.01, 0.10], where κ is the step size.

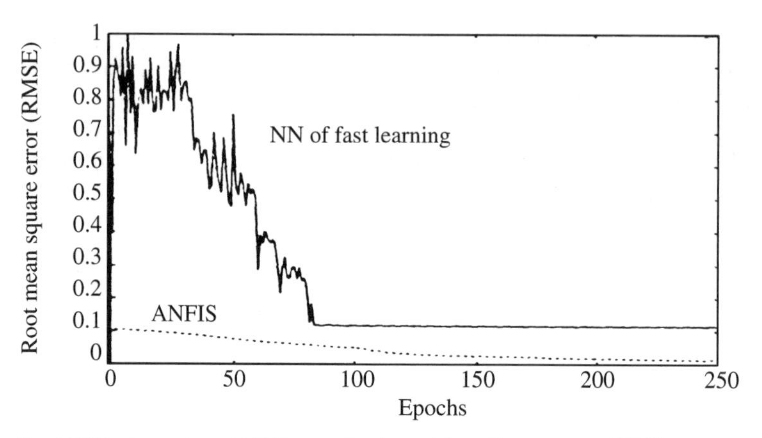

Fig. 2.2.12. Comparison of RMSE curves for the multilayer Perceptron and ANFIS

From Fig. 2.2.12 we can observe that the ANFIS system performance is a lot better than the back propagation algorithm, something which is possibly owed to the trapping of the NN in a local minimum.

Figure 2.2.13 shows the training data and the surface that were reconstructed in different epochs with the ANFIS system. Finally, Fig. 2.2.14 shows the initial and final membership functions of x and y, from which we can see the steep changes of the surface of training data around the origin, have as a result the shifting of the membership functions towards the origin (theoretically the membership functions of x and y must be symmetrical as per the origin).

Example 2.2.2.2 (Modeling of a non linear function of three inputs)
The ANFIS/Takagi-Sugeno method will be implemented here for approximating the following non linear function of three inputs (x, y, z) with homogenous data in the area $[1,6] \times [1,6] \times [1,6]$.

$$\text{Output } f = (1 + x^{0,5} + y^{-1} + z^{-1,5})^2 \tag{2.2.68}$$

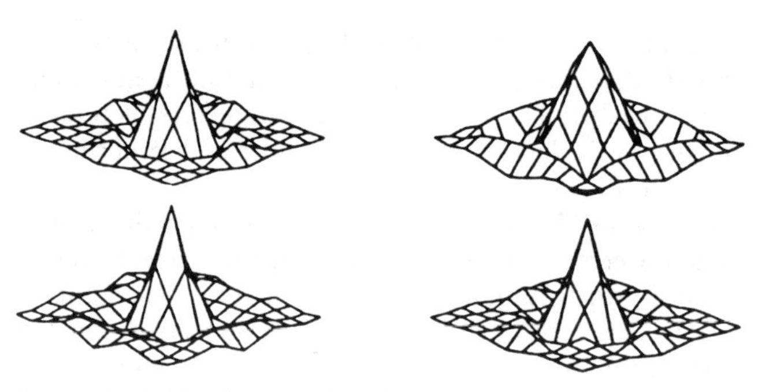

Fig. 2.2.13. Training data (top left) and surfaces that result after 1 (top right), 100 (bottom left) and 250 epochs (bottom right)

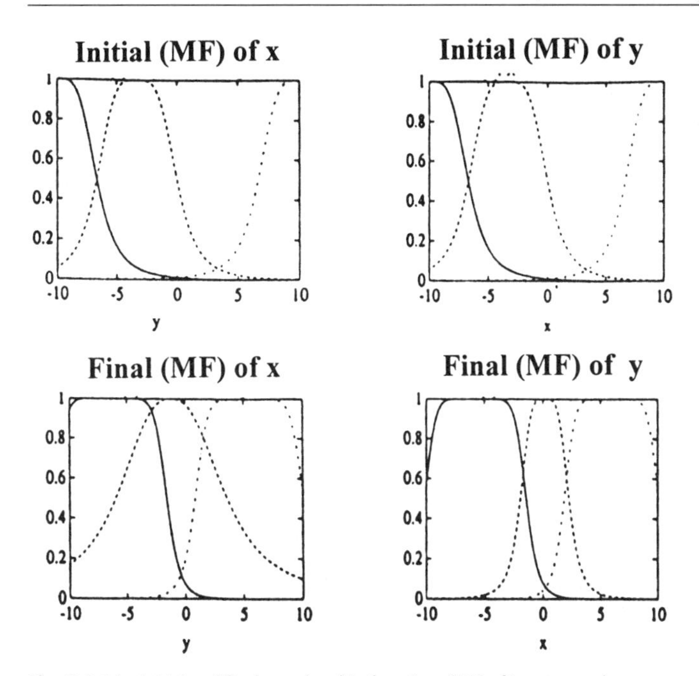

Fig. 2.2.14. Initial and final membership functions (MF) of inputs *x* and *y*

Solution
This application was implemented in MATLAB [1] using the topology of Fig. 2.2.15, where we have eight rules with two membership functions for each input variable. We used 216 training pairs in the above region and for evaluation purposes, 125 verification pairs were used in the region $[1.5, 1.5] \times [1.5, 1.5] \times [1.5, 1.5]$.

As index of the quality of the modeling, the average percentage error (APE) $E_{\text{ME}\Sigma}$ was used:

$$E_{\text{ME}\Sigma} = \frac{1}{N}\sum_{i=1}^{N}\frac{|f_d(i)-\hat{f}(i)|}{|f_d(i)|}100\% \tag{2.2.69}$$

where N is the number of pairs and $f_d(i)$, $\hat{f}(i)$ is the i target output and the predicted output, respectively.

Figure 2.2.16 shows the membership functions before and after the training and the error curves $E_{\text{ME}\Sigma}$ for different step sizes ($k = 0.01$ until 0.09) are given in Fig. 2.2.17, which shows that if the step size k is not very large, it does not have a big impact on the quality of the approach that is given by the ANFIS.

The training and verification error after 200 epochs is:

$E_{\text{ME}\Sigma}(\text{training}) = 0.043\%$ and $E_{\text{ME}\Sigma}(\text{verification}) = 1.066\%$

which are much smaller than the equivalent errors that arise with other "pure fuzzy" models. For example, in such a case the errors that we got were:

$E_{\text{ME}\Sigma}(\text{training}) = 1.5\%$ and $E_{\text{ME}\Sigma}(\text{verification})=2.1\%$

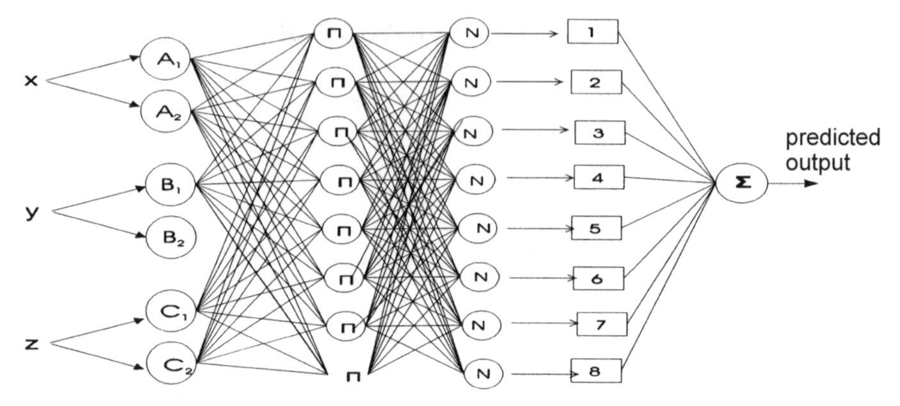

Fig. 2.2.15. ANFIS model/Takagi-Sugeno for function (2.2.68) (the connections from the inputs to layer 4 were not included)

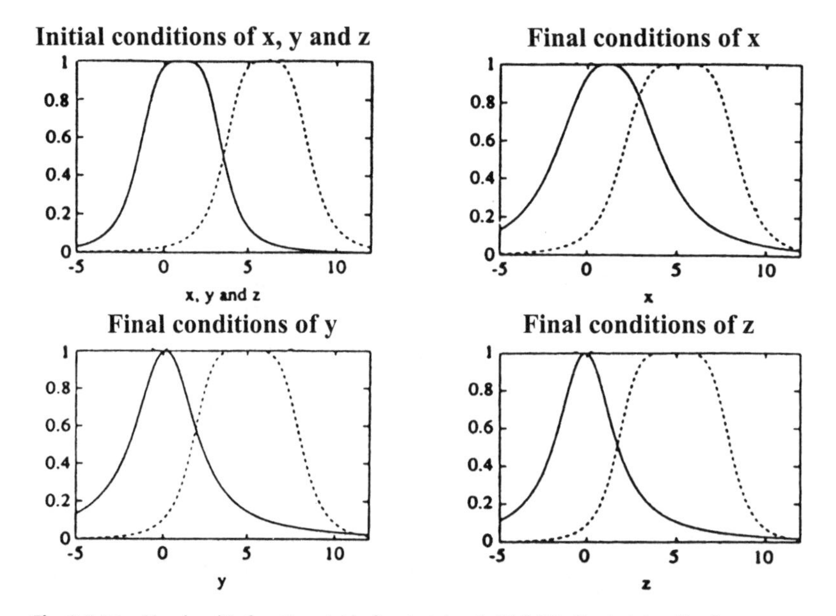

Fig. 2.2.16. Membership functions (a) before training (initial) (b) after training (final)

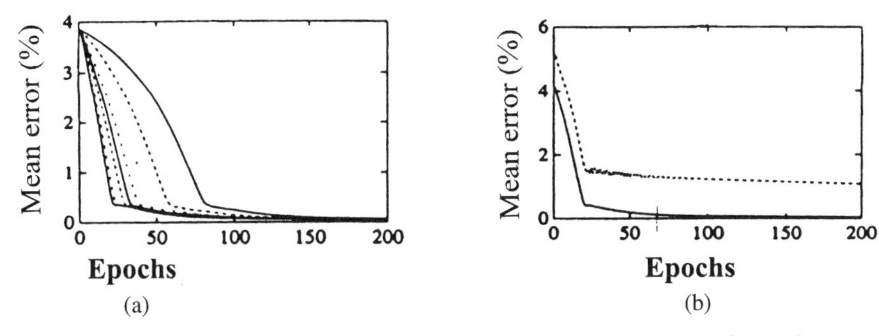

Fig. 2.2.17. Curves of "mean percentage error" $E_{ME\Sigma}$. (a) Nine error curves of training for initial step size from 0.01 (far right curve) upto 0.09 (far left curve); (b) Error curve of training (continuous line) and verification (dotted line) for the initial step size $k = 0.1$

Example 2.2.2.3
(Application of the GARIC system in the Inverse pendulum balance problem) [1]
The GARIC system was applied to the balance problem of an inverse pendulum which is placed in a vehicle as shown in Fig. 2.2.18. Here, we will describe the basic elements of the solution through GARIC.

Solution
We define the following parameters of the "vehicle-pendulum":
$x, \dot{x}$ = horizontal position and velocity of the vehicle, respectively
$\theta, \dot{\theta}$ = angle and angular velocity and per the vertical line of the inverse pendulum.
f = the applied force to the vehicle

$$\ddot{x} = \frac{f + ml\,[\dot{\theta}^2 \sin\theta - \ddot{\theta}\cos\theta] - \mu_K \operatorname{sgn}(\dot{x})}{m_K + m} \tag{2.2.70a}$$

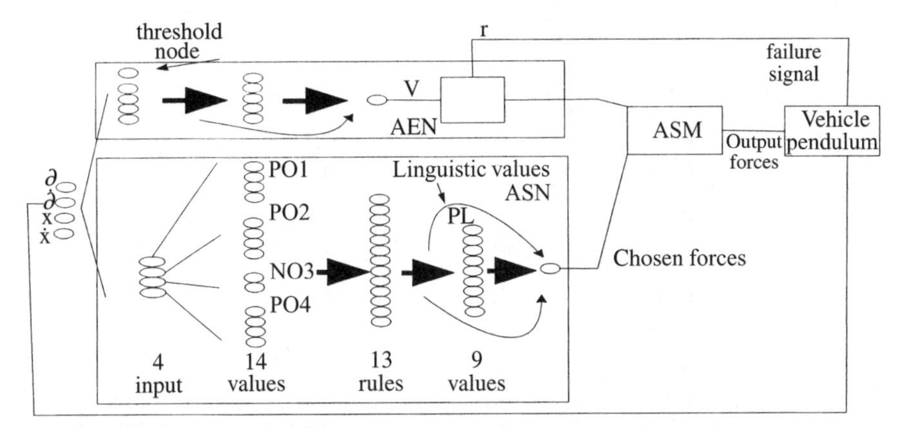

Fig. 2.2.18. GARIC System applied to an inverse pendulum on a moving platform (vehicle). Action variable, F. For every w_j, layer 4 performs calculations using the mean of local maxima, the fuzzification rule

$$\ddot{\theta} = \frac{g\sin\theta + \cos\theta\left[\frac{-f - ml\dot{\theta}^2\sin\theta + \mu_K\,\mathrm{sgn}(\dot{x})}{m_K + m}\right] - \frac{\mu_p\dot{\theta}}{ml}}{l\left[\frac{4}{3} - \frac{m\cos^2\theta}{m_K + m}\right]} \tag{2.2.70b}$$

where g = the acceleration due to gravity, m_K = the pendulum mass, $2l$ = the length of the pendulum, μ_K = the friction coefficient of the vehicle with the floor and μ_p = the coefficient of the pendulum with the vehicle.

The Eqs. (2.2.70a, b) were simulated by the Euler method and with a time step of Δt = 20 msec.

The training system knows only its state vector in each time step. Eqs. (2.2.70a, b) are not known (the system is a "black box" to the training mechanism). We consider there to be a failure when: $|\theta| > 12°$ and $|x| > 24\ m$ (even if the system practically learned and functioned successfully, even with smaller limits). The Action Evaluation Network (AEN) has four input nodes, one threshold node, five hidden nodes and one output node. The input vector is normalized so that the position of the vehicle and the pendulum are within the interval [0,1]. Similarly, the speeds are also normalized, without though being limited to a certain interval. The initial values of the 35 network weights are selected randomly in the interval [−0.1, +0.1] and the learning rate is γ = 0.3. The external signal (that is the failure signal) r, is received by AEN and is used in the calculation of the internal reinforcing signal $\hat{r}$, according to (2.2.39) with coefficient α = 0.9. The Action Selection Network (ASN), has four input nodes, 14 nodes in the second layer, 13 nodes in the third layer (that means 13 rules), 9 nodes in the fourth layer (that is 9 decisions) and one output node which calculates the force applied to the vehicle. The 13 rules with the 9 decisions/conclusions of the force are given in Table 2.2.2 and the initial membership function definitions (14 for the left members and 9 for the right members of the rules) are given in Table 2.2.3.

Table 2.2.2. The 13 rules of the ASN with the 9 values of the force

PO1	PO2	ZE	ZE	PL
PO1	ZE2	ZE	ZE	PM
PO1	NE2	ZE	ZE	ZE
ZE1	PO2	ZE	ZE	PS
ZE1	ZE2	ZE	ZE	ZE
ZE1	NE2	ZE	ZE	NS
NE1	PO2	ZE	ZE	ZE
NE1	ZE2	ZE	ZE	NM
NE1	NE2	ZE	ZE	NL
VS1	VS2	PO3	PO4	PS
VS1	VS2	PO3	PS4	PVS
VS1	VS2	NE3	NE4	NS
VS1	VS2	NE3	NS4	NVS

Table 2.2.3. The triangular membership functions (14 for the conditions and 9 for the conclusions)

Linguistic value	Center	Left open	Right open	Linguistic value	Center	Left open	Right open
PO1	0.3	0.3	−1	PL	20.0	5.0	−1.0
ZE1	0.0	0.3	0.3	PM	10.0	5.0	6.0
NE1	−0.3	−1	0.3	PS	5.0	4.0	5.0
VS1	0.0	0.05	0.05	PVS	1.0	1.0	1.0
PO2	1.0	1.0	−1.0	NL	−20.0	−1.0	5.0
ZE2	0.0	1.0	1.0	NM	−10.0	6.0	5.0
NE2	−1.0	−1.0	1.0	NS	−5.0	5.0	4.0
VS2	0.0	0.1	0.1	NVS	−1.0	1.0	1.0
PO3	0.5	0.5	−1.0	ZE	0.0	1.0	1.0
NE3	−0.5	−1.0	0.5				
PO4	1.0	1.0	−1.0				
NE4	−1.0	−1.0	1.0				
PS4	0.0	0.01	1.0				
NS4	0.0	1.0	0.01				

A spreal the size of −1.0 corresponds to ∞. The parameters of the membership functions define the initial weights of layers 2 and 4 of the action selection network. The state variables $x, \dot{x}, \theta, \dot{\theta}$ take on four fuzzy (linguistic) values:

PO: Positive, ZE: Zero, VS: Very Small and NE: Negative

Force f takes nine linguistic values:

PL (positive large), PM (Positive medium), PS (Positive small), PVS (Positive very small), NVS (negative very small), NS (Negative small), NM (Negative medium), NL (Negative large), ZE (Zero).

The above linguistic values are given in Fig. 2.2.19.

From the 13 rules, the nine are used for the vertical balance of the pendulum and the remaining four for the horizontal position of the vehicle.

Experimental results

In each experiment the system "vehicle-pendulum" began from an initial state and ended up either with the indication of "failure" or "successful control", after a long time period (in this case 100.000 steps or 33 mins of real time). After each failure, the initial state was altered arbitrarily or randomly.

Figure 2.2.20 shows the controller behavior throughout the training duration. It appears here that the membership functions move to the right position with the training.

In this experiment we moved the center (0) of the linguistic value ZE in −5 N and the system learned to restore it approximately to 0 after 322 trials. Similar results occured by changing the linguistic values of the other variables, something that shows the vigorous behavior of the GARIC controller.

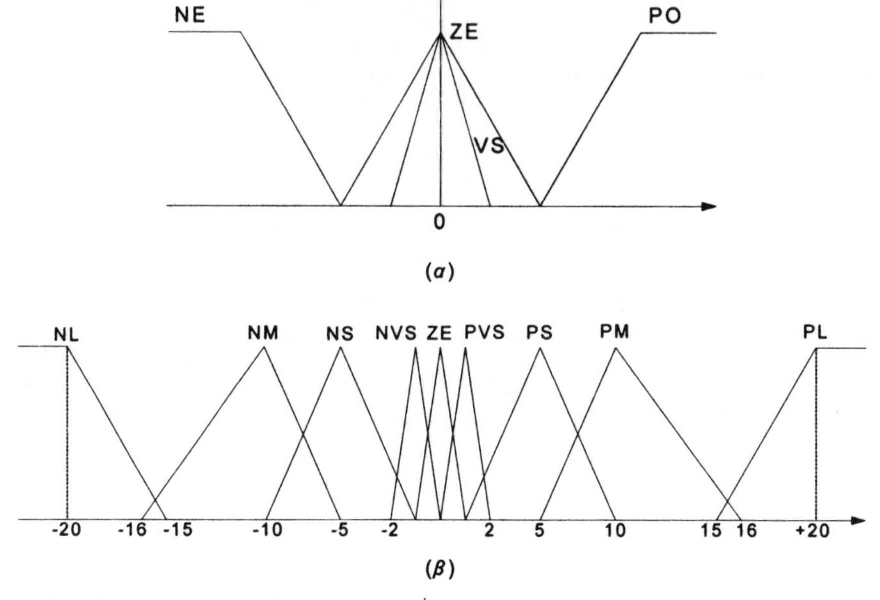

Fig. 2.2.19: (a) The four linguistic values of $x, \dot{x}, \theta, \dot{\theta}$; (b) The nine linguistic values of force f

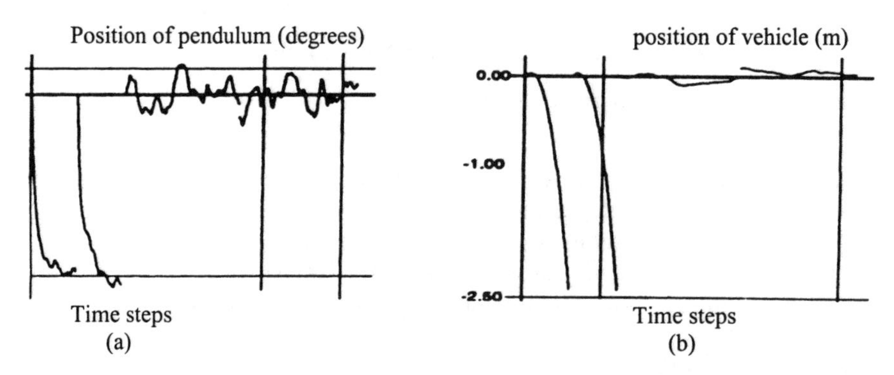

Fig. 2.2.20. Value ZE has moved to +5. The system has restored it to 0, after training with 322 trials. (a) position of pendulum, (b) position of vehicle

Example 2.2.2.4
("softmin" operator and defuzzification of "Local Mean of maxima")

The following two rules are given:

R_1: IF x is A_1 AND y is B_1 THEN z is C_1
R_2: IF x is A_2 AND y is B_2 THEN z is C_2

with membership functions A_1, A_2, B_1 and B_2:

$$\mu_{A_1}(x), \mu_{A_2}(x), \mu_{B_1}(y) \text{ and } \mu_{B_1}(y)$$

(a) If x and y have the particular values x_0 and y_0 it is asked to find the "activation forces" of the two rules based on the "softmin" operator that were proposed for the GARIC system [1].

(b) Also, it is asked to find the crisp value z^* that results wholly from the system of two rules by applying the defuzzification method of "Local Mean of Maxima" (LMOM) that was applied to the previous example of "vehicle/inverse pendulum".

Solution

The forces of activation of the two rules that result when the fuzzy sets "softmin" operator is applied, are (Fig. 2.2.21):

Rule R_1:

$$w_1 = \frac{\mu_{A_1}(x_0)e^{-k\mu_{A1}(x_0)} + \mu_{B_1}(y_0)e^{-k\mu_{B1}(y_0)}}{e^{-k\mu_{A1}(x_0)} + e^{-k\mu_{B1}(x_0)}}$$

Rule R_2:

$$w_2 = \frac{\mu_{A_2}(x_0)e^{-k\mu_{A2}(x_0)} + \mu_{B_2}(y_0)e^{-k\mu_{B2}(y_0)}}{e^{-k\mu_{A2}(x_0)} + e^{-k\mu_{B2}(y_0)}}$$

The values w_1 and w_2 of the above equations approach those that result from the "min" rule (the "hardmin") for $k \to \infty$, that is in the values:

$$w_{1k} = \min\{\mu_{A_1}(x_0), \mu_{B_1}(y_0)\}, w_{2h} = \min\{\mu_{A_2}(x_0), \mu_{A_2}(y_0)\}$$

b) If z_1 and z_2 are the output values (control/action) of the rules R_1 and R_2, we have:

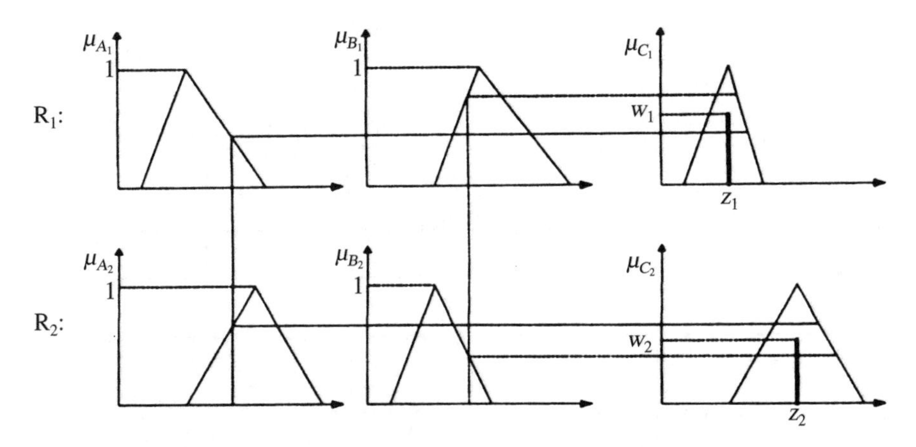

Fig. 2.2.21. fuzzy sets intersection with "softmin"

z_1 and z_2 depend on the defuzzyfication method, that is:

$$z_1 = \mu_{C_1}^{-1}(w_1) \text{ and } z_2 = \mu_{C2}^{-1}(w_2)$$

The resulting (crisp) value z^* of the two rules is given by the usual process of averaging using weights w_i $(i = 1, 2)$:

$$z^* = \frac{\sum_i w_i z_i}{\sum_i w_i} \tag{2.2.71}$$

The z_1 and z_2 that result from the defuzzification of "local mean of maxima" (LMOM) are clarified in Fig. 2.2.22a for the triangular membership functions.

The value $z = \mu_F^{-1}(w_r)$ that corresponds to w_r for a triangular membership function (Fig. 2.2.22a) with peak value c_F, left region from s_{F_a} and right region from s_{F_b}, is found as the average value of α and β. The α and β are found by applying the similarities of the triangles left of c_F and right of c_F.

$$\frac{(c_F - a)}{1 - w_r} = \frac{s_{F_a}}{1} \text{ and } \frac{(\beta - c_F)}{1 - w_r} = \frac{s_{F_b}}{1}$$

Solving as per α and β we find:

$$\begin{aligned} z = \mu_F^{-1}(w_r) &= \frac{\alpha + \beta}{2} \\ &= \frac{c_F - (1 - w_r)s_{F_a} + c_F + (1 - w_r)s_{F_b}}{2} \\ &= c_F + \frac{1}{2}(s_{F_b} - s_{F_a})(1 - w_r) \end{aligned} \tag{2.2.72}$$

If the membership function is monotonous (Fig. 2.2.22 b) then the $\mu_F^{-1}(w_r)$ is the typical mathematical inverse $z = \mu_F^{-1}(w_r)$, so $w_r = (1/c_F)z$ from which:

$$z = \mu_F^{-1}(w_r) = c_F w_r \tag{2.2.73}$$

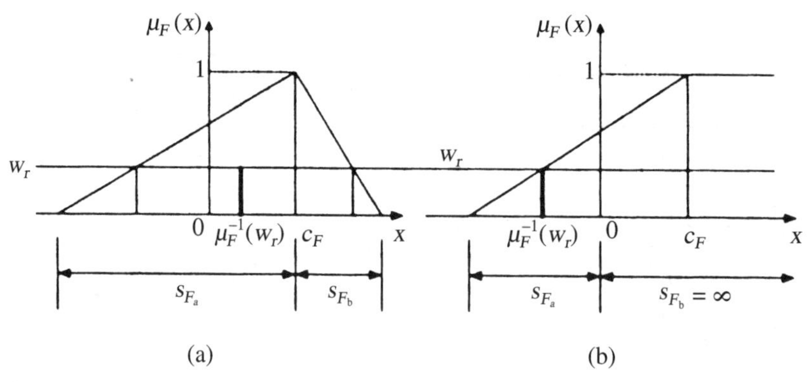

Fig. 2.2.22. Defuzzification LMOM

In the case where $w_r = 0$, we use the limit value given by (2.2.72) for $w_r = 0$:

$$z = \mu_F^{-1}(w_r = 0) = c_F + \frac{1}{2}(s_{F_b} - s_{F_a}) \tag{2.2.74}$$

Example 2.2.2.5
(Predicting Mackey-Glass Time Series with the Fuzzy ART System) [1]
The fuzzy ART system was applied for the prediction of the chaotic time series Mackey-Glass which is described by the equation:

$$\dot{x}(t) = \frac{0.2x(t-\tau)}{1+x^{10}(t-\tau)} - 0.1x(t) \tag{2.2.75}$$

where τ is a time delay. This time series is used as a classic example of evaluating prediction methods. Here, we will describe the application method and we will compare the results to another group of methods.

The fuzzy rules that were used are of the Takagi-Sugeno type, while their number was standard for every experiment. The free parameters that were specified as the weights of NN are: the parameters that define the hypercubes (categories) of the input variables (input weights), the gradients of the membership functions and the parameters of the right member (linear polynomials) of the rules. Two new attributes of the algorithm that were used are the following: breakage of the rule in two, with the worse local behavior (in difficult regions) and the addition of a new rule, when the output error exceeds a certain high limit (threshold).

In the beginning there are N non-committed rules, where each rule corresponds to a hypercube (category) and a fuzzy subset. As input vectors are presented, categories (hypercubes) are created and the corresponding nodes are committed. These categories are defuzzyfied in order to form the fuzzy conditions (fuzzy sets) of the left member of the rules. As a behavior indicator, the mean square error (MSE) was used where the error is: $e = y_d - y$(yd = desired output, y = real output). The error update is done with the rule:

$$MSE^{new} = 0{,}9995\, MSE^{old} + 0{,}0005\, e^2$$

Only the activated rules took part in the update of the weights. The learning in all cases was done with the algorithm δ.

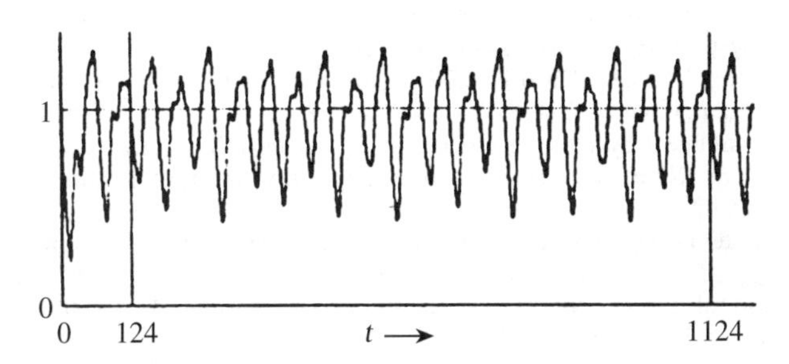

Fig. 2.2.23. The Mackey-Glass time series from $t = 0$ upto $t = 1200$

Table 2.2.4. Results of fuzzy ART compared to six other methods (each epoch of training includes 500 repetitions for the presentation of 500 pairs of education)

Methods	Non-dimensional error indicator
Fuzzy ART (6 rules, 4 M(x), 100.000 epochs)	0.00687
Fuzzy ART (6 rules, M(x), 100.000 epochs)	0.00691
Fuzzy ART (8 rules M(x), 40.000 epochs)	0.00572
Fuzzy ART (12 rules, M(x), 6.000 epochs)	0.00586
Fuzzy ART (12 rules, M(x), 40.000 epochs)	0.00398
Fuzzy ART (16 rules, M(x), 40.000 epochs)	0.00359
ANFIS 16 rules	0.007
Auto-regressive model (AR)	0.19
NN of cross-correlation (serial)	0.06
ND of back propagation	0.02
Polynomial approach sixth degree	0.04
Method of linear prediction (2000 training data)	0.55

Table 2.2.4 shows the results of the method for various numbers of rules and epochs and also, for comparative reasons, the results of six other methods. The error that appears in this table is the non-dimensional size of the square root of the mean square error divided by the formal divergence of the time-series (which is roughly 0.227 for the training data and 0.226 for the verification data).

The Fuzzy ART/Takagi-Sugeno method is superior (in all cases) to all the other methods, including the ANFIS method.

References

[1] Rocha A F (1982) Basic properties of neural circuits, Fuzzy Sets and Systems 7:109–121.

[2] Kosko B (1991) Neural Networks and Fuzzy Systems, Prentice-Hall, Englewood Cliffs, NJ.

[3] Takagi H, Hayashi I (1991) NN-driven fuzzy reasoning, International Journal of Approximate Reasoning, 5(3):191–213 (Special Issue of IIZUKA'88).

[4] Yamakawa T, Furukawa M A, Design algorithm of membership functions for a fuzzy neuron using example-based learning, Proceedings of the First IEEE Conference on Fuzzy Systems, San Diego, pp 75–82.

[5] Erickson R P, Lorenzo P M D, Woodbury M A (1994) Classification of taste responses in brain stem: Membership in fuzzy sets, Journal of Neurophysiology 71(6): 2139–2150.

[6] Furukawa M, Yamakawa T (1995) The design algorithms of membership functions for a fuzzy neuron, Fuzzy Sets and Systems, 71(3):329–343.

[7] Chen C H (1996) Fuzzy Logic and Neural Network Handbook, McGraw-Hill, New York.

[8] Cox E (1994) The Fuzzy Systems Handbook, Academic Press, New York.

[9] Lewis F L, Campos J, Selmic R (2002) Neuro Fuzzy Control of Industrial Systems with Actuator Nonlinearities, Siam, Philadelphia.

[10] Lin C T, Lee G (1995) Neural Fuzzy Systems, Prentice Hall, Englewood Cliffs, New Jersey.

[11] Nie J, Linkens D (1995) Fuzzy Neural Control: Principles, Algorithms and Applications, Prentice Hall, Englewood Cliffs, N.J.

[12] Sutton R, Barto A (1998) Reinforcement Learning – An Introduction, MIT Press, Cambridge, M.A.

[13] Tzafestas S G, Venetsanopoulos A N (1994) Fuzzy Reasoning in Information Decision and Control Systems, Kluwer, Dordrecht/Boston.

[14] Tzafestas S G, Borne P, Tzafestas E S (2000) Soft Computing Methods and Applications, Mathematics and Computers in Simulation (Special Issue), 51.
[15] Wasserman D P (1993) Advanced Methods in Neural Computing, Van Nostrand Reinhold, New York.
[16] Welsted S T (1994) Neural Networks and Fuzzy Logic Applications in C/C^{++}, John Wiley, New York.
[17] Wesley Hines J W (1997) Matlab Supplement to Fuzzy and Neural Approaches in Engineering, John Wiley, New York/Chichester.
[18] Zimmerman H J (1985) Fuzzy Set Theory and Its Applications, Kluwer, Boston, M.A.

3 Neuro-Fuzzy Applications in Speech Coding and Recognition

Francesco Beritelli, Marco Russo, Salvatore Serrano

3.1 Introduction

The recent growth of multimedia mobile communications based on man-machine interaction has increased the demand for advanced speech processing algorithms capable of providing good performance levels, even in adverse acoustic noise conditions (car, babble, traffic noise, etc.), with as low a computational load as possible. Robust speech classification represents, in fact, a crucial point both in speech coding and recognition, two fundamental applications in modern multimedia systems. In particular, in the field of speech coding, an accurate speech classification is fundamental in selecting the appropriate coding model and in maintaining a high perceived quality of the decoded speech. On the other hand, in the field of speech recognition, a robust signal classification is fundamental in obtaining a good word recognition rate, also in the presence of high background noise levels.

For some time now, more sophisticated methods based on Neural Networks (NNs) have been introduced for intelligent signal processing. Lately, various applications of Fuzzy Logic (FL) in the field of telecommunications have also been developed. Recently, the authors of this chapter presented Genetic Algorithms (GAs) as an emerging optimization algorithm for signal processing, pointing out that their use together with neural networks and fuzzy logic, would arouse increasing attention in the future.

In particular, this chapter, through a series of applications in speech and signal classification, shows that a neurofuzzy pattern recognition approach is both more efficient and noise robust, than traditional solutions. More specifically, the signal classifier is based on the extraction of a salient set of acoustic parameters and on the modelling of each phonetic class through a set of fuzzy rules obtained automatically by hybrid neurofuzzy techniques.

Comparative results with other methods show that the impending neuro-fuzzy systems represent a computationally simple and noise robust alternative to the matching techniques of a traditional pattern recognition approach.

The chapter is organized as follows. Section 3.2 presents a brief overview on soft computing. The hybrid learning tool used for fuzzy rule extraction is described in Section 3.3. Section 3.4 presents an overview on speech recognition and coding tecniques. In Section 3.5 we point out that to deal with the problem of speech classification, it is more convenient to use methodologies like fuzzy logic. Section 3.6 presents a series of neurofuzzy applications in the field of speech coding and recognition. Section 3.7 concludes the chapter.

3.2 Soft Computing

The term Soft Computing (SC) was coined by Lotfi A. Zadeh [1]. In the paper in which he uses this new term, he introduces the concept of Machine Intelligence Quotient (MIQ) and states that the current trend in research is towards producing machines with an increasingly higher MIQ. An increase in MIQ is, in fact, being achieved in two totally separate directions. On the one hand, there are machines based on classical logic, and thus on precision, which are evolving towards increasingly faster machines, with a rising degree of parallelism. On the other hand, new forms of logic are being developed, such as Fuzzy Logic (FL), Neural Networks (NNs), and Probabilistic Reasoning (PR) (comprising Genetic Algorithms (GAs)) their strength, by contrast, lying in their capacity to deal with uncertainty and imprecision. In the former case, solutions are found for problems which are very precise and consequently have a high computational cost. This is why Zadeh speaks of Hard Computing (HC). In the latter case, on the other hand, he speaks of SC, as imprecise or uncertain solutions can be found at a much lower cost in terms of calculation effort.

There are a number of cases in which excessive precision is quite useless, so a non-traditional approach is preferable. In some problems this is the only way, as the computational complexity of a classical approach would be prohibitive.

Figure 3.1 shows, at least as far as FL, NNs and GAs are concerned, how the various components of SC can be approximately ordered on a time scale and on a scale relating to their learning capacity.

The time scale shown in Fig. 3.1, is ordered according to the learning time. FL, as conceived of by Zadeh, is not capable of learning anything. NNs and GAs however, have this capacity, although it can be said that on average, pure GAs

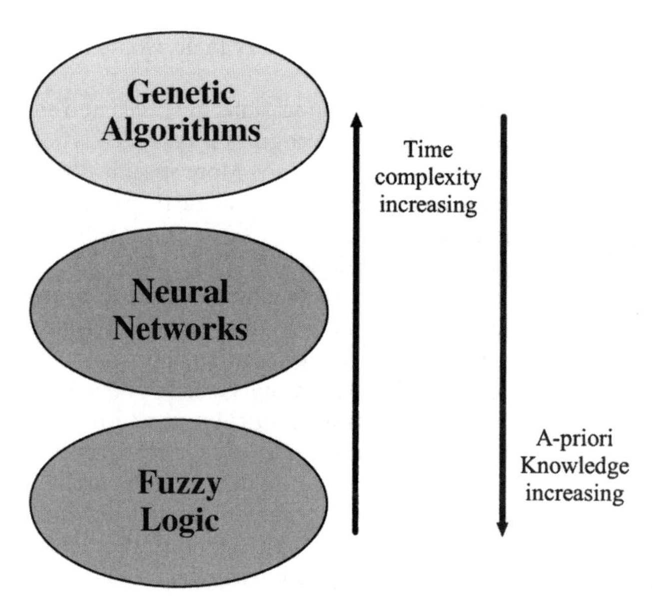

Fig. 3.1. Soft Computing

generally need a longer learning time. From another viewpoint, the order is inverted. GAs, in fact, need no a priori knowledge, NNs need very little and FL at times needs quite detailed knowledge of the problem to be solved.

In effect, each of these three areas of SC has its advantages and disadvantages. FL does not share the inherent concept of NNs, i.e. automatic learning. So it is impossible to use it when experts are not available. It does, however, have a great advantage over the other two techniques: expressed according to fuzzy canons, the knowledge base is computationally much less complex and the linguistic representation is very close to human reasoning.

NNs, at least in regards to the typical features of gradient descent learning networks, are quite different. Firstly, they were conceived of specifically for learning. They are therefore fundamental, when all that is available are some significant examples of the problem to be solved, rather than a solution algorithm. There are two evident disadvantages in using NNs. In general they can learn correctly from examples, but what is learnt is not easy for humans to understand, i.e. the knowledge base extracted from them does not have such intuitive representation as that provided, for example, by FL. Secondly, the type of functions that can be used in NNs must possess precise regularity features and the derivative of these functions has to be known a priori.

Similar considerations hold for GAs, with certain clarifications. Their learning speed is usually slower. However, they have two great advantages over NNs. The functions that can be used in GAs can be much more general in nature and knowledge of the gradient of the functions is not usually required. Lastly, as these algorithms explore several directions at the same time, they are affected by the problem of local extremes much less than NNs. That is, a GA is more likely (compared to an NN) to find a global extreme, rather than a local one. Even if the extreme found is not a global one, it is likely to correspond to a less significant learning error.

On the basis of these considerations, techniques which make use of a combination of SC aspects – i.e. GAs, NNs and FL – are an interesting prospect. Hybrid techniques, in fact, generally inherit all the advantages, but not the less desirable features of the single SC components.

3.3 FuGeNeSys: a Neuro-Fuzzy Learning Tool for Fuzzy Modeling

The rules governing a fuzzy system are often written using linguistic expressions which formalize the empirical rules by means of which a human operator is able to describe the process in question, using their own experience. Fuzzy rules are therefore generally obtained manually, according to heuristic criteria derived from knowledge of the problem. This approach, however, may have great drawbacks when no a priori knowledge is available about the input-output link the fuzzy system has to provide, or when it is highly complex or nonlinear. In addition, it requires a long refinement phase in the definition of the term sets of each input variable which optimize the system (e.g. low, medium, high).

In this chapter, we have adopted the FuGeNeSys (Fuzzy Genetic Neural System) tool [2, 3]. It represents a new fuzzy learning approach based on genetic algorithms and neural networks, which obtains the fuzzy rules in a supervised manner. The advantage of this is that the fuzzy knowledge base is created automatically. The tool also makes it possible to obtain the minimum set of fuzzy rules for optimization of the problem, discarding any rules that are of no use.

The main features of FuGeNeSys are:

1) The number of rules needed for learning is always very low. In the various applications developed, the number has always been below ten.
2) The learning error is comparable to, if not better than, that of other techniques described in literature.
3) Simplified fuzzy knowledge bases can be generated, i.e. the tool is capable of eliminating the unnecessary antecedents in any rule.
4) Significant features are correctly identified.

The tool can be used in both classification and interpolation problems.

Below, we will give a brief outline of the basic procedures followed by the learning tool, used for fuzzy rule extraction.

3.3.1 Genetic Algorithms

GAs have been deeply investigated for two decades as a possible approach to solving many search and optimization problems [3]. These algorithms are a sort of artificial evolution of virtual individuals, selected by means of a Fitness Function.

In practice, GAs are a robust way of searching for the global optimum of a generic function with several variables; they are very flexible and not sensitive to the classical problem of local optimum. The flow of our GA can be simply explained as follows:

$t = 0$;
$P(0) = (P_1, P_2, \ldots, P_N)$;
While not (End Condition) do
Generate P_R from $P(t)$, applying reproduction operator;
Generate P_C from P_R, applying crossover operator;
Generate P_M from P_C, applying mutation operator;
Generate P_H from P_M applying hill descending operator;
$P(t+1) = P_H$
$t = t + 1$;
end while.

The operations performed by the genetic operators, are described in Table 3.1.

We maintain the best solution in the population because it is proven that canonical GAs will never converge to the global optimum. In order to enhance performance in terms of execution time and minimization of GA error, an

Table 3.1. Genetic operators used

Operator	Description
Reproduction	The individuals in the population P_R come from the probabilistic selection of the individuals in the population $P(t)$ with the greatest fitness.
Crossover	Applied to two chromosomes (parents) it creates offspring using the genes of both. In most cases a point called crossover is chosen and two offspring are created. The first will have a chromosome with the father's chromosome in the first part and the mother's in the second. The sibling will have the opposite.
Mutation	Each single bit of a generic sibling will be inverted with a pre-established probability of p_m.
Hill descending	Every time there is best fitness enhancement we call a backpropagation procedure until there are further enhancements.

equivalent NN was associated to the processing of the fuzzy rules. The resulting network is a three-layered one. This is the non-standard genetic operator mentioned above, as the hill descending operator. There is an autonomous mechanism to adjust the learning speeds in this equivalent network, establishing different learning rates at different layers. Therefore, the mixture of GAs and NNs generates a fuzzy inference, i.e. fuzzy rules and tuned membership functions.

3.3.2 The Fuzzy Inferential Method Adopted and its Coding

Each individual in the genetic population is a set of R fuzzy rules. Each rule comprises I inputs (antecedents) and O outputs (consequents), represented by an equal number of fuzzy sets connected by the fuzzy operator AND (i.e. minimum). The membership functions are Gaussians, which can thus be characterized by a center and a variance.

Every element is therefore an individual, whose genetic heritage is made up of the sequence of I inputs and O outputs for all the R rules, giving a total of $(I+O)R$ fuzzy sets.

As we used this tool in classification, we adopted the Weighted Mean (WM) defuzzification method which offers better results for this kind of problem.

Let us illustrate the equations used to calculate the fuzzy inferences in greater detail. Each rule has a maximum of I antecedents. For the generic r-th rule, the i-th antecedent P_{ir} has the following form:

$$P_{ir} = (X_i \text{ is } FS_{ir}) \tag{3.1}$$

X_i is the i-th input and is crisp. That is, X_i is the fuzzy set which is always null with the exception of the numerical value x_i, for which the degree of membership assumes the value of 1. FS_{ir} is a fuzzy set. It has a Gaussian membership function $\mu_{ir}(x)$, univocally determined by a centre c_{ir} and a variance parameter

γ_{ir} according to the following equation:

$$\mu_{ir}(x) = e^{-\gamma_{ir}^2 (x - c_{ir})^2} \tag{3.2}$$

Considering the crisp nature of the inputs, the degree of truth α_{ir} of a generic antecedent P_{ir} is:

$$a_{ir} = e^{-\gamma_{ir}^2 (x_i - c_{ir})^2} \tag{3.3}$$

The connector between the various antecedents is always AND, as calculated by Zadeh [1]. So, the degree of truth, θ_r, of the r-th rule, is the minimum of the degrees of truth α_{ir} for the antecedents belonging to the rule:

$$\theta_r = \min_{i=1}^{I} \alpha_{ir} \tag{3.4}$$

For the conclusion of the rules, each rule has an associated numerical output value z_r. Using WM defuzzification means that the higher the degree of truth of a rule, the closer the defuzzified value y will be to z_r.

$$y = \frac{\sum_{r=1}^{R} \theta_r z_r}{\sum_{r=1}^{R} \theta_r} \tag{3.5}$$

where z_r are crisp values (output singletons).

Figure 3.2 gives a practical example of calculating a fuzzy inference, made up of only two rules. The inference has two rules, two inputs X_1 and X_2 and an output Y.

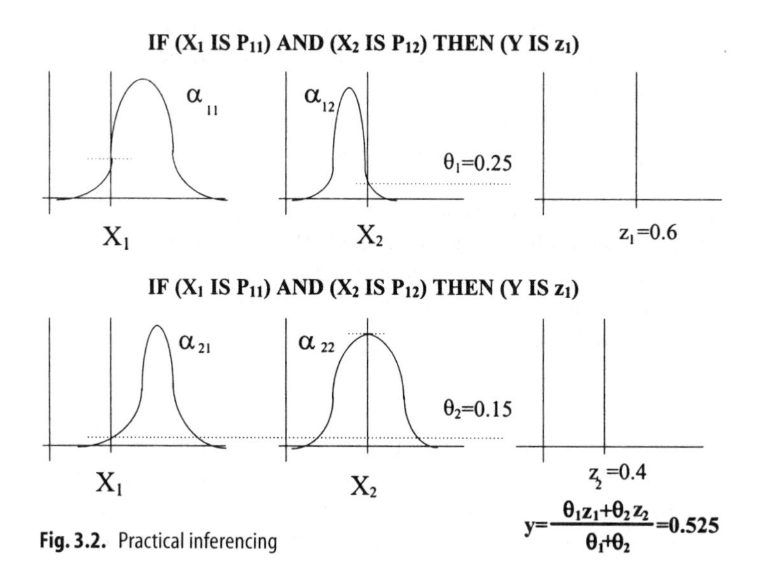

Fig. 3.2. Practical inferencing

3.3.3 Fuzzy Inference Complexity

The computational load of a fuzzy system is very low. For a software implementation, it is possible to compute the degrees of truth of a premise θ simply through look-up tables. Therefore, using a fuzzy system with R rules, I inputs and O outputs, to compute the degree of activation of the rules, $R \cdot I$ memory accesses are needed to calculate the degree of truth of the antecedents, R minimum operations among I values to calculate the degree of truth of the premises of the rules θ_r, and $R \cdot O$ products, $2(R \cdot O - 1)$ sums and O divisions for the defuzzification process.

3.4 Conventional Speech Coding and Recognition Techniques

3.4.1 Speech Recognition

Figure 3.3 depicts a typical block scheme representing the working principle of a voice command recognition (VCR) system.

The endpoint detector identifies the boundaries of the spoken word or voice command, giving the system the possibility of rejecting the segment of signal not including voice. Due to the high degree of redundancy in a speech signal and variations in the pronunciation of the same phonetic units even by a single individual, it is necessary to use a features extraction block to obtain a reduced set of acoustic parameters from the dense information content.

The parameters obtained, generally fall into two classes: the first is called "acoustic features", including formants, pitch, energy, nasality, vocalisation etc. through which it is possible to characterise phonetic segments. The second category includes sets of parameters that can be directly obtained from processing the spectrum of the signal and do not necessarily have a physical meaning: the most frequently adopted are the cepstral coefficients [19].

These represent the Fourier series coefficients of the envelope of the logarithm of the power spectrum and can be obtained either directly from frequency analysis with an array of uniformly spaced filters (Cepstral), or by analysis with an array of filters spaced according to the Mel scale (Mel Frequency Cepstral or MFCC), or by linear prediction coding (LPC Cepstral or LPCC). Frequently, the cepstral coefficients are used in combination with their first- and second-order time derivatives. These sets are known as Delta-Cepstral and Delta-Delta-Cepstral. Lastly, in addition to these, the energy of the speech signal is often used as a recognition parameter [19].

In choosing the parameters, it is also necessary to take into account the subsequent block in the system, i.e. the one dealing with analysis and classification of the speech signal. Basically, its task is to establish which phonetic unit is present in the segment of speech signal being analysed and so it varies according to the minimum unit taken into consideration. If it is a word, the block has to establish the vocabulary element it belongs to, whereas if it is part of a word (a

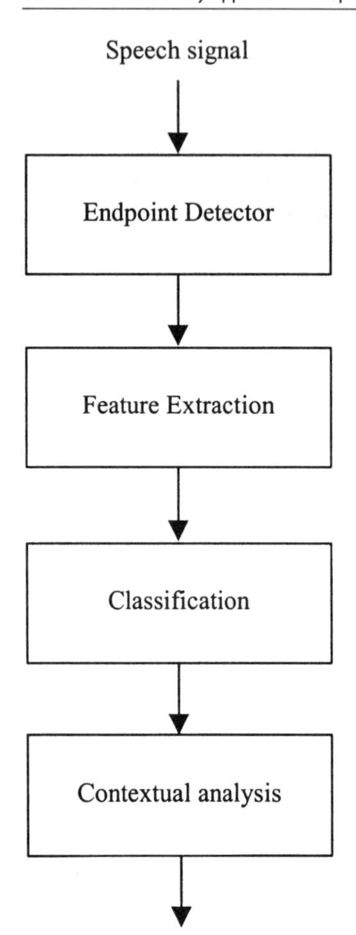

Fig. 3.3. Block scheme of a voice command recognition system

phoneme, syllable etc.), the decision will refer to one of the possible phonetic units. The main techniques used in making this decision, are divided into two approaches: acoustic-phonetic and pattern recognition [4, 19].

The last block, i.e. the contextual analysis block, is not necessarily present in all systems: its task, in fact, is to examine the result provided by the classifier, taking into account the grammar and syntax rules it knows, and at the same time analyse the result of the last classifications made. Even though it has a fundamental function, especially in the recognition of connected, continuous speech, the contextual analysis block can also be successfully used in isolated word recognition systems.

The overall accuracy of a VCR system depends on the performance of the single processes in the blocks shown in Fig. 3.3, so each of these phases is still the subject of intense research activity. There is still a large number of problems to be solved to hope for an all-inclusive approach and consequently, there are various categories of VCR systems, according to the features that are considered to be most important for specific applications.

The most important classification to be made is without doubt between speaker-dependent and speaker independent recognition systems [19]. In the former case, before using the system, the user will have to train it to recognise the words in the vocabulary, according to his timbre and pronunciation. In the latter, a certain amount of accuracy is sacrificed, to make the system immediately usable by several users.

Recently, an intermediate speaker adaptation technique has been introduced, which dynamically adapts the parameters of the system to the speaker. This adjustment can either be initial or continuous, while the system is being used.

3.4.1.1 Acoustic-phonetic approach

According to the acoustic-phonetic theory, a speech signal contains a certain number of phonetic segments that can be identified via a set of parameters known as "acoustic features" [4]. This approach recognises speech by identifying these segments in a voice message through segmentation of the whole signal and classification of the single segments. The basic scheme of an acoustic-phonetic recognition system includes analysis of the speech signal (array of filters, LPC), and extraction of a certain number of "features" (formants, nasality, pitch, etc.). By analysing these parameters, the next block produces hypotheses regarding the segmentation of the signal and proposes the phonetic units that are most likely to be contained in each segment. Although this approach is an attractive one and theoretically unexceptionable, it encounters difficulty in obtaining a sufficiently accurate phonetic segmentation and so has not been very successful in real applications, at least as a "stand-alone" solution.

3.4.1.2 Pattern recognition techniques

The pattern recognition approach [14] is certainly the most successful in practical applications. This is for several reasons, including its relative conceptual simplicity and its versatility, which make it suitable for a wide range of speech recognition tasks, from isolated to connected speech, and from speaker-dependent to speaker-independent and speaker-adapted systems [19]. Its recorded performance and potential for growth are both very good.

All pattern recognition methods are based on a common structure: in an initial phase called "training" the system is presented with one or more repetitions per elementary phonetic unit (word, phoneme, syllable, etc.) to be recognised. During this phase, prototypes for each of the phonetic units are developed. The actual recognition phase consists of comparing an unknown sequence with each of the prototypes created during the training stage. At this point, each prototype is assigned a score on the basis of which one the decisional logic establishes which to be "closest" to the input speech signal, thus indicating which word (or phoneme) has been uttered.

There are two categories of pattern recognition techniques: deterministic and statistical, the main difference lying in the kind of prototypes created during the training stage [4]. In deterministic systems, the results of the training are called "templates" and comprise the sequence of parameters that is most representative of the relative phonetic unit. In statistical techniques, on the other hand, the training phase returns models that contain the statistical properties of the speech signal. This is a significant step forward with respect to the deterministic approach, because it enables even highly complex models to be created, according to the requirements of specific applications. The downside, is that to obtain reliable speech signal statistics, the speech database used for training has to be much larger than the one used in deterministic models.

The differences between the two approaches are not, however, confined to the kind of models used. The pattern matching stage is also different. The most common deterministic models use the Vector Quantization (VQ) technique or Dynamic Time Warping (DTW) as a matching measure [4]. Once the distortion measure has been established, a "template" is matched with a speech sequence by calculating the distance between the two patterns (the unknown and reference patterns). From this point of view, statistical techniques have an advantage in that there are no reference patterns on which time matching has to be performed, statistical models are represented by stochastic processes and the score attributed to each of them in the matching phase is proportional to the probability that the process will produce a sequence equal to the unknown one. The most commonly used statistical speech recognition models are based on Hidden Markov Models (HMM) [4]. Recently, Artificial Neural Networks (ANNs) [31] have been widely applied to the classification of static speech patterns, obtained by some kinds of spectral analysis of isolated words [32, 33]. For small vocabularies, the results obtained are almost comparable with those given by advanced HMM-based recognizers.

A. Vector quantization

Vector Quantization (VQ) is a pattern matching technique which encodes each input vector by comparison with a set of reference codewords stored in a codebook. In particular, in the training phase, N vectors obtained via vector quantization are memorised. The codebooks are obtained from the input sequence using a generalised version of the Lloyd algorithm for vector quantization. During the recognition phase, the distortion between the input sequence and the word models obtained via vector quantization, is calculated.

In the traditional technique, the distortion is obtained by summing up the distance of each vector belonging to the input sequence and the vector of the closest quantized model. Indicating the codebook $\aleph r = (x1, x2, x3, \ldots, xN)$ and the pattern to be recognised for each vector ti as $\tau = (t1, t2, t3, \ldots, tI)$, the index

$$j_i \operatorname{argmin}_{h=1}^{N} d(\mathbf{x}_h, \mathbf{t}_i) \tag{3.6}$$

is determined and therefore the global distance from the model is equal to:

$$D^r = \frac{1}{I}\sum_{i=1}^{I} d(\mathbf{x}_{j_i}, \mathbf{t}_i) \tag{3.7}$$

B. Dynamic time warping

In comparing sequences of spectral vectors, it is necessary to bear in mind that a speaker rarely utters the words in his vocabulary at the same speed. It is consequently necessary to prevent differences in the duration or the rhythm at which a word is pronounced by calculating the distance between spectral sequences when the comparison is made. The comparison is generally made between sequences with different durations, which we will indicate here as:

$$X=(x_1,x_2,\ldots,x_{T_x}) \text{ and } Y=(y_1,y_2,\ldots,y_{T_x}) \tag{3.8}$$

The durations *Tx* and *Ty* are therefore not identical. In addition *xi* and *yi* are vectors. In general, the distance between two vectors $d(x_i, y_i)$ will be indicated for reasons of notational simplicity as $d(i_x, i_y)$ where $i_x = 1, \ldots, Tx$ and $i_y = 1, \ldots, Ty$. Due to the difference in the duration of the spectral sequences, the indexes identifying the pair of vectors to be compared, have to meet certain conditions. If, for example, time linear normalisation is used, the indexes have to meet the following condition:

$$i_y = \frac{T_y}{T_x} \cdot i_x \tag{3.9}$$

The distance between the sequences can be defined as:

$$d(X,Y)=\sum_{i_x}^{T_x} d(i_x,i_y) \tag{3.10}$$

Linear alignment implicitly, assumes that there are no variations in the speed at which the speaker speaks and so the sounds are all uttered more slowly or more quickly. This constraint is obviously not capable of modelling what happens in reality.

It is therefore necessary to normalise these fluctuations, so that the comparison is made under the best possible conditions. This is in part achieved by an algorithm called DTW (Dynamic Time Warping) [4].

A more general time alignment procedure needs to use two warping functions connected to the same time axis *k*, i.e.:

$$i_x = \phi_x(k) \quad \text{for } k=1,\ldots,T \tag{3.11}$$

$$i_y = \phi_y(k) \quad \text{for } k=1,\ldots,T \tag{3.12}$$

Figure 3.4 illustrates some typical local continuity constraints. The distance between two sequences is defined, introducing a "weight" factor $m(k)$ and a normalisation factor M_ϕ as:

$$d_\phi(X,Y)=\frac{1}{M_\phi} d[\phi_x(k),\phi_y(k)] \cdot m(k) \tag{3.13}$$

Both the onerous computational load of DTW and the amount of memory required to store the whole reference example, limit the use of time warping to applications with a hardware scenario of medium-high complexity.

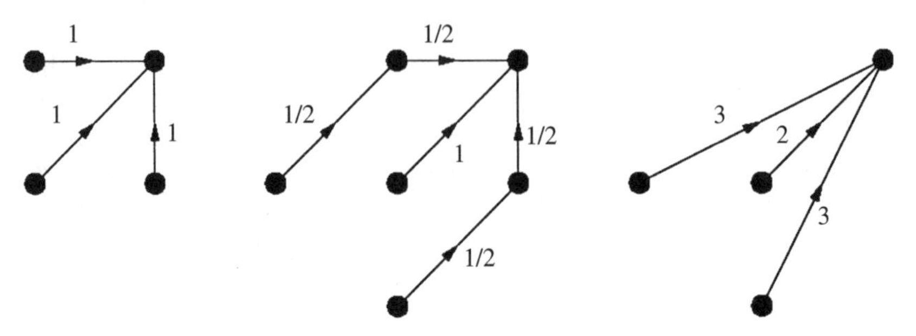

Fig. 3.4. Typical local continuity constraints

C. Hidden Markov models

An HMM for speech recognition is consists of a non-observable "hidden" process (a Markov chain), and an observation process which links the acoustic vectors extracted from the speech signals to the state of the hidden process. Since there are strong temporal constraints in speech, left-to-right HMMs are generally used. An HMM can model a specific speech unit, such as a word. Three problems must be solved for the model to be practical [4, 19]:

1) The evaluation problem: given a model and a sequence of observation on the speech signal, what is the probability of observation sequence, conditioned on the model. An efficient solution can be found with the forward pass of the forward-backward algorithm.
2) The learning problem: given an HMM and a sequence of observation, how to adjust the model parameters to maximize the probability of generating the observations. The observation sequence is called the training sequence.
3) The decoding problem: given a model and a sequence of observation, what is the state sequence in the model that best explains the observation. The solution to this problem requires an optimality criterion to find the best possible solution. Typically, the Viterby algorithm [4] is used.

HMMs were first introduced in speech recognition by considering the observation as a discrete symbol belonging to a finite alphabet. In this case, it is possible to use a discrete probability density within each state of the model. Since observations are often continuous in nature, various methods have been used to transform continuous signals into sequences of discrete symbols, especially Vector Quantization methods based on clustering techniques like the K-Means algorithm [4]. However, the discretization of continuous signals, introduces distortions that may, in some cases, significantly reduce recognition performance. To overcome the limitations of discrete HMMs, Continuous Density HMMs have been proposed. In this case, some restrictions must be defined on the Probability Density Function (pdf). The most general pdf model, for which a consistent re-estimation method has been designed, is a finite mixture of P Gaussian functions.

3.4.1.3 Endpoint Detection

The Endpoint Detection (EPD) block, is one of the elements that most affect the accuracy of an IWR system, as shown in [4]. The problems affecting correct evaluation of the endpoints of a word are essentially due to the speaker and the acoustic environment they are in. A speaker, in fact, also utters sounds that are not part of a word but produced by breathing or smacking the lips and so on.

In addition, in real situations, a word can be uttered in the presence of background noise, caused for example by fans or machinery, non-stationary noise such as the shutting of a door or a car horn, or interference due to other speech signals coming from a television, a radio or a conversation.

Another aspect to be borne in mind is computational simplicity, which is still a limit, mostly for applications in which it is necessary to use low-cost or low calculating power DSPs, as in mobile devices.

3.4.2 Speech Coding

3.4.2.1 Analysis by Synthesis Coding

In the scenario of modern communications, voice services still represent the main business of telecommunication providers. In view of the future development of wireless networks integrating multimedia in a mobile environment, a lot of effort is presently devoted to the study of more efficient speech coding techniques providing high perceptive quality at a very low bit rate [6, 11].

The low-rate speech coding systems which are most widely studied and offer the best performance are based on closed-loop linear prediction Analysis-by-Synthesis techniques. These systems offer the good speech quality typical of waveform coders with the compression efficiency of coding techniques based on models of the speech generation process.

More specifically, Analysis-by-Synthesis techniques enhance the performance of traditional Linear Prediction Coding (LPC) vocoders, by better modeling of the excitation at the synthesis filter, thus overcoming the main limitation of assuming that speech signals are either voiced or unvoiced [8]. The general model belonging to this class of coders, in fact, uses a single excitation signal carefully optimized by closed-loop analysis, which allows coding of the excitation frame at a very low bit rate, while maintaining a high level of speech quality. Figure 3.5 shows a general model for Analysis-by-Synthesis coding.

It consists of one short term predictor $A(z)$ to model the short-time spectral envelope of the speech waveform and one long term predictor $P(z)$ for the fine structure of the speech spectrum. The excitation frame at the LPC filter input is optimized, by minimizing the perceptually weighted difference between the original speech signal $s(n)$ and the reconstructed one $s\wedge(n)$. The weighting filter has the transfer function $W(z) = A(z)/A(z/\gamma)$, where γ is a bandwidth expansion factor. As a large part of the perceived noise in a coder comes from the frequency regions where the signal level is low, the filter $W(z)$ gives less weight to

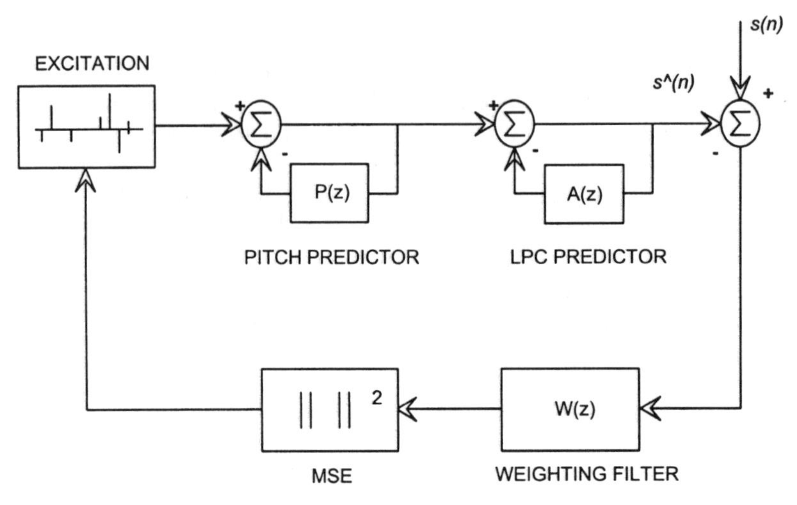

Fig. 3.5. General model for Analysis-by-Synthesis coding

the residual error in the formant regions where the noise is better masked by the speech signal with high energy.

A new generation of cost-effective digital signal processing algorithms has led to the development of several Constant Bit Rate (CBR) speech coding standards, such as ITU-T Rec. G. 728 (16 kbit/s LD-CELP) and ITU-T Rec. G. 729 (8 kbit/s CS-ACELP and its extension to 12 kbit/s) [12], for high-quality wired communications, and the ETSI standards for the European GSM digital cellular system, mainly the 6.5-kbit/s Half-Rate and the new 13-kbit/s Enhanced Full-Rate [13] speech coding schemes. All these new-generation coders use the linear predictive-based analysis-by-synthesis coding technique, which is commonly known as Code Excited Linear Prediction (CELP) coding [11].

3.4.2.2 Variable Rate Coding

To improve the efficiency of transmission systems, speech coders are applied in conjunction with DSI (Digital Speech Interpolation) devices, more recently indicated as VADs (Voice Activity Detectors) [11]. In this case, CBR source coding is used during the talkspurt periods whilst no packets are transmitted during silence periods. So the resulting source emits Variable Bit Rate (VBR) traffic and, an the receiver side, a comfort noise algorithm is used to reconstruct the background noise during the inactivity periods.

Although the approach based on DSI has been widely and successfully applied to several types of digital circuit multiplication equipment (DCME), voice storage and packetized voice networks, the emerging request for the provision of high-quality personal communication services able to support integrated multimedia applications has recently stimulated investigation into more efficient VBR speech compression algorithms, based on multimode coding [26, 27], such as the so-called Variable Rate CELP (VR-CELP). A multimode VBR coder typically improves the spectral efficiency by exploiting, not only the silence periods, but also the diverse characteristics of the speech waveform. Moreover, it takes into account the different peculiarities of background noise, thus allowing an improvement in signal reconstruction, as compared to simple comfort noise systems. Finally, multimode VBR coding provides greater system versatility, as it allows the quality and bit rate to be controlled according to the user's requirements, the conditions of the communication channel and the network load. On the basis of these considerations, multimode VBR speech coding today appears to be the most promising coding approach for the integration of voice in the service scenarios foreseen for third-generation radio systems.

A multimode speech coder exploits the large amount of silence during a conversation and the great variations in the characteristics of active speech, using an appropriate coding model for each phonetic class considered. Naturally, to achieve a toll quality, it is necessary to have both signal classification algorithms (which are robust to the background noise typical of mobile environments) and efficient signal coding models.

Typically, a multimode speech coder is based on a set of operating modes for background noise coding and another set for talkspurt segment coding. The frame classifier has a multilevel architecture and it uses efficient algorithms based on both a simple threshold comparison and a pattern recognition approach.

3.4.2.3 Phonetic Classification

Phonetic classification is undoubtedly one of the most delicate issues in variable speech coding [11]. The coding of a speech segment by means of an inappropriate model causes degradation in quality, which in some cases reaches unacceptable values. A phonetic classifier architecture presents several levels of classification. Usually, at the first level, we have a robust Voice Activity Detector (VAD) to distinguish activity segments (talkspurts) from non-activity segments (silence or background noise).

At the second level of classification, if the speech segment is active, a voicing detection algorithm discriminates between unvoiced and voiced sounds. In the category of voiced sounds the fully voiced speech segments are identified by a backward cross-correlation algorithm. If the frame is classified as silence or noise, it is initially separated into stationary and non-stationary background noise. This is done by comparing the threshold of the spectral distortion measure of the current and previous frames. Both stationary and non-stationary frames are closed-loop classified as acoustic noise segments, requiring a noise-like or codebook LPC excitation. Finally, within stationary frames, by means of a simple changing level control the speech segments that maintain their energy level unchanged with respect to the previous frame are distinguished from those that have undergone a variation.

3.5 A Soft Computing-Based Approach in Speech Classification

Indicating the zero crossing rate as a useful parameter for the voicing decision, Rabiner and Schafer in [4] emphasize that: "*[The statement] if the zero-crossing rate is high, the speech signal is unvoiced, otherwise, if the zero-crossing rate is low, the speech signal is voiced, is very imprecise, because we have not said what is high and what is low and of course, it really is not possible to be precise*".

This consideration, which holds for any parameter typically used by a voicing discrimination algorithm, suggests that it is more convenient to deal with the problem of audio classification using methodologies like fuzzy logic, that are suitable for problems requiring approximate rather than exact solutions, and that can be represented through descriptive or qualitative expressions, in a more natural way than mathematical equations [1]. Fuzzy logic also allows the various inputs to be interpreted as linguistic variables (low, medium, high), exploits the concepts of degree of membership and fuzzy set, and requires a very low computational load. Finally, there is the advantage that the contribution of each single parameter to the final decision is weighted in a non-linear fashion.

For some time now, more sophisticated methods based on NNs have been introduced for intelligent signal processing [5]. Recently in [3] the author presented GAs as an emerging optimization algorithm for signal processing, pointing out that their use with neural networks and fuzzy logic would generate increasing attention in the future. The aim of this chapter is to demonstrate the advantages of the impending SC-hybrid systems, as compared with traditional signal processing methods, especially in the field of speech classification. Figure 3.6 shows the general functional scheme of the implemented classification systems.

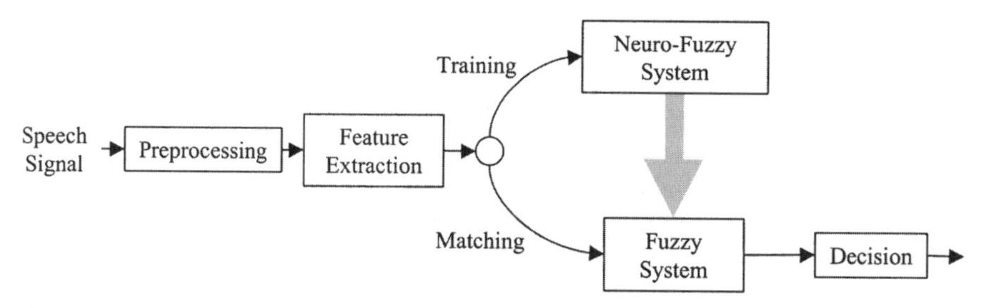

Fig. 3.6. Functional scheme of the implemented classification systems

3.6 Neuro-Fuzzy Applications in Speech Coding and Recognition

3.6.1 Voiced/Unvoiced Classification

3.6.1.1 Introduction

Voicing detection represents a crucial point in the perceived quality and naturalness of a very low bit-rate speech coder [6–8]. Voiced (V) to Unvoiced (UV) misclassifications degrade the speech quality as a noise-like excitation to the LPC filter is not suitable for reconstruction of voiced frames, whereas UV to V errors involve a higher coding rate and also degrade the speech quality when the voiced LPC excitation model used is unsuitable for synthesizing UV sounds. The main cause of degradation in the performance of a voicing detector is the presence of background noise which alters the characteristics of the speech waveform [8]. Therefore, in order to develop more efficient very low bit-rate speech coding algorithms for mobile communications, in which the level of background noise may be very high, new robust speech classification methodologies are required.

In this Section, we present a simple, efficient pattern recognition approach to voicing decision based on fuzzy logic. The Fuzzy Voicing Detector (FVD) proposed, uses a matching phase based on a set of fuzzy rules extracted by FuGeNeSys, the hybrid learning tool proposed in Section 3.3, which integrates the advantages of neural networks, genetic algorithms and fuzzy logic. Through a series of comparisons with a traditional solution, we outline the advantages of this approach in terms of great robustness to the background noise typical of mobile environments.

3.6.1.2 The Fuzzy Voicing Detector

Pre-processing of the speech signal consists of a pre-emphasis procedure [4], performed by a first-order FIR filter, and a 150-Hz high-pass filtering of the speech waveform, in order to reduce the effect of the car noise. The feature extraction module obtains an efficient pattern of acoustic parameters needed for the matching phase. The five parameters used, chosen from those typically used for voicing detection [4, 7], are calculated in a window of 80 speech samples (a 10-ms frame, sampling at 8 kHz) and are as follows: the normalized first autocorrelation coefficient (ρ_1), the first coefficient of the 10th-order LPC filter (a_1), the zero average crossing rate (ZCR), the energy level in dB (E) and the inverse of the short term prediction gain (I_g). The matching phase, performed here by a fuzzy system, has the task of mapping the pattern of input parameters onto a scalar value, ranging between 0 and 1, which indicates the degree of membership in the V/UV classes. Finally, the decision module, using a threshold comparison, assigns a membership class to the analysis frame.

The fuzzy system was trained off-line using a database of 60 speech phrases (in English and Italian) spoken by 36 mother-tongue speakers, 18 males and 18 females, lasting for a total of 260 seconds and recorded in a quiet environment. More specifically, the database was subdivided into a learning database (10354 patterns) and a testing database (15670 patterns), which naturally contains different phrases and speakers from the first one. The patterns extracted from the speech frames labeled silence or mixed, were discarded.

After the training phase, we obtained a knowledge base of only 3 fuzzy rules. In Fig. 3.7, in the rows we have the rules, in the first 5 columns the inputs of the fuzzy system, and in the last column the crisp output.

As can be seen from Fig. 3.7, the membership functions chosen for the fuzzy sets are Gaussian. Each of the fuzzy sets represented has its universe of discourse corresponding to the relative input on the abscissa, and the truth values, on the ordinates. The connectors in the premise are ANDs calculated through the minimum operator. Finally, the defuzzification method used, the weighted mean, only requires a crisp value for each output (the numerical value that is present in the sixth column of Fig. 3.6). The output of the fuzzy system is given by the weighted mean of these crisp values, where the weights are the degree of activation of the fuzzy rules. We say that 0 is UV and 1 is V. Linguistic interpretation of the rules is easy. For example, rules two and three can be written as:

Rule 2 IF (a_1 is negative) AND (ZCR is very low)
AND (E is very high) AND (I_g is low)
THEN (speech frame is V)

Rule 3 IF (a_1 is positive) AND (ZCR is medium)
AND (E is low) AND (I_g is very high)
THEN (speech frame is UV)

As the reader can see, from a linguistic perspective the fuzzy inference evaluation follows human common sense, exactly. Moreover, the FVD output does not depend on whether the single inputs have exceeded a threshold or not, but on an overall and non-linear evaluation of the values they have taken. So, even though the possibility of linear separation of the parameters between the two classes V/UV decreases as the noise level increases, the single fuzzy sets handle the uncertainty in the input pattern well, guaranteeing great robustness to the variations the parameters undergo, in the presence of background noise.

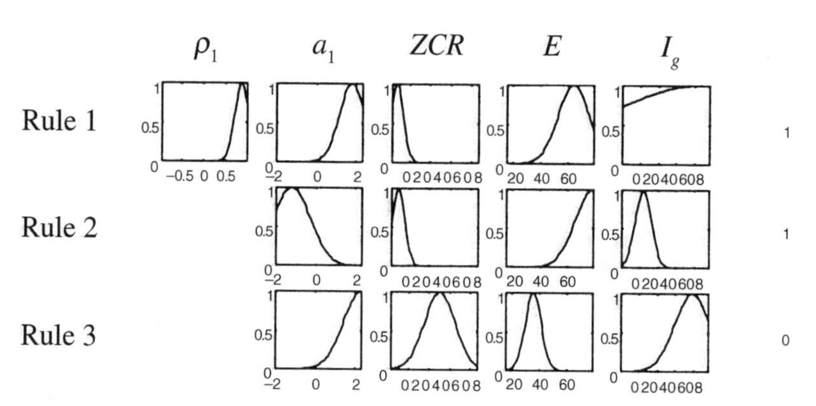

Fig. 3.7. The three fuzzy rules of the FVD

3.6.1.3 Experimental Results

To test the validity of the classifier proposed, the performance of the FVD was compared with that of a voicing detector based on Fisher's discriminant analysis, adopted in the LPC-10E vocoder [9], and many recent low bit-rate speech coders. For the Fisher analysis the set of weights considered, in the order in which the five parameters used are presented, was [0.41, 0.22, –0.01, –0.95, –0.42], with an offset constant equal to 0.01 and a detection threshold of 0.5. For the FVD the threshold value which minimizes the function resulting from the sum of the V to UV and UV to V learning misclassifications is 0.32. Moreover, in order to evaluate the performance of the FVD in noisy environments typical of mobile communications, the following four types of background noise were digitally added to the testing speech database: white, car, traffic and babble noise. Table 3.2 shows the results of the comparison, in terms of percentage of misclassification with varying types of noise and SNR. The last column in the table, indicates the improvement in the total error percentage. The results show the advantage of the fuzzy-based approach, as compared with the traditional technique. More specifically, the performance is quite similar in the clean case (for which the Fisher method performs well), but when the noise level increases, FVD performance improves considerably as compared with the traditional solution. It should also be noted that the improvement mainly concerns the reduction in V to UV misclassifications, i.e. errors which degrade the speech quality most. Further, by reducing the effect of car noise, the high-pass pre-processing

Table 3.2. Fisher-FVD comparison in terms of percentage of misclassification (W = White, C = Car, B = Babble, T = Traffic, $\Delta E_n = [E_{\text{Fisher}} - E_{\text{FVD}}]/E_{\text{Fisher}}$)

SNR (dB)	Noise	V → UV (%)		UV → V (%)		ΔE_n (%)
		Fisher	FVD	Fisher	FVD	
Clean		1.30	1.33	0.94	0.77	6.25
20	W	9.32	9.10	0.21	0.19	2.52
10	W	26.44	21.47	0.02	0.04	18.70
5	W	45.21	31.50	0.01	0.02	30.30
0	W	68.18	41.55	0.00	0.22	38.74
20	C	1.27	1.33	1.02	0.79	7.42
10	C	1.19	1.24	1.39	1.13	8.14
5	C	1.15	1.15	2.08	1.98	3.10
0	C	1.12	1.05	3.57	3.37	5.76
20	B	5.13	5.32	0.48	0.37	-1.43
10	B	12.52	9.07	0.43	0.84	23.47
5	B	18.19	9.08	1.16	3.11	37.00
0	B	23.27	6.39	2.52	7.06	47.85
20	T	3.48	3.68	1.21	0.92	1.92
10	T	7.67	4.89	3.05	3.92	17.82
5	T	9.79	3.51	4.42	6.75	27.80
0	T	12.09	1.73	6.42	9.37	40.03

filter guarantees optimal performance in the presence of this kind of noise, even at low SNRs.

3.6.2 Voice Activity Detection

3.6.2.1 Introduction

A Voice Activity Detector (VAD) aims to distinguish between speech and several types of acoustic background noise, even with low signal-to-noise ratios (SNRs). Therefore, in a typical telephone conversation, a VAD, together with a comfort noise generator (CNG), achieves a silence compression. In the field of multimedia communications, silence compression allows the speech channel to be shared with other information, thus guaranteeing simultaneous voice and data applications [10]. In a cellular radio system that uses the Discontinuous Transmission (DTX) mode, such as the Global System for Mobile communications (GSM), a VAD reduces co-channel interference (increasing the number of radio channels) and power consumption, in portable equipment. Moreover, a VAD is vital in reducing the average bit rate in future generations of digital cellular networks, such as the Universal Mobile Telecommunication Systems (UMTS), which provide variable bit-rate (VBR) speech coding. Most of the capacity gain is due to the distinction of speech activity and inactivity.

The performance of a speech coding approach based on phonetic classification, however, strongly depends on the classifier which must be robust to every type of background noise [8]. As is well known, for example, above all with low SNRs, the performance of a VAD is critical for the overall speech quality. When some of speech frames are detected as noise, intelligibility is seriously impaired due to speech clipping in the conversation. If, on the other hand, the percentage of noise detected as speech is high, the potential advantages of silence compression are not obtained. In the presence of background noise it may be difficult to distinguish between speech and silence, so for voice activity detection in wireless environments, more efficient algorithms are needed [11].

The activity detection algorithm proposed in this Section is based on a pattern recognition approach in which the feature extraction module uses the same set of acoustic parameters adopted by the VAD recently standardized by ITU-T in Rec. G.729 annex B [12], but basing the matching phase on fuzzy logic. Through a series of performance comparisons with the ITU-T G.729 Annex B VAD and the VAD standardized by ETSI for the Full Rate GSM codec [13], varying the type of background noise and the signal-to-noise ratios, we outline the validity of the new methodology in terms of both communication quality improvement and bit-rate reduction, as compared with the traditional solution.

3.6.2.2 The Fuzzy VAD Algorithm

The functional scheme of the Fuzzy Voice Activity Detector (FVAD) is based on a traditional pattern recognition approach [14, 15]. The four differential parameters used for speech activity/inactivity classification are the same as those used in G.729 Annex B and are: the full-band energy difference ΔE_f, the low-band energy difference ΔE_l, the zero-crossing difference ΔZC and the spectral distortion ΔS. The matching phase is performed by a set of fuzzy rules obtained automatically by means of the hybrid learning tool presented in Section 3.3. As is well known, a fuzzy system allows a gradual, continuous transition rather than a sharp change between two values. So, the Fuzzy VAD proposed returns a continuous output ranging from 0 (Non Activity) to 1 (Activity), which does not depend on whether the single inputs have exceeded a threshold or not, but on an overall evaluation of the values they have assumed. The FVAD translates several individual parameters into a single continuous value which, in our case, indicates the degree of membership in the Activity class and the complement of the degree of membership in the Non Activity class. The final decision is made by comparing the output of the fuzzy system, which varies in a range between 0 and 1, with a fixed threshold experimentally chosen by minimizing the sum of Front End Clipping (FEC), Mid Speech Clipping (MSC), OVER, Noise Detected as Speech (NDS) [16] and the standard deviation of the MSC and NDS parameters. In this way, we found an appropriate value for the hangover module that satisfies the MSC and NDS statistics, reducing the total error. The hangover mechanism chosen is similar to that adopted by the GSM [13].

3.6.2.3 Speech Database

The speech database used to obtain the learning and testing patterns contains sequences recorded in a non-noisy environment (Clean sequences, SNR = 60 dB), sampled at 8000 Hz and linearly quantized at 16 bits per sample. It consists of 60 speech phrases (in English and Italian) spoken by 36 native speakers, 18 males and 18 females. The database was then subdivided into a learning database and a testing database, which naturally contains different phrases and speakers from the first one. The two databases were marked manually as active and non-active speech segments. In order to have satisfactory statistics in regards to the languages and the speakers, the male and female speakers and the languages were equally distributed between the two databases. Further, to respect the statistics of a normal telephone conversation (about 40% of activity and 60% of non-activity), we introduced random pause segments, extracting from an exponential population, the length of talkspurt and silence periods.

In order to evaluate the effects of changes in the speech level, we considered 3 different levels in the testing database: 12, 22, 32 dB Below Codec Overload (BCO), i.e. from the overload point of 16 bit word length, whereas the effects of background noise on VAD performance was tested by digitally adding various types of stationary and non-stationary background noise (Car, White, Traffic and Babble) the clean testing sequence at different signal-to-noise ratios (20, 10, 0 dB). The learning database consists of only clean sequences, so the trained fuzzy system used for the matching phase is independent of any type of background noise.

To summarize, the learning database comprises clean speech sequences at 22 dB below codec overload, lasting about 4 minutes, whereas the testing database includes clean speech and noisy sequences corresponding to about 342 minutes of signal, divided in 57 files of 6 minutes each (6 types of superimposed noise, white, car, street, restaurant, office and train noise, with 3 different SNRs and 3 different levels, plus 3 clean files at different levels).

3.6.2.4 Fuzzy Rules

After the training phase, we obtained a knowledge base of only six fuzzy rules. Fig. 3.8 shows the six fuzzy rules the tool extrapolated from the examples. In the rows we have the rules, and in the first four columns the four fuzzy system inputs. Each of the fuzzy sets represented, has the Universe of Discourse corresponding to the relative input on the abscissa and the truth values on the ordinates. The crisp value of the output singleton is presented in the last column. More specifically, we say that 0 is inactivity and 1 is activity. We adopted the Weighted Mean defuzzification method, which offers better results in classification problems.

If we neglect very large fuzzy sets, we can give a linguistic representation of the six fuzzy rules:

Rule 1 : IF (ΔS is medium-low) THEN (Y is active)
Rule 2 : IF (ΔE_f is very high) THEN (Y is inactive)
Rule 3 : IF (ΔE_l is low) AND (ΔS is very low) AND (ΔZC is high) THEN (Y is active)

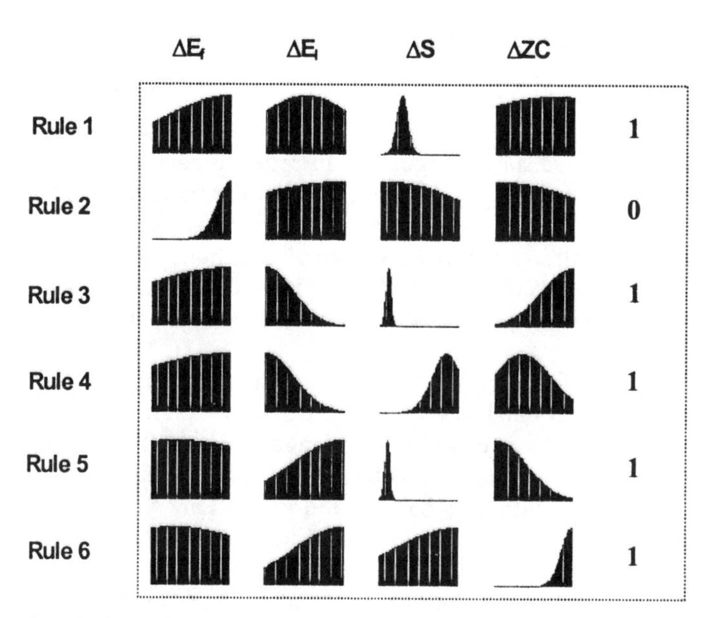

Fig. 3.8. FVAD rules

Rule 4 : IF (ΔE_l is low) AND (ΔS is high) AND (ΔZC is medium) THEN (Y is active)

Rule 5 : IF (ΔE_l is high) AND (ΔS is very low) AND (ΔZC is low) THEN (Y is active)

Rule 6 : IF (ΔE_l is high) AND (ΔS is not low) AND ΔZC is very high) THEN (Y is active)

Of course, the output of the fuzzy system, which indicates the degree of membership in the Activity/Inactivity classes, depends on an overall evaluation of the input parameter values by means of the defuzzyfication process. For example, we have a high output (i.e. the frame is detected as active) if ΔE_f is not high and ΔS is medium-low, whereas we have a low output (i. e. the frame is detected as inactive) if ΔS is medium and ΔE_f is very high and ΔZC is not high; in this last case, in fact, only the degree of truth of rule 2 is high, while the degree of truth of the other rules is low.

3.6.2.5 Decision Module

In order to establish an optimal threshold value with which to compare the fuzzy system output, we analyzed the total misclassification error with respect to a threshold value, F_{th}, ranging between 0 and 1. The threshold was chosen in such a way, as to achieve a trade-off between the values of the four parameters FEC + MSC + OVER + NDS. Although some of them (specifically MSC and FEC) can be improved by introducing a successive hangover mechanism which delays the transitions from 0 to 1, the presence of a hangover block makes the values of the OVER and NDS parameters worse. The latter were therefore given priority over MSC and FEC, in choosing the threshold.

The minimum total error is achieved with about $F_{th} = 0.21$. We chose $F_{th} = 0.25$, so as to reduce the value of OVER and NDS; as mentioned previously, the corresponding increase in FEC and MSC can be solved by introducing a hangover mechanism. The threshold F_{th} was also chosen so as to minimize the variance of the parameters affected by the hangover: this then allows us to design a suitable hangover for our VAD. We used a VAD hangover to eliminate mid-burst clipping of low speech levels. The mechanism is similar to the one used by the GSM VAD.

3.6.2.6 Experimental Results

In this subsection, we compare the performances of the ITU-T G.729 standard VAD, the Full Rate GSM VAD and the FVAD [17]. All results were averaged on the six types of background noise: white, car, street, restaurant, office and train noise. The results were analyzed considering the percentage of FEC and MSC in active voice frames, to calculate the amount of clipping introduced, and the percentage of OVER and NDS in non-active voice frames, to calculate the increase in activity.

Figure 3.9(a–b) shows a performance comparison in the case of a signal level of 22 dB below codec overload. Both in terms of clipping introduced (FEC +

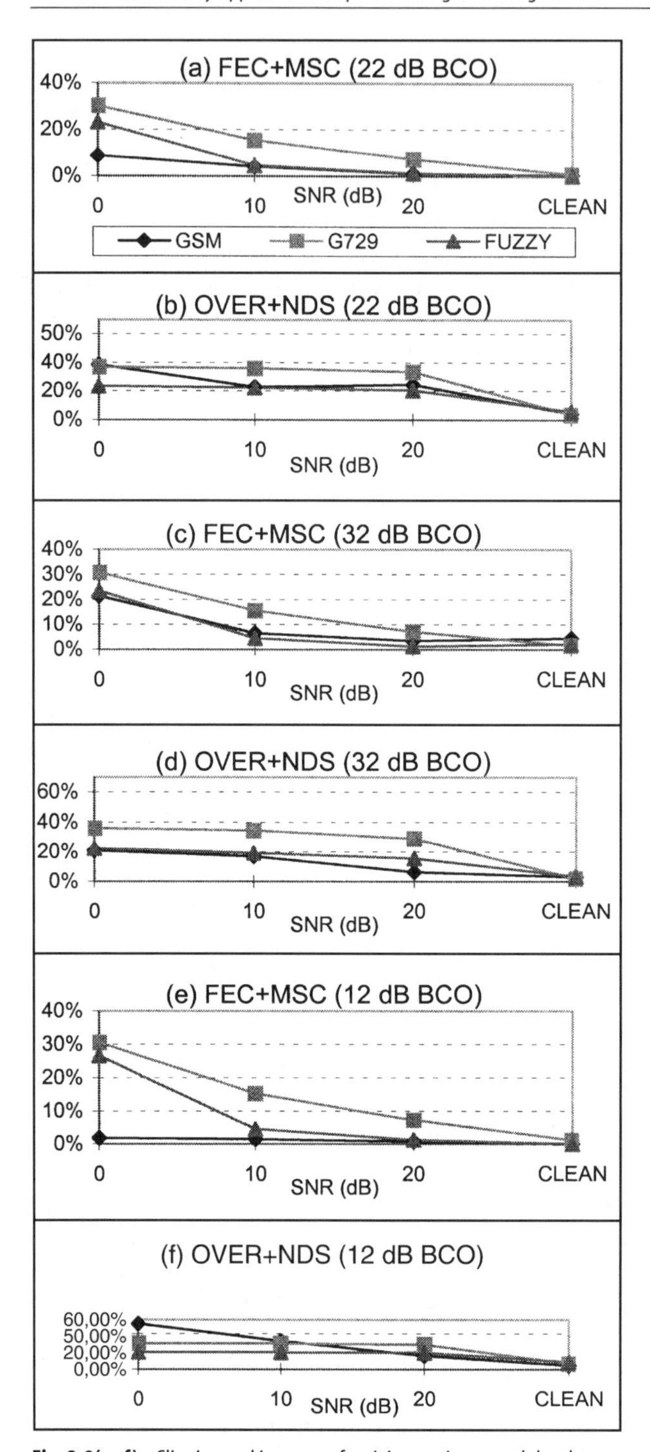

Fig. 3.9(a–f). Clipping and increase of activity varying speech level

MSC) and in terms of increase in activity (OVER + NDS), the FVAD performs better than the G.729, except in the clean case for which performance is similar. At SNR = 10 dB, for example, we halved both misclassification errors. We also observed that on average, the FVAD performance is similar to that of the GSM VAD, which in turn performs better than the ITU-T standard. In Fig. 3.9(c–d), we compare performance in the case of a signal level of 32 dB below codec overload. The performance of the FVAD and G.729 VAD, are substantially unchanged, whereas we observed an improvement in the performance of the GSM VAD in terms of the activity factor but a deterioration in terms of clipping, above all with very high and very low SNRs. Finally, Fig. 3.9(e–f) shows a comparison in the case of a signal level of 12 dB below codec overload. In terms of FEC+MSC, the FVAD still performs better than the G.729 (in fact the performance is substantially unchanged with respect to the 22 dB BCO case), whereas we observed a slight improvement in the GSM VAD performance, in terms of clipping.

In terms of OVER + NDS, the GSM VAD gives a worse performance, when the SNR is below 10 dB, due to the high signal level. We observed a deterioration in the performance of both the FVAD and the G.729 in the clean case, whereas below SNR = 20 dB, the FVAD performance is better than that of both the G.729 and the GSM VAD.

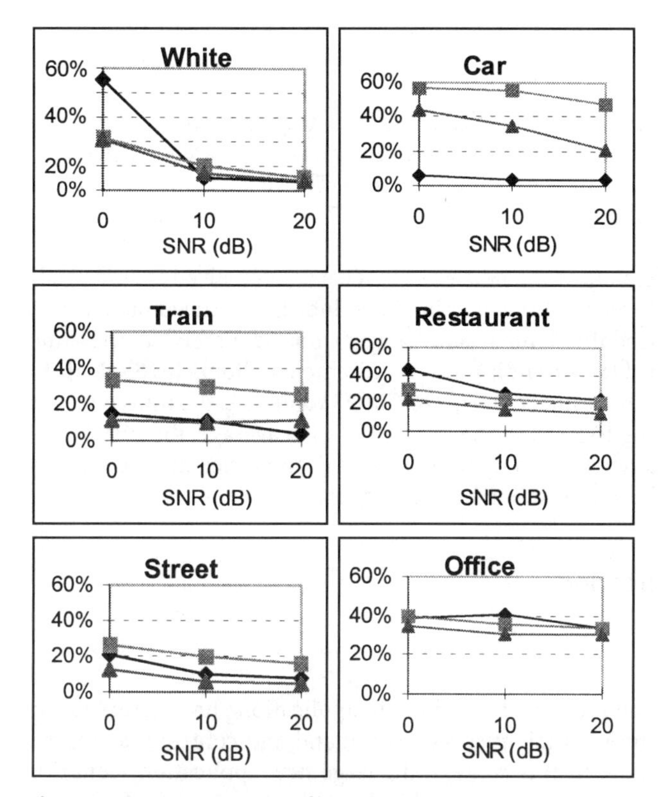

Fig. 3.10. Results varying types of background noise

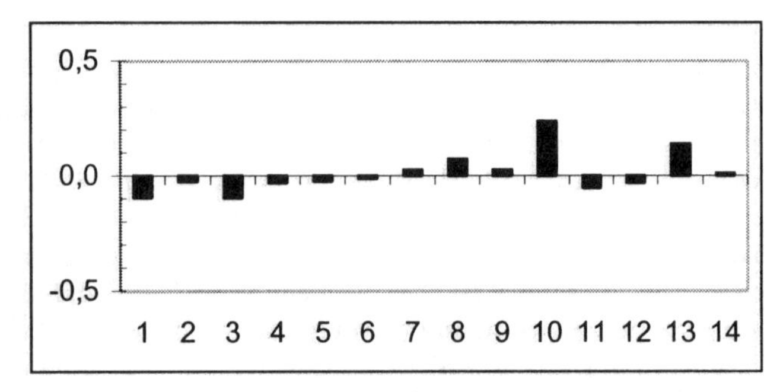

Fig. 3.11. Comparison Mean Opinion Score varying acoustic condition (a minus sign means that VAD proposed is worse than the ITU-T VAD). 1 – Clean; 2 – Car SNR =20 dB; 3 – Babble SNR = 20 dB; 4 – Traffic SNR = 20 dB; 5 – Office SNR = 20 dB; 6 – Car SNR =10 dB; 7 – Babble SNR = 10 dB; 8 – Traffic SNR = 10 dB; 9 – Office SNR = 10 dB; 10 – Car SNR =0 dB; 11 – Babble SNR = 0 dB; 12 – Traffic SNR = 0 dB; 13 – Office SNR = 0 dB; 14 – Mean value

A performance evaluation in terms of FEC + MSC + OVER + NDS with varying types of background noise, is shown in Fig. 3.10. The FVAD results are always better than those of the G.729. More specifically, we have a significant improvement in the case of car, train and street noises. Further, for non-stationary background noise, the FVAD performance is also better than that of the GSM VAD, whereas for stationary noise, performance is similar except for the car noise case.

We also made comparisons considering several sequences of modern and classical music, sampled at 8 kHz. More specifically, we calculated the percentage of clipping introduced by the 3 different VADs. The results indicate that the GSM VAD introduces about 5% of clipping, the G.729 VAD 20% and the FVAD 14%.

To evaluate the efficiency of the new VAD, in terms of perceived speech quality and the effect on listeners of the clipping introduced, we carried out a series of listening tests. We used the Comparison Category Rating method, in the same conditions adopted in [18] extending the requirements about the SNR up to 0 dB. Figure 3.11 gives the results in terms of CMOS values, i.e. the differences in MOS scores between the FVAD and the ITU-T VAD. On average, the FVAD presents similar performance to the G.729 VAD. For car and office noise at SNR = 0 dB FVAD performs better by about 0.2 MOS scores.

3.6.3 Endpoint Detection

3.6.3.1 Introduction

In the last few years, the growth of multimedia applications has increased the demand for new and more efficient speech command and control systems for man-machine interaction. In this context, the large new application scenarios will require systems and methodologies able to guarantee good performance levels, even in adverse acoustic noise conditions, with as low a computational load as possible [19]. As the recognition rate of an isolated word recognizer strongly depends on the accuracy of the End Point Detector (EPD) [19, 20], for the effective automatic recognition of speech, a robust word boundary detection algorithm is essential [21].

In this Section, we propose a new fuzzy logic-based boundary detection algorithm that meets the requirements of both computational simplicity and robustness to background noise. The Fuzzy End Point Detector (FEPD) [22] uses a set of four simple differential parameters and a matching phase based on a set of fuzzy rules. Experimental evaluation shows that the FEPD outperforms traditional word detection methods.

3.6.3.2 Description of the Algorithm

The architecture of the endpoint detector proposed, is based on a pattern recognition approach. More specifically, it consists of pre-processing the speech signal, extracting its significant features, a matching phase, a post-processing module and finally a decision block. Pre-processing of the speech signal consists of 140-Hz high-pass filtering, in order to eliminate the undesired low-frequency components.

In order to guarantee a robust word boundary detection in the presence of high noise levels, rather than using absolute parameters like energy, correlation and zero crossing, we propose a different approach, based on a set of parameters differentiated with respect to a local average. More specifically, the four differential parameters used for speech/noise classification are calculated in a window of 80 speech samples (a 10-ms frame, sampling at 8 kHz) and are: the full-band energy difference ΔE_f, the low-band energy difference ΔE_l, the zero-crossing difference ΔZC and the spectral distortion ΔS [23].

For the matching phase a new methodology is used having the advantage of exploiting all the information in the input pattern by means of a set of six fuzzy rules automatically extracted by FuGeNeSys tool. The fuzzy system has the task of mapping the pattern of input parameters onto a scalar value, ranging between 0 and 1, which indicates the degree of membership in the voice inactivity/activity classes [17].

In order to reduce sharp variations in the fuzzy system output, it is post-processed by a 7-th order median filter. Finally, the decision module, by means of a threshold comparison, returns the start and the ending point. Experimentally we chose a threshold value of $T_h = 0.9$ in that, observing the output of the post-processing module for several types of background noise and signal-to-noise ratios, it was rarely observed to exceed 0.9 in segments of pure noise, whereas inside word boundaries it is always close to 1. Further, in order to avoid false alarms, due to non-stationary noise, or premature endpoints due to possible intra-word pauses or unvoiced sounds, we inserted a simple control mechanism with two windows W_1 and W_2 frames in width, similar to the one proposed in [24]. To determine the start point, the window slides along the post-processed signal; if the number of elements that exceed the threshold value T_h is greater than k^*W_1, it is a start point; otherwise it is a false alarm. The same procedure is followed to determine the endpoint, which is detected if the signal is below the threshold k^*W_2 times. We experimentally fixed $k = 0.8$, $W_1 = 5$, $W_2 = 40$.

3.6.3.3 Experimental Results

In this section we evaluate the performance of new endpoint detection algorithm analyzing the results of a series of tests carried out in various acoustic noise and speech level conditions. The testing database consists of isolated words sampled at 8 kHz, linearly quantized at 16 bits per sample, normalized at −20 dBmO and preceded and followed by about 0.5 seconds of silence. More specifically, we chose as test words, a ten-digit Italian vocabulary pronounced by 20 male and 20 female speakers, different from those used in the training phase. In order to evaluate the FEPD robustness to noisy environments we digitally added three typical background noises (car, babble, traffic) at different signal-to-noise ratio values (30, 15, 5, 0 dB). Figure 3.12(a)–(d) shows a comparison between the FEPD performance and that of three traditional methods in terms of weighted accuracy (WA), an evaluation parameter, expressed in frames, which gives greater weight to the effect of word clipping than to widening, as suggested in [20]. In the comparison, to each method, we applied the same window mechanisms for the false alarm and intra-word silence problem. Further, in the graphs the WA values were saturated at a value of 50 frames. In regards to the performance of traditional methods, the first two reference endpoints detectors, proposed by Tsao et al. [24] and by Lamel et al. [25], present quite good performance only with SNRs greater than 20 dB, but they have the advantage of a low complexity. The third method, recently proposed by Junqua et al. [21], performs better in noisy environments but is computationally more complex. FEPD performance is better than traditional methods with all SNRs and types of background noise. More specifically, the results demonstrate not only an improvement with high SNRs, but also optimum FEPD behaviour in very noisy conditions. At SNR = 0 dB, for example, the WA parameter is halved with respect to traditional word boundaries detectors. In the presence of babble noise, we can observe that the WA tends to saturate when the background noise level increases and is reduced with traffic noise at SNR = 0 dB. This anomalous behaviour is due to the combined action of two factors: a) a saturation of the fuzzy output with high noise levels, which causes a widening of the endpoints (thus reversing the trend occurring up to SNR = 5 dB); b) the criterion used to define the WA parameter, which gives greater weight to clipping than to widening.

In all conditions, we observed a great reduction in the amount of clipping and widening greater than 180 ms (18 frames) in length, i.e of the errors that most degrade the performance of a speech recognition algorithm.

Finally, one of the problems of an energy-based speech activity detector, is the large dynamic range of speech signal levels. Above all in the presence of noise, the performance of a word boundary detector deteriorates as the speech level decreases. This problem is usually dealt with by using an activity gain con-

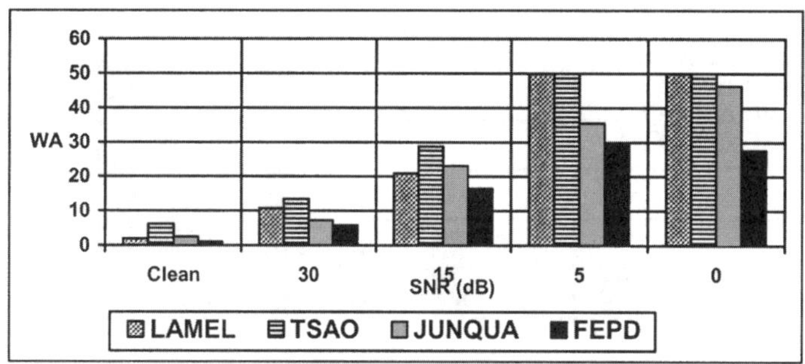

(a) All three types of noise

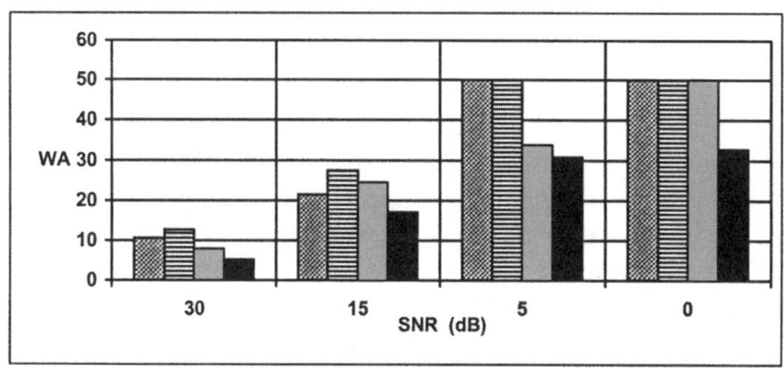

(b) Babble noise

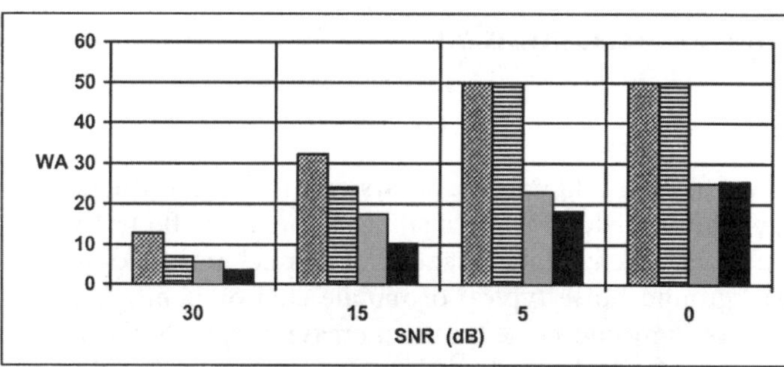

(c) Car noise

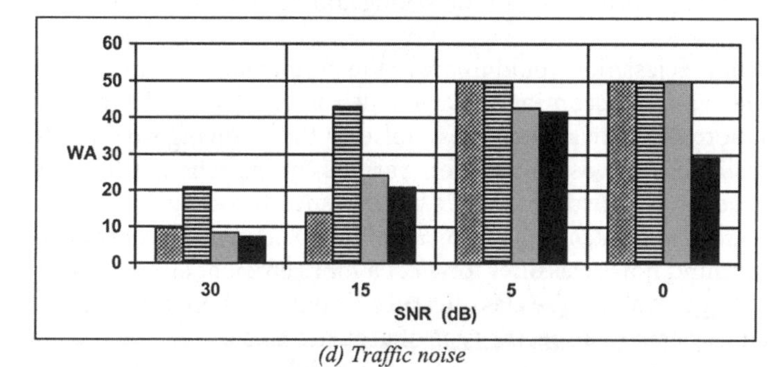

(d) Traffic noise

Fig. 3.12(a–d). Weighted accuracy versus signal-to-noise ratio for different types of background noise

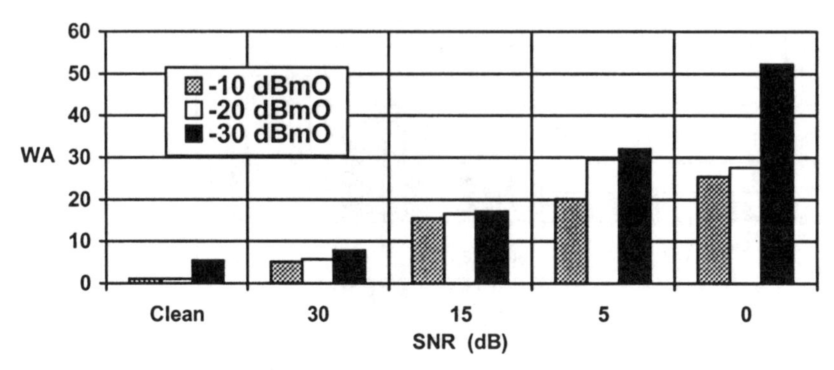

Fig. 3.13. FEPD performance at varying speech levels

trol (AGC) system. Thanks to the efficient set of differential parameters together with a robust, non-linear methodology for the pattern matching, the performance of the FEPD is quite independent of speech level changes. Figure 3.13 gives the weighted accuracy with varying SNRs and signal levels (± 10 dB as compared with the nominal level). Except for the 0 dB/-30 dBmO case, performance is similar to the reference nominal case of –20 dBmO, so the FEPD is robust to level variations and does not need an AGC system.

3.6.4 Background noise classification

3.6.4.1 Introduction

One important aspect of recent digital cellular systems is the robustness of the speech coding algorithms needed for the channel to be used efficiently: they have to be robust not only to channel degradation (channel noise, fading, etc.), but also to the background noise typical of mobile environments. A case in point is the effect of background noise on the increase in speech cutting percentages or the activity of a VAD, or the percentage of misclassification of a voicing algorithm, which would lead to the wrong coding model for the frame being analyzed.

A background noise classifier would improve the performance of these algorithms and, in the case of discontinuous transmission (DTX), would make it possible to use a more sophisticated comfort noise on the receiving side for the synthesis of long silence periods. In addition, various coding schemes have recently been proposed in the field of variable bit-rate speech coding, based on a phonetic classification approach [26, 27] or an object-oriented [28] approach, which use a background noise classifier to select a more efficient and appropriate coding model. In general, a noise classification module is useful for adapting a speech processing system to a specific type of acoustic noise.

In this Section, a background noise classifier for mobile environments is presented. The methodology adopted is based on a pattern recognition approach,

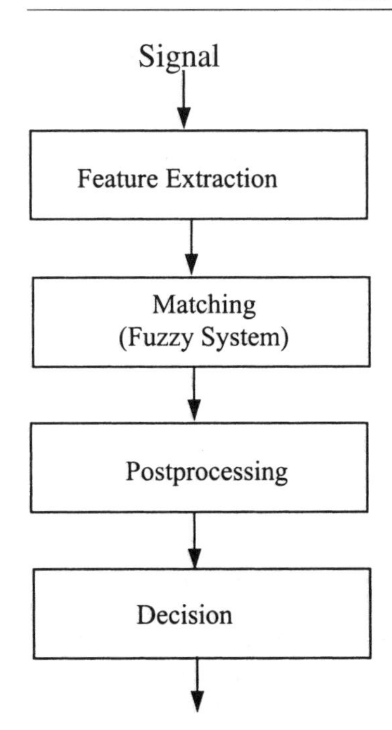

Fig. 3.14. Noise classifier architecture

in which the matching phase is performed using a fuzzy system. The fuzzy rules are automatically extracted from examples by the hybrid learning tool presented in Section 3.3.

The noise classifier operates at two levels: a) discrimination between steady-state and transient background noise; b) discrimination between white Gaussian and car noise and discrimination between babble and traffic noise.

3.6.4.2 Noise Classifier Architecture

Figure 3.14 shows the first level of the architecture of the background noise classifier proposed in [29]. Each 10-ms of waveform the feature extraction module computes the set of parameters adopted. The matching phase is performed by a set of fuzzy rules obtained through supervised learning using FuGeNeSys. The fuzzy system output is postprocessed by a median or sliding average filter. The final decision is made by a threshold comparison. Several background noises typical of mobile environments, were taken into consideration at the first level of classification and grouped into two classes according to their steady-state or transient features. The second level comprises two more classifiers, whose feature extraction and matching modules have been optimized for the subclassification of steady-state noise into white Gaussian (W) and car (C) noise, and transient noise into babble (B) and traffic (T) noise.

In addition, we present: 1) a criterion to group a large range of environmental noise into a reduced set of classes of noise with similar acoustic characteristics; 2) a larger set of background noise together with a new multilevel classification architecture; 3) a new set of robust acoustic parameters. The improved version of the Fuzzy Noise Classifier (FNC) has been assessed in terms of misclassification percentage and compared with a Quadratic Gaussian Classifier (QGC) recently introduced in literature [30].

3.6.4.3 Noise Database

The database contains a large number of different types of background noise. The learning and testing patterns contain sequences recorded in various noisy mobile environments, sampled at 8000 Hz and linearly quantized at 16 bits per sample. The training set consists of the following 4 examples of stationary noise: Bus, Car, Train, and Dump; and the following 8 examples of non-stationary noise: Office, Restaurant, Street, Factory, Construction, Shopping, Rail station, and Pool.

The set of examples thus built, comprises a total of 20000 patterns, equally distributed between the classes, giving a total of 200 seconds of signal. A different set of examples was constructed to validate the results obtained. It is made up in part of noise with the same features as that used in the training set, but also included types of noise not considered previously. The testing set comprises 24000 patterns, again equally distributed between the two classes, corresponding to four minutes of signal.

3.6.4.4 Noise classes Selected and Classification Tree

In order to group the 12 environmental noises into a reduced set of classes of noise with similar acoustic characteristics, we measured the linear separability among each pair of noises, by means of the Fisher Discriminant Ratio (FDR) [14, 15] over a large initial set of acoustic parameters (more than 100). In particular, two different types of noise belong to the same class if the maximum value of FDR varying the parameters is less than 1, i. e. when the FDR of all features is very low. On the basis of this criterion, we obtained the following seven noise classes listed in Table 3.3.

Table 3.4 lists the pairs of noises with a maximum FDR of less than one, the value of the maximum FDR and the relative parameter. Figure 3.15 shows the classification tree of the Fuzzy Noise Classifier (FNC) proposed in this contribution. The noise classifier operates at four levels and the FNC discriminates:

Table 3.3. Classes of noise with different characteristics

Stationary	Non-Stationary
Car	Street
Train	Factory
Bus-Dump	Construction
	Babble (Pool, Office, Restaurant, Shopping, Rail station)

Table 3.4. Pairs of noises with FDR max less than one

Noises	FDR max	Parameter
Dump-Bus	0.88	Inverse of the short prediction gain
Office-Rest	0.59	8° Cepstral parameter
Office-Shop	0.74	1° Cepstrum parameter
Office-Pool	0.60	1° LPC parameter
Office-Rail station	0.70	1° Cepstrum parameter
Shop-Pool	0.69	2° Cepstral parameter
Shop-Rest	0.71	2° Cepstral parameter
Shop-Rail station	0.98	2° Cepstral parameter
Pool-Rest	0.87	2° LPC parameter
Pool-Rail station	0.73	3° LPC parameter

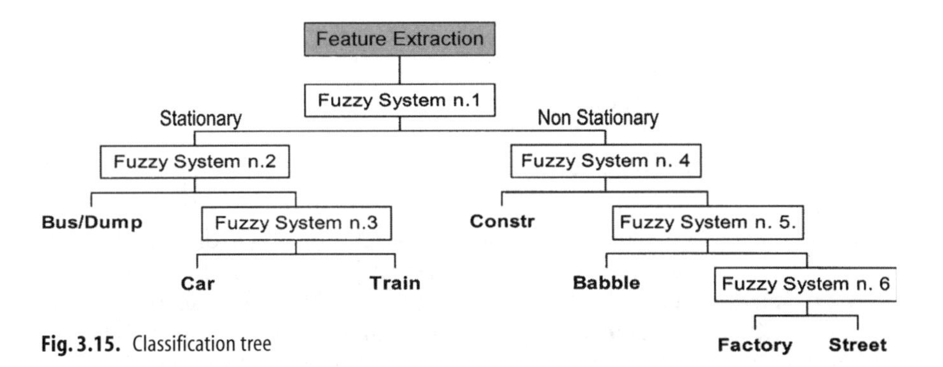

Fig. 3.15. Classification tree

1) at the first level between stationary and non-stationary noise; 2) at the second level between Bus/Dump and Car/Train noise for stationary noise sub-classification, and between Construction and Babble/Factory/Street noise for the non-stationary noise sub-classification; 3) at the third level between Car and Train noise for the stationary noise sub-classification, and between Babble and Factory/Street noise for the non-stationary noise sub-classification; 4) finally, at the fourth level between Factory and Street noise for the non-stationary noise sub-classification.

For each classification level we have an automatically trained fuzzy system discriminating between the two acoustic classes. In total, we have six fuzzy systems for each of which the appropriate number of fuzzy rules was 10.

3.6.4.5 Feature Selection

Feature selection represents one of the most important tasks in a pattern recognition approach. It allows for a low computational load, without increasing the misclassification error.

The goal is to obtain an efficient, small vector of acoustic features which represent the input pattern for the fuzzy system being trained. Each fuzzy system

Table 3.5. Features selected for each fuzzy system

Input Parameters	Fuzzy System					
	1	2	3	4	5	6
1) Differential power	*					*
2) Differential variance	*					
3) Differential left variance	*					
4) Differential short term prediction gain					*	
5) Inverse of the short time prediction gain		*				
6) Norm of cepstrum		*				
7) Norm of LAR coefficients					*	
8) Norm of LPC coefficients			*			
9) Norm of 10 cepstral coefficients			*			
10) First normalized autocorrelation coefficient				*		*
11) First cepstral coefficient			*			
12) First LPC coefficient		*	*	*		*
13) Regularity				*		
14) Right variance					*	
15) Normalized energy in the band 0–900 Hz						*

provides a model for the acoustic parameter space of each category of background noise.

The method adopted to identify the optimal set from a total of more than 100 parameters, is based on the Fukunaga criterion [14]. A total of 15 parameters were extracted, even though the six fuzzy systems only use a subset of these parameters. We have, in fact, a maximum of 4 inputs for each system. Table 3.5 shows the set of 15 parameters selected, highlighting the optimal subsets of input features for the six fuzzy models. All parameters require a very low computational load.

3.6.4.6 Results

A series of tests were performed to evaluate the performance of the classifier in terms of misclassification matrices and error rate. Moreover, the results were compared with those obtained using a matching model based on the Quadratic Gaussian Classifier (QGC) and recently proposed in [30], as the best algorithm among those tested. The QGC uses Line Spectral Frequencies (LSFs) as input patterns. In order to maintain similar test conditions, in the comparison between the QGC and the FNC, we only considered the five types of background noises used in [30]. Further, since the QGC presents a frame length of 20 ms, in order to have the same processing delay, we used a 3rd-order median filter as the post-processing module, increasing the FNC delay to 20 ms.

3.6.4.7 Stationary/Non-Stationary Classification

The first level of classification represents a crucial point in the performance of all classifiers. The testing results relating to the first level of classification be-

Table 3.6. Stationary/non-stationary noise confusion matrix

		OUTPUT			
		Stationary		Non-Stationary	
INPUT		QGC (%)	FNC (%)	QGC (%)	FNC (%)
	Stationary	92.5	99.3	7.5	0.7
	Non-Stationary	14.4	0.3	85.6	99.7

Table 3.7. QGC-FNC comparison in terms of total stationary/ non-stationary noise misclassification

Type of classifier	Total error (%)
QGC	10.95
FNC	0.5

tween stationary and non-stationary noise are summarized in Table 3.6. In this case, only 3 parameters are needed for discrimination (refer to Table 3.5). As can be seen, the misclassification between stationary and non-stationary noise is very low and, consequently, this guarantees a low propagation error.

Table 3.7 compares the two classifiers in terms of the total misclassification error, in the various cases examined. As can be seen, the performance of the fuzzy method, in term of stationary/non-stationary misclassification, is clearly better than that of the traditional technique.

3.6.4.8 Performance at the Sub-Classification Levels

The results relating to the whole FNC classification tree, are given in Table 3.8. These results are good in that, on average, a 90% discrimination is achieved.

Table 3.8. Confusion matrix

		OUTPUT (%)						
		Bus-Dump	Train	Car	Constr	Street	Fact	Babble
INPUT	Bus-Dump	88.2	2.4	8.3	0	0	0	1.1
	Train	5	83.6	10.4	0	0	0	1
	Car	13	10.6	76.4	0	0	0	0
	Cons	0	0	0	99	0	1	0
	Street	0.1	0	0	0	58.3	28.4	13.2
	Fact	0	0	0	8.2	1.5	88.5	1.8
	Babble	0.3	0.2	0.3	1.2	10.4	9.8	77.8

Table 3.9. Confusion matrix for QGC classifier

	Car	Bus	Factory	Street	Babble
Car	99.6	0.2	0.2	0	0
Bus	0	85.2	2.2	3.8	8.8
Factory	0	5.6	93.2	0.2	1.0
Street	0.0	24.8	2.0	71.4	1.8
Babble	0.0	12.8	2.0	5.4	79.8

Table 3.10. Confusion matrix for FNC classifier

	Car	Bus	Factory	Street	Babble
Car	94.3	5.7	0	0	0
Bus	9.5	90.4	0	0	0.1
Factory	0	0	93.5	4.9	1.6
Street	0	6.5	3.4	79.3	10.8
Babble	5.1	2.7	3.1	5.1	84.0

Comparing the QGC and FNC classifiers, as can be seen in Tables 3.9 and 3.10, on average, we again have an improvement in the recognition rate. The improvement is for all types of background noise except for car noise. We have a total error of 14.16% for the QGC classifier and of 11.7% for the FNC classifier.

3.7 Conclusions

In this chapter, we have presented a series of applications in speech and signal classification, based on a neurofuzzy pattern recognition approach. In particular, first we have presented an efficient fuzzy logic-based method for V/UV speech classification, which is language- and speaker-independent and computationally simple. Successively, we have presented a voice activity detector based on a set of six fuzzy logic and than we have presented a computationally simple and noise robust endpoint detection algorithm. Finally, a simple criterion for grouping a large range of noise into a reduced set of background noise classes has been proposed. A series of tests have demostrated for each application that a pattern classification based on the neurofuzzy approach is both more efficient and noise robust than traditional solutions.

References

[1] Zadeh L A (1994) Fuzzy Logic, Neural Networks and Soft Computing, Comm. of the ACM, vol. 37, no. 3, pp. 77–84.

[2] Russo M (1997) FuGeNeSys: Sw and Hw Implementation, Proc. of IEEE First Int. Conf. on Conventional and Knowledge-Based Intelligent Electronic System, (Adelaide, Australia), pp. 209–218.

[3] Russo M (1998) FuGeNeSys: A Fuzzy Genetic Neural System for Fuzzy Modeling, IEEE Transaction on Fuzzy Systems , Vol. 6, pp. 373–388.
[4] Rabiner L R, Schafer R W (1978) Digital Processing of Speech Signal, Prentice-Hall, Inc., Englewood Cliffs, New Jersey.
[5] Tang K S, Man K F, Kwong S, He Q, (1996) Genetic Algorithms and their Applications, IEEE Signal Processing Magazine, vol. 13, n. 6, pp. 22–37.
[6] Cox R V, Kroon P (1996) Low Bit-Rate Speech Coders for Multimedia Communication, IEEE Communication Magazine, pp. 34–41.
[7] Beritelli F, Casale S (1997) Robust Voiced/Unvoiced Speech Classification using Fuzzy Rules, Proc. IEEE Workshop on Speech Coding, Pennsylvania, USA, pp. 5–6.
[8] Hess W J (1992) Pitch and Voicing Determination, Advances in Speech Signal Processing edited by Sadaoki Furui, M. Mohan Sondhi, pp. 3–48, Dekker.
[9] Campbell J P, Tremain T E (1986) Voiced/Unvoiced Classification of Speech with Applications to the U.S. Government LPC-10E Algorithm, Proc. of IEEE Int. Conf. on Acoustics, Speech and Signal Processing, Tokyo, pp.473–6.
[10] Sririam K, Varshney P K, Shanthikumar J G (1983) Discrete-Time Analysis of Integrated Voice/Data Multiplexers with and without Speech Activity Detectors, IEEE J-SAC vol. 1, n. 6, pp. 1124–1132.
[11] Gersho A (1996) Advances in Speech and Audio Compression, IEEE Proc., vol. 82, no. 6, pp. 900–918.
[12] Rec. ITU-T G.729 Annex B, 1996.
[13] ETSI GSM 06.32 (ETS 300-580-6), September 1994.
[14] Fukunaga K (1990) Introduction to Statistical Pattern Recognition, Academic Press, San Diego, California.
[15] Kil D H, Shin F B (1996) Pattern recognition and prediction with applications to signal characterization, AIP (American Institute of Physics) Press, Woodbury, New York.
[16] Southcott C B et al. (1989) Voice Control of the Pan-European Digital Mobile Radio System, ICC '89, pp. 1070–1074.
[17] Beritelli F, Casale S, Cavallaro A (1998) A Robust Voice Activity Detector for Wireless Communications Using Soft Computing, IEEE Journal on Selected Areas in Communications (JSAC), special Issue on Signal Processing for Wireless Communications, Vol. 16, No. 9, pp. 1818–1829.
[18] Pascal D (1996) Results of the Quality of the VAD/DTX/CNG of G.729 A (CCR Method), ITU-T contribution, Geneva.
[19] Junqua J C, Haton J P (1996) Robustness in Automatic Speech Recognition: Fundamentals and Applications, Kluwer Academic Publishers.
[20] Rabiner L, Juang B H (1993) Fundamentals of speech recognition, Prentice Hall International.
[21] Junqua J C, Mak B, Reaves B (1994) A robust algorithm for word boundary detection in the presence of noise, IEEE Trans. on Speech and Audio Proc., Vol. 2, No.3, pp. 406–412.
[22] Beritelli F (2000) Robust Word Boundary Detection Using Fuzzy Logic, Electronics Letters, Vol. 36, No. 9, pp. 846–8.
[23] Benyassine A, Shlomot E, Su H Y, Massaloux D, Lamblin C, Petit J P (1997) ITU Reccomendation G.729 Annex B: A Silence Compression Scheme for Use with G.729 Optimized for V.70 Digital Simultaneous Voice and Data Applications, IEEE Comm. Magazine, Vol. 35, No. 9, pp. 64–73.
[24] Tsao C, Gray R M (1984) An endpoint detector for LPC speech using residual error look-ahead for vector quantization applications, Proc. of IEEE Int. Conf. on Acoustic, Speech and Signal Processing (ICASSP'84), pp. 18B.7.1–18B.7.4.
[25] Lamel L F, Rabiner L R, Rosemberg A E, Wilpon J G (1981) An improved endpoint detector for isolated word recognition, IEEE Trans. on Acoustic, Speech and Signal Processing, Vol. 29, No. 4, pp. 777–785.
[26] Paksoy E, Srinivasan K, Gersho A (1994) Variable Bit-Rate CELP Coding of Speech with Phonetic Classification, European Trans. on Telecommunications, vol. 5, pp. 591–601.

[27] Cellario L, Sereno D (1994) CELP Coding at Variable Rate, European Trans. on Telecommunications, vol. 5, pp. 603–613.
[28] Chiariglione L (1997) MPEG and Multimedia Communications, IEEE Trans.on Circuits and Systems for Video Technology, vol. 7, No. 1, pp. 5–18.
[29] Beritelli F, Casale S, Ruggeri G (2000) New Results in Fuzzy Pattern Classification of Background Noise, International Conference on Signal Processing (ICSP2000).
[30] El-Maleh K, Samouelian A, Kabal P (1999) Frame-Level Noise Classification in Mobile Environments, ICASSP'99, Phoenix, Arizona.
[31] Haykin S (1998) Neural Networks: A Comprehensive Foundation, Prentice Hall, 1998.
[32] Farell K R, Mammone R J, Assaleh K T (1994) Speaker recognition using neural networks and conventional classifiers, IEEE Transactions on speech and audio processing, Vol. 2, No. 1, Part II, pag. 194–203.
[33] Karnianadecha M, Zahorian S A (1999) Signal modeling for isolated word recognition, International Conference on Acoustic, Speech, ans Signal Processing, Phoenix, Arizona.

4 Image/Video Compression Using Neuro-Fuzzy Techniques

Wan-Jui Lee, Chen-Sen Ouyang and Shie-Jue Lee

4.1 Introduction

The phenomenal increase in generation, transmission, and use of digital images and video in various applications, is placing an enormous demand on storage space and communication bandwidth. Data compression is a viable approach to alleviate such storage and bandwidth demands and enables applications such as digital video broadcasting, video streaming over internets, and mobile videophones, which were impracticable only a few years ago [6, 18, 22, 55, 71]. Image/video compression concerns transmitting image/video data with as much high quality and low bandwidth as possible, within a processing time acceptable by the user. Two types of compression are possible, i.e., lossless and lossy. With lossless compression, there is no loss of information and therefore the image can be perfectly recovered. Lossy compression, on the other hand, is a trade-off, which results in less amount of data with a sacrifice to fidelity.

4.1.1 Image Compression

A number of image compression techniques have been developed. Generally, there are four directions in image compression: vector quantization, transform coding, predictive coding, and entropy coding. Vector quantization (VQ) tries to reduce spatial redundancy in an image [49]. Representative blocks are chosen from the image to be compressed, as code-words. Each block of the image is compared with the code-words to find out the most similar one and labeled with the index of the corresponding code-word. As a result, the underlying image is represented by a series of indices, which are then transmitted through the communication channel. At the receiving end, the code-words corresponding to the received indices are used to reconstruct the original image. Since only table lookups are involved at the receiving end, VQ is particularly suitable for real-time decoding. This simplicity also results in low complexity of the decoder which is attractive for low power systems and applications.

Transform coding compresses an image by converting the image into a small number of coefficients. Discrete Cosine Transform (DCT) and wavelet transform, are two well-known methods. DCT transforms a block of pixels into a matrix of coefficients, which represent spatial frequencies. Because most high-frequency coefficients are nearly zero, compression is attained. However, the underlying block-based scheme generally degrades the performance at low bitrates since correlation across the block boundaries is not eliminated. Recently, the wavelet transform coding (also referred to as sub-band coding) has emerged as a cutting edge technology in the field of image compression. The basic idea of wavelet transform is to represent any arbitrary function as a superposition of a set of wavelets or basis functions, which are obtained by scaling and shifting from the mother wavelet. Wavelet-based coding is more robust under transmission and decoding errors, and also facilitates progressive transmission of images. Because of their inherent multi-resolution nature, wavelet coding schemes are especially suitable for applications where scalability and tolerable degradation are important.

Predictive coding works on the basis that adjacent pixels in an image are generally very similar, and thus the redundancy between successive pixels can be removed and only the residual between the actual and predicted pixel values is encoded. Differential Pulse Code Modulation (DPCM) is one particular example of predictive coding. Entropy coding relies on uneven distribution of values to be encoded, and the code length of a value associated inversely with the probability the value occurs in the image. Two popular entropy coding schemes, are Huffman coding and arithmetic coding. Huffman coding aims to produce variable-length codes for symbols to represent data at a rate of its entropy. From a given probability function for each value of image pixels, a tree is constructed by summing the two lowest probabilities iteratively, until no merge can be done. Then each symbol is encoded by tracing the corresponding path in the tree. In this way, Huffman coding, maps fixed length symbols to variable length codes. Arithmetic coding converts a variable number of symbols into a variable length code-word. A sequence of symbols is represented by an interval with length equal to its probability. The interval is specified by its lower and upper boundaries. The code-word of the sequence is the common bits in binary representations of its lower and upper boundaries.

4.1.2 Video Compression

Video data usually contains streams of image frames, each stream consisting of a set of sequential image frames. Transmitting a video stream requires a large bit rate and needs a high compression ratio for real applications. Two types of approaches, block-based and object-based, have been developed for video compression. In block-based approaches, e.g., MPEG-2 [2], an image frame is encoded as a set of fixed-size blocks, and redundancies are removed by attempting to match and reuse the blocks from the previous frames of the current frame. The most basic form of block-based approaches checks if a block in the current frame is identical to a block around the same place in the previous frame. If they are the same, the data of the underlying block is not encoded. Otherwise, the best matched block in the previous frame is found and the difference among them is encoded. The area in which matching is searched, affects the quality of the reconstructed image frame. The larger the area in which matching is searched, the larger the chance that matching can be found. However, most matched blocks are found around the place of the original block. Besides, increasing the search area also increases the computation required.

Object-based approaches, e.g., MPEG-4, treats an image frame as a set of objects. Instead of coding each individual block of an object, we can represent the whole object by a simple code and thus transmission efficiency can be increased. The same objects between adjacent image frames can be described by the same identity plus the difference between them. Furthermore, object-based approaches allow more sophisticated content-based interactivity, such as tuning compression parameters for different objects and manipulation of various objects. In an image frame, a region is defined as a contiguous set of pixels that are homogeneous in terms of certain features, i.e., texture, color, motion, or shape. A video object is a collection of regions which have been grouped together under some criteria across several image frames. For instance, a shot of a person walking can be segmented into a collection of adjoining regions by different criteria, but all the regions may exhibit consistency in their motion attributes. Therefore, finding video objects from image frames is the key issue for object-based compression approaches.

4.1.3 Fuzzy Theory and Neural Networks

Clearly, many compression techniques require a mechanism for clustering or prediction. For example, VQ may apply a clustering mechanism to derive code-words. The blocks of an image can be separated into clusters, and one representative is decided for each cluster and is used as a code-word. A similar clustering mechanism can be applied in object-based video compression for determining various objects from a video stream. The predictive coding method for image compression may use a predicting mechanism to predict the value of the current pixel from the values of the previous pixels.

Quantitative approaches based on conventional mathematics can be used to group similar elements into categories, but they are not suitable when the underlying application is complex, ill-defined, or uncertain. Fuzzy theory [69] was proposed to deal with the applications in which the categorical boundaries are not crisp. Bezdek et al. [9] proposed the fuzzy c-means (FCM) algorithm, which generalizes the hard c-means algorithm [20] to produce a soft partition for a given dataset. Lin et al. [42] obtained fuzzy partitions by iteratively cutting each dimension of the input space into two parts. Wong and Chen [64] proposed another idea for clustering. Reference vectors attract one another and form different clusters according to a similarity measure. Juang and Lin [33, 34] obtain fuzzy clusters from a given dataset via an aligned clustering-based algorithm and a projection-based correlation measure.

Neural networks have learning capabilities and can learn clusters from given training data. Adaptive resonance theory (ART) [13, 14, 28] is a network for data clustering. The stored prototype of a category is adapted when an input training pattern is sufficiently similar to the prototype. When an input training pattern is not sufficiently similar to any existing prototype, a new category is formed with the input training pattern as the prototype. Kohonen clustering network (KCN) [38], also known as Kohonen self-organizing map

(SOM), is a fully connected linear network. The output generally is organized in a one- or two-dimensional arrangement of neurons. The weights connecting the input to the output perform association between weights and inputs. Not only the winner of the competition, but also its neighbors have their weights updated according to the competitive rule. The HEC network [43] performs a partitional clustering using the regularized Mahalanobis distance. It consists of two layers. The first layer employs a number of principal component analysis subnetworks to estimate the hyper-ellipsoidal shapes of currently formed clusters, and the second layer performs a competitive learning using the cluster shape information provided by the first layer. The spiking neural network [11], consists of a fully connected feedforward network of spiking neurons. Each data-point is translated into a multidimensional vector of spike-times in the input layer. If the distance between clusters is sufficiently small, the winner-takes-all competition tunes output neurons to the spike-time vectors associated with the centers of the respective clusters. The cluster-detection-and-labeling (CDL) network [21] consists of two layers and a threshold calculation unit. The first layer performs similarity matching, whereas the second layer implements cluster assignments.

Recently, neuro-fuzzy approaches for data clustering have attracted a lot of attention [35, 53, 64, 66, 67]. Such approaches, combine advantages of both fuzzy theory and neural networks. Fuzzy ART [15] is capable of learning categories in response to arbitrary sequences of analog or binary input patterns. Input vectors are normalized according to a complement coding process which makes the MIN operator and the MAX operator of fuzzy theory complementary to each other. The fuzzy min-max clustering neural network proposed in [57] adopts hyperbox fuzzy sets. Learning is done by creating and expanding/contracting hyperboxes in the pattern space. The fuzzy Kohonen clustering network (FKCN) [59] combines the fuzzy c-means algorithm and KCN. It offers automatic control on the learning rate distribution and the extent of topological neighborhood using fuzzy membership values. The fuzzy bidirectional associative clustering network (FBACN) [62] is composed of two layers of recurrent networks, performing fuzzy-partition clustering according to the objective-functional method. The first layer of FBACN is implemented by a Hopfield network, while the second layer is implemented by a multi-synapse neural network with added stochastic elements of simulated annealing. The self-constructing fuzzy neural network (SCFNN) [41] is able to partition a given dataset into a set of clusters based on similarity tests. Membership functions associated with each cluster are defined according to statistical means and variances of the data points included in the cluster. Besides, parameters can be refined to increase the precision of the resulting clusters.

4.2 Neuro-Fuzzy Techniques

As indicated earlier, a lot of neuro-fuzzy techniques have been developed. In this section, we describe three techniques: FKCN, Fuzzy-ART, and SCFNN.

FKCN is a non-sequential fuzzy neural network, i.e., all the training data are considered together. Fuzzy-ART and SCFNN, on the other hand, are sequential ones, in which data are considered one at a time.

4.2.1 Fuzzy Kohonen Clustering Networks (FKCN)

Assume that we have N training vectors each of dimension n. The task of FKCN is to find out c reference vectors to partition these N training vectors (or patterns) into c clusters properly. A FKCN network consists of three layers: an input layer, a distance layer, and an output (membership) layer, having n, c, and c neurons, respectively, as shown in Fig. 4.1. The input layer receives the input training patterns and broadcasts them forward to the distance layer. The distance layer calculates distances between input patterns and reference vectors, and transmits the distances to the output layer. The output layer computes the membership degrees that the input pattern belongs to each cluster. A matrix of trainable weights, V,

$$V = \begin{bmatrix} \boldsymbol{v}_1 \\ \boldsymbol{v}_2 \\ \vdots \\ \boldsymbol{v}_c \end{bmatrix} = \begin{bmatrix} v_{11} & v_{12} & \cdots & v_{1c} \\ v_{21} & v_{22} & \cdots & v_{2c} \\ \vdots & \vdots & \vdots & \vdots \\ v_{n1} & v_{n2} & \cdots & v_{nc} \end{bmatrix} \tag{4.1}$$

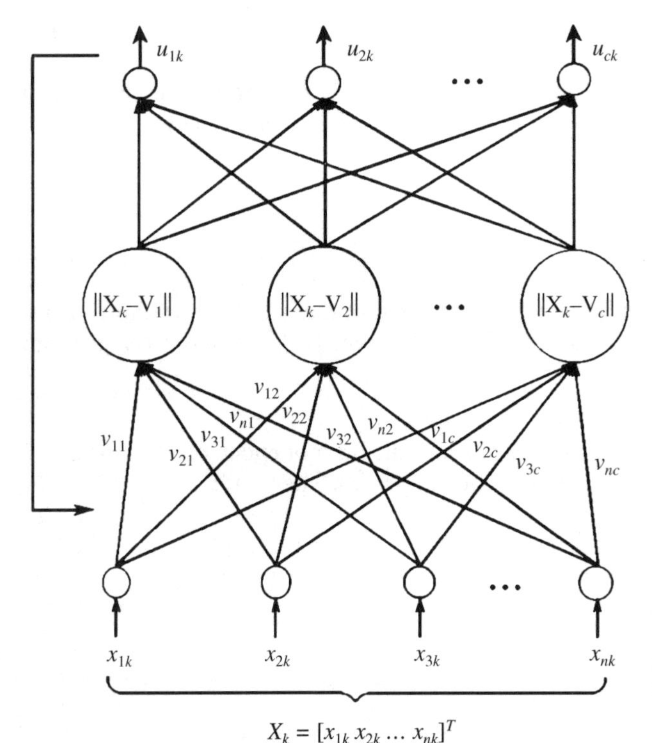

Fig. 4.1. Architecture of FKCN networks

exists between the input layer and the distance layer, where v_{ij} is the weight of the connection between the input node i and the distance node j. Basically, V is obtained by minimizing the following objective function:

$$\sum_{k}\sum_{j}(u_{kj})^m \| \mathbf{x}_k - \mathbf{v}_j \|^2 , \tag{4.2}$$

where u_{ki} is the membership degree of $\mathbf{x}_k$ belonging to cluster j and m is a controlling exponent to be calculated later.

During the training, each input pattern $\mathbf{x}_k$, where $\mathbf{x}_k = [x_{1k}, \ldots, x_{nk}]$ and $1 \le k \le N$, is presented to the input layer. Each node j, $j = 1, 2, \ldots, c$, in the distance layer calculates the distance d_{ki} between $\mathbf{x}_k$ and $\mathbf{v}_j$ *as follows:*

$$d_{kj} = \|\mathbf{x}_k - \mathbf{v}_j\|^2 = (\mathbf{x}_k - \mathbf{v}_j)^T(\mathbf{x}_k - \mathbf{v}_j). \tag{4.3}$$

Then all the weights are adjusted by:

$$\mathbf{v}_j(t) = \mathbf{v}_j(t-1) + \left[\sum_{k=1}^{N} \eta_{kj}(\mathbf{x}_k - \mathbf{v}_j(t-1))\right] / \sum_{k=1}^{N} \eta_{kj} \tag{4.4}$$

where η_{kj} is the learning rate defined by

$$\eta_{kj} = (u_{kj})^m \tag{4.5}$$

$$m = m_0 - z\Delta m , \tag{4.6}$$

$$\Delta m = \frac{m_0 - 1}{z_{max}} \tag{4.7}$$

with m_0 being a positive constant greater than 1, z being the current epoch count, z_{max} being the limit of the epoch count, and u_{kj} being calculated by:

$$u_{kj} = \frac{1}{\sum_{i=1}^{c}\left(\frac{d_{kj}}{d_{ki}}\right)^{\frac{2}{m-1}}}. \tag{4.8}$$

The process iterates until either that the weight matrix V in two consecutive epochs are close enough or the epoch count z exceeds z_{max}. The FKCN algorithm can be summarized below:

procedure FKCN
  Set the number of clusters to be c and the limit of epoch count to be z_{max};
  Initialize the weight matrix V and choose m_0, where $m_0 > 1$;
  for $z = 1, 2, \ldots, z_{max}$
  Compute all learning rates $\eta_{kj}, j = 1, 2, \ldots, c, k = 1, 2, \ldots, N$, with Eq. (4.5);
  Update all weight vectors $\mathbf{v}_j, j = 1, 2, \ldots, c$, with Eq. (4.4);
  Compute $E = \|V^z - V^{(z-1)}\|^2 = \Sigma_j \|\mathbf{v}_j^z - \mathbf{v}_j^{(z-1)}\|^2$;
  if $E \le \epsilon$ *break*;
  else $z = z + 1$;
  return with c clusters;
end FKCN

When the process terminates, $v_j, j = 1, 2, \ldots, c$, is treated *as* the center or mean of cluster j. A pattern $\boldsymbol{x}_k$ belongs to cluster p if u_{kp} of node p is the highest output at the output layer.

4.2.2 Fuzzy-ART Networks

Fuzzy-ART [15] networks are similar to ART networks [13, 14], but they accept continuous inputs between 0 and 1 that represent fuzzy membership values.

A fuzzy-ART network consists of three layers: an input layer, a choice layer, and a match layer, as shown in Fig 4.1. Like FKCN networks, a matrix $\boldsymbol{V}$ of weights exists between the input layer and the choice layer. However, unlike FKCN networks, clusters are generated as needed and the number of clusters is determined automatically. Initially, the network contains only the input layer

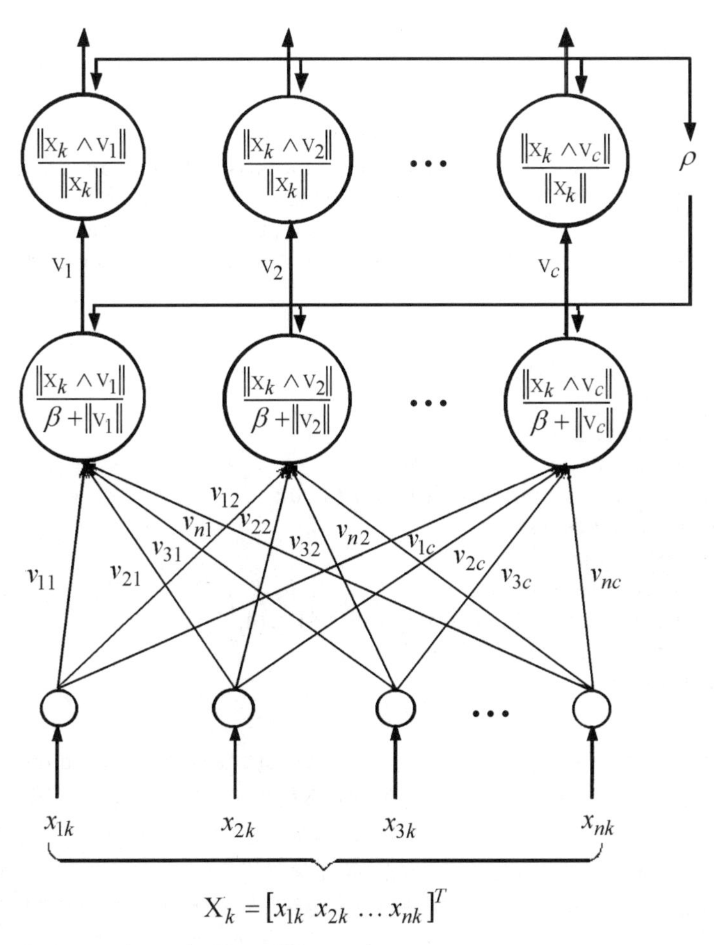

Fig. 4.2. Architecture of Fuzzy-ART networks

and the number of clusters, J, is set to 0. When an input vector $\boldsymbol{x}_k$ is presented to the network, we compute distance d_j by:

$$d_j = \frac{\| \boldsymbol{x}_k \wedge \boldsymbol{v}_j \|}{\beta + \| \boldsymbol{v}_j \|} \tag{4.9}$$

for each node j in the choice layer, where $\wedge$ is the fuzzy min operator and β is a constant insuring that the denominator is larger than 1. Let $\boldsymbol{v}_{j*}$ be the reference vector of node j^* having the largest distance, and be called the winner reference vector. Then we perform a vigilance test which checks the degree of similarity between $\boldsymbol{v}_{j*}$ and $\boldsymbol{x}_k$, as follows:

$$\frac{\| \boldsymbol{x}_k \wedge \boldsymbol{v}_j \|}{\| \boldsymbol{x}_k \|} > \rho \tag{4.10}$$

where ρ is the vigilance parameter specified by the user. Note that the test is done at the corresponding node, node j^*, at the match layer. If $\boldsymbol{v}_{j*}$ passes the vigilance test, it is adapted to $\boldsymbol{x}_k$ by:

$$\boldsymbol{v}_j(t) = (1 - \eta)\boldsymbol{v}_j(t - 1) + \eta(\boldsymbol{x}_k \wedge \boldsymbol{v}_j(t - 1)). \tag{4.11}$$

Otherwise, the current winner node is deactivated and the next winner is chosen. This process repeats until either a reference vector passes the vigilance test or none passes the test. If the latter case occurs, we increase J by one, i.e., $J = J + 1$, create a new subnet for $\boldsymbol{x}_k$, as shown in Figure 4.3, and add it to the network. Note that $\boldsymbol{v}_J$ is set to $\boldsymbol{x}_k$ in this subnet. The Fuzzy-ART algorithm can be summarized below:

procedure Fuzzy-ART
  Set parameters, η, β, and ρ. Initialize $J = 0$;
  for $k = 1, 2, \ldots, N$
  Input the training vector, $\boldsymbol{x}_k$;
  Choose the winner node j^* by Eq. (4.9);
  Do vigilance test of the winner node by Eq. (4.10);
  while fails the vigilance test, *do*
    Find the next winner node;
  if one node passes the vigilance test
  then update the weights of the winner node by Eq. (4.11);
  else $J = J + 1$ and create a new subnet J for $\boldsymbol{x}_k$;
 return with J clusters;
end Fuzzy-ART

When the process terminates, $\boldsymbol{v}_j, j = 1, 2, \ldots, J$, is treated as the center or mean of cluster j. A pattern $\boldsymbol{x}_k$ belongs to cluster p if node p has the highest output at the match layer.

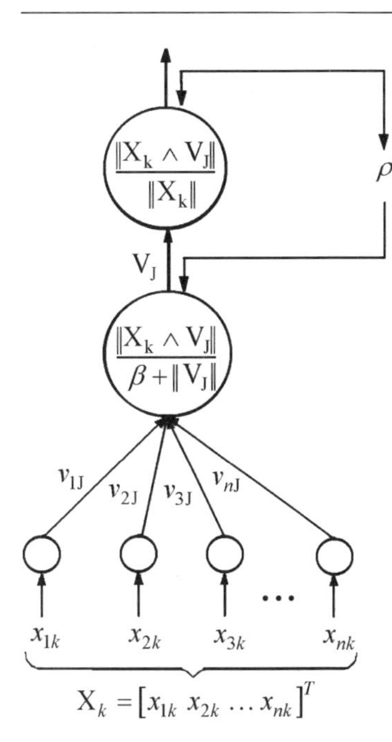

Fig. 4.3. Subnet *J* is created for Fuzzy-ART

4.2.3 Self-Constructing Fuzzy Neural Networks (SCFNN)

The task of SCFNN is to partition the given data set into fuzzy clusters, with the degree of association being strong for data within a cluster and weak for data in different clusters. Let $\boldsymbol{x} = [x_1, x_2, \ldots, x_n]$ be an input vector of n dimensions. A fuzzy cluster C_j is defined as a Gaussian function of the following form:

$$C_j = G(\boldsymbol{x}; \boldsymbol{v}_j, \sigma_j) = \prod_{i=1}^{n} g(x_i; v_{ij}, \sigma_{ij}) \tag{4.12}$$

where $g(x_i; v_{ij}; \sigma_{ij})$ is

$$g(x_i; v_{ij}, \sigma_{ij}) = \exp\left[-\left(\frac{x_i - v_{ij}}{\sigma_{ij}} \right)^2 \right]. \tag{4.13}$$

where $\boldsymbol{v}_j = [v_{1j}, \ldots, v_{nj}]$ denotes the mean vector and $\sigma_j = [\sigma_{1j}, \ldots, \sigma_{nj}]$ *de*notes the deviation vector for C_j. Gaussian functions are adopted in SCFNN for representing clusters because of their superiority over other functions in performance [68].

Like Fuzzy-ART networks, clusters are generated as needed and the number of clusters is determined automatically in SCFNN networks. A SCFNN network consists of three layers: an input layer, a fuzzification layer, and a competition layer, as shown in Figure 4.4 [41, 47]. The input layer contains n nodes. It receives input patterns and broadcasts them to the fuzzification layer. The fuzzification layer contains J groups each of which contains n nodes. The correspond-

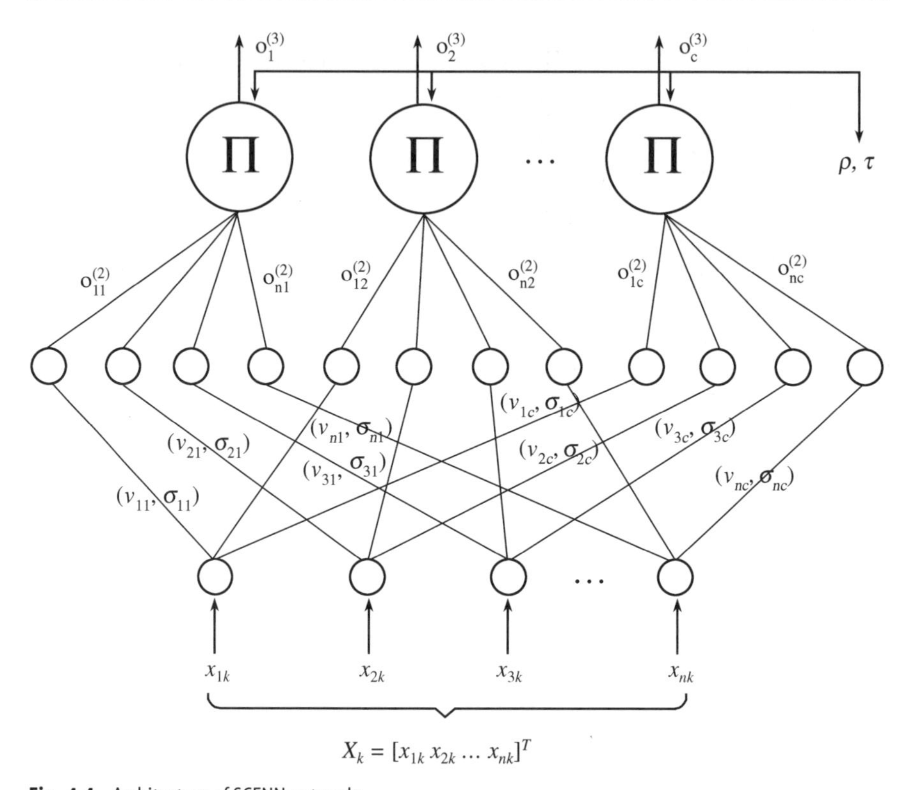

Fig. 4.4. Architecture of SCFNN networks

ing weight vector $[v_{1j}, v_{2j}, \ldots, v_{nj}]$ of each group j represents the prototype of cluster j. The ith node of group j calculates the Gaussian function value $g(x_i; v_{ij}, \sigma_{ij})$. The competition layer contains J nodes. The output of node j of this layer is the product of all its inputs from the previous layer, i.e.,

$$\prod_{i=1}^{n} g(x_i; v_{ij}, \sigma_{ij})\,. \tag{4.14}$$

Note that the weights $\{(v_{ij}, \sigma_{ij}) | 1 \le i \le n, 1 \le j \le J\}$ between the input layer and the fuzzification layer are adjustable, and the other weights are fixed to 1.

Let $\boldsymbol{x}_k$ be a training pattern. We define that $\boldsymbol{x}_k$ belongs to cluster C_j if $\boldsymbol{x}_k$ contributes to the distribution of patterns in C_j, i.e., $\boldsymbol{v}$ and σ_j have to be recalculated due to the addition of $\boldsymbol{x}_k$. Let S_j indicate the size of C_j, i.e., the number of patterns that belong to C_j. Also, we define an operator, *comb*, to combine a cluster C_j and $\boldsymbol{x}_k$ to result in a new cluster C_j', as follows:

$$C_j' = comb(C_j, \boldsymbol{x}_k) = G(\boldsymbol{x}_k; \boldsymbol{v}_j', \sigma_j') \tag{4.15}$$

$$= \prod_{i=1}^{n} g(x_i; v_{ij}', \sigma_{ij}') \tag{4.16}$$

where the mean and deviation vectors, v'_j, and σ'_j, associated with C'_j are computed by:

$$v'_{ij} = \frac{S_j v_{ij} + x_{ik}}{S_j + 1}, \tag{4.17}$$

$$\sigma'_{ij} = \left[\frac{(S_j - 1)(\sigma_{ij} - \sigma_0)^2 + S_j (v_{ij})^2 + (x_{ik})^2}{S_j} - \frac{S_j + 1}{S_j} \left(\frac{S_j v_{ij} + x_{ik}}{S_j + 1} \right)^2 \right]^{1/2} + \sigma_0 \tag{4.18}$$

for $1 \le i \le n$, with σ_0 being a user-defined constant.

Let J be the number of existing fuzzy clusters. Initially, J is 0 since no cluster exists at the beginning. For a training pattern $\boldsymbol{x}_k$ being applied to the network, we first find the winner node j^* at the competition layer

$$j^* = \arg\max_j \left\{ \prod_{i=1}^{n} g(x_i; v_{ij}, \sigma_{ij}) \right\}, 1 \le j \le J. \tag{4.19}$$

Then we check if $\boldsymbol{x}_k$ is similar enough to cluster C_{j*} by the following similarity test:

$$\prod_{i=1}^{n} g(x_i; v_{ij*}, \sigma_{ij*}) \ge \rho \tag{4.20}$$

where ρ, $0 \le \rho \le 1$, is a predefined threshold. If $\boldsymbol{x}_k$ passes the similarity test on cluster C_{j*}, we further check the variance of the resulting cluster $C'_{j*} = comb(C_{j*}, \boldsymbol{x}_k)$ induced by the addition of $\boldsymbol{x}_k$. We say that pattern $\boldsymbol{x}_k$ passes the variance test on cluster C_{j*} if

$$\|\sigma'_{j*}\| \le \tau \tag{4.21}$$

where σ'_{j*} is computed by Eq. (4.18) and τ is another user-defined threshold. If either Eq. (4.20) or Eq. (4.21) fails, the current winner node is deactivated and the next winner is chosen.

Two cases may occur. First, there are no existing fuzzy clusters on which pattern $\boldsymbol{x}_k$ has passed both the similarity test and the variance test. For this case, pattern $\boldsymbol{x}_k$ is not close enough to any existing cluster and a new fuzzy cluster C_J, $J = J + 1$, is created with

$$\boldsymbol{v}_J = \boldsymbol{x}_k, \sigma_J = [\sigma_0, \sigma_0, \ldots, \sigma_0] \tag{4.22}$$

as shown in Figure 4.5. Note that the new cluster C_J contains only one member, pattern $\boldsymbol{x}_k$. The reason that σ_J is initialized to a non-zero vector is to avoid the null width of a singleton cluster. Of course, the number of clusters is increased by 1 and the size of cluster C_J should be initialized, i.e.,

$$J = J + 1, S_J = 1. \tag{4.23}$$

On the other hand, if there is a winning node j^* on which pattern $\boldsymbol{x}_k$ has passed both the similarity test and the variance test, pattern $\boldsymbol{x}_k$ is close enough to clus-

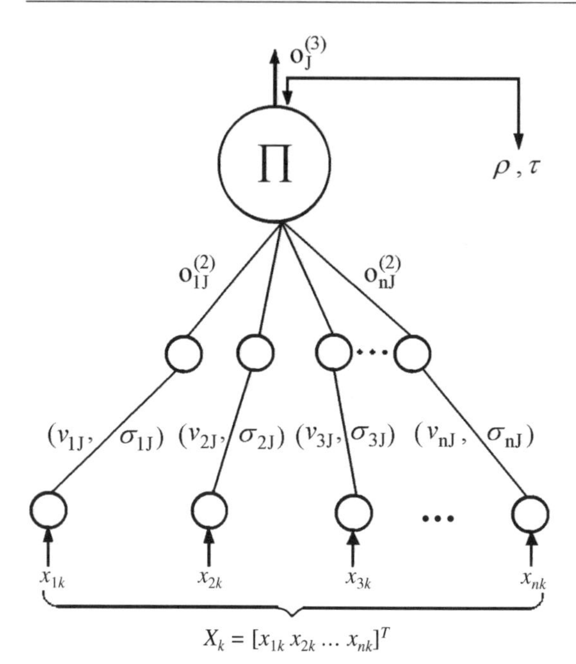

Fig. 4.5. Subnet J is created for SCFNN

ter C_{j*} and the weights of the winning node $j*$ are modified to include pattern $\boldsymbol{x}_k$ by Eq. (4.18) and Eq. (4.17), and the size of C_{j*} is increased by 1, i.e.,

$$S_{j*} = S_{j*} + 1 . \tag{4.24}$$

Note that J is not changed in this case.

The above process is iterated until all the training patterns are processed. The SCFNN algorithm can be summarized below:

procedure SCFNN
  Set parameters, σ_0, ρ and τ;
  for $k = 1, 2, \ldots, N$
    Input the training vector $\boldsymbol{x}_k$;
    Choose the winner node $j*$ by Eq. (4.19);
    Perform similarity and variance tests of the winner node;
    while fails either similarity test or variance test, *do*
      Find the next winner node;
    if one node passes both tests
    then add $\boldsymbol{x}_k$ into the winner node;
    else $J = J + 1$ and create a new subnet J for $\boldsymbol{x}_k$;
  return with J clusters;
end SCFNN

When the process terminates, v_j, $j = 1, 2, \ldots, J$, which is the mean vector of cluster C_j. Furthermore, the variance, σ_j, is also provided for each cluster C_j by the

SCFNN algorithm. A pattern x_k belongs to cluster p if node p has the highest output at the competition layer.

Because of the similarity and variance tests, SCFNN can generate compact and dense clusters, and capture the real distribution of the training vectors. Thus, the clusters generated can represent training vectors appropriately. Besides, the network obtained by SCFNN can be tuned when applied to the supervised recognition problem. This is useful in identifying objects for video compression, to be presented later.

4.3 Neuro-Fuzzy Based Vector Quantization for Image Compression

As mentioned, vector quantization (VQ) is attractive for image compression due, to its simplicity in decoding at the receiving end. One of the key issues for VQ is the generation of code-words based on which image blocks are encoded and decoded. Many methods have been proposed for generating code-words for VQ [10, 32]. The LBG algorithm [50, 51] is one of the most famous methods. It starts with a set of randomly selected code-words which form initial clusters. According to the Euclidean distances from code-words, a training pattern is clustered to the code-word nearest to it. Data in the same cluster form a new code-word which replaces the old one. The process repeats, until the variation of average distortion of all clusters is smaller than a predefined threshold. Modified ART2 [30, 60] is another approach for VQ based on the ART algorithm. Codewords are constructed gradually as the data are fed one by one. When the first block of image comes in, the first code-word is set up. Incoming blocks are compared with existing code-words. If a code-word is similar enough to an incoming block, the block is categorized to the code-word and the code-word is modified to include the block. The similarity threshold increases in each iteration. The algorithm terminates when the number of code-words reaches the desired one, or the threshold reaches the predefined upper-bound.

VQ incorporate well with neuro-fuzzy clustering methods in order to derive the code-words for image compression. Any neuro-fuzzy clustering method presented in Section 4.2, can do the job. Obviously, the code-book size is identical to the number of clusters and all the v_j vectors form the desired code-words. In this section, we describe how the SCFNN clustering method is used to generate code-words for vector quantization. The fuzzy clusters obtained by SCFNN have a high-degree of intra-cluster similarity and a low-degree of inter-cluster similarity. The mean vector of each obtained fuzzy cluster becomes naturally a code-word. The advantages of using SCFNN include that the fuzzy clusters generated are compact and dense, the real distribution of image content can be captured, and image content can be represented by code-words more appropriately.

4.3.1 VQ Encoding/Decoding

For simplicity, we only focus on gray-level images. Extension to color images is obvious. Given an original image, $\bar{I}$, of N_x by N_y pixels to be transmitted, namely,

$$\begin{bmatrix} I_{11} & I_{12} & \cdots & I_{1N_x} \\ I_{21} & I_{22} & \cdots & I_{2N_x} \\ \vdots & \vdots & \vdots & \vdots \\ I_{N_y 1} & I_{N_y 2} & \cdots & I_{N_y N_x} \end{bmatrix}$$

where I_{ij} represents the gray value *of* the pixel located at position (i, j), we divide $\bar{I}$ into non-overlapping blocks each of which contains p by p pixels. Usually, N_x and N_y are both multiples of p. Therefore, $\bar{I}$ can be divided into N_xN_y/p^2 blocks which are numbered 1, 2, …, N_xN_y/p^2 from left to right and top to bottom, as shown in Figure 4.6. Similarly, the pixels in a block are numbered 1, 2, …, p^2 from left to right and top to bottom, as shown in Figure 4.7.

The VQ-based compression/decompression system we are concerned with is shown in Figure 4.8. The system consists of three major components: encoder, decoder, and code-book. Usually, the size of the code-book is 2^b, i.e., it contains 2^b code-words. Each code-word is a block of pixels. We obtain the code-book using SCFNN from the blocks of the underlying image. Then at the transmitting end, each block of the image is encoded by comparing it with the code-words of the code-book. The code-word which is most similar to the block is chosen and the index of the code-word is used to represent the block and is transmitted to the receiving end through the communication channel. At the receiving end, a received index is decoded by checking against the code-book, which is the same as that used by the encoder at the transmitting end. The corresponding code-word is recalled to become the block reconstructed. When we are done with all the indices received, the whole image is reconstructed which is an approximate version of the original image at the transmitting end.

1	2	$\cdots$	$\frac{N_x}{p}$
$\frac{N_x}{p}+1$	$\frac{N_x}{p}+2$	$\cdots$	$\frac{N_x \times 2}{p}$
$\vdots$	$\vdots$	$\ddots$	$\vdots$
$\frac{N_x}{p}\times\left(\frac{N_y}{p}-1\right)+1$	$\frac{N_x}{p}\times\left(\frac{N_y}{p}-1\right)+2$	$\cdots$	$\frac{N_x \times N_y}{p^2}$

Fig. 4.6. Block numbering in an image

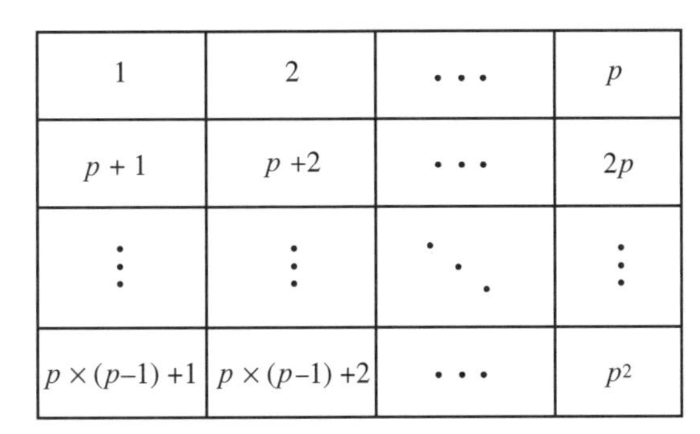

Fig. 4.7. Pixel numbering in a block

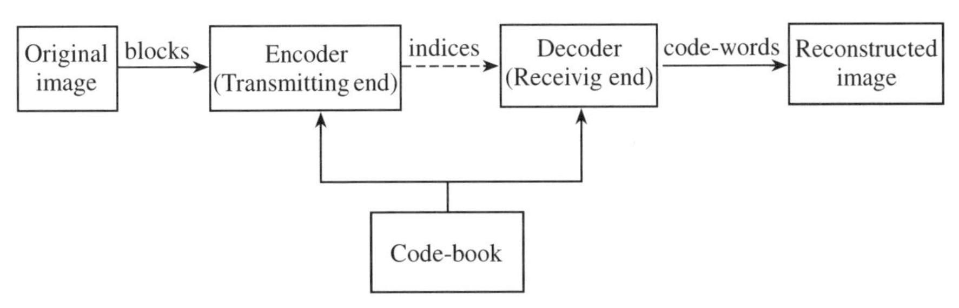

Fig. 4.8. The VQ-based compression system

4.3.2 Clustering by SCFNN

As mentioned earlier, the pixels of a given image are divided into N_xN_y/p^2 blocks and each block contains p^2 pixels. We represent each block as a vector of size p^2. Let $N = N_xN_y/p^2$. *Therefore,* we have N training vectors, $\boldsymbol{x}_1, \boldsymbol{x}_2, \ldots \boldsymbol{x}_N$, and each training vector $\boldsymbol{x}_k$ has p^2 dimensions, i.e., $n = p^2$. *These* training vectors are given to the SCFNN algorithm of Section 4.2.3. Suppose J clusters are obtained. Then the mean vector of each cluster obtained becomes a code-word for encoding/decoding.

Note that SCFNN determines the number of clusters by itself. However, it is desirable that the size of the code-book be 2^b. Therefore, we have to make the number of clusters obtained as close to 2^b as possible, i.e., $J \simeq 2^b$. This is achieved by adjusting iteratively the values of the two involved parameters, ρ and τ. Suppose we would like the number of final clusters to be $B = 2^b$. Initially, each training pattern is treated as a cluster and the number of clusters is N. We randomly choose some values for ρ and τ and perform fuzzy clustering. Let the number of clusters generated be L. If $L > B$, then we decrease ρ and increase τ by an amount proportional to $(L - B)/(N - L)$. If $L < B$, then we increase ρ and decrease τ by an amount proportional to $(B - L)/(N - L)$. This process iterates until L is close enough to B. Therefore, a code-book obtained by SCFNN can be summarized below.

procedure Code_Book_SCFNN
    Let $B = N$;
    while B is not close to 2^b *do*
        Adjust ρ and τ appropriately;
    $B = 0$;
    for each training block $\boldsymbol{x}_k$, $1 \leq k \leq N$
        $W_1, = \{C_j | G(\boldsymbol{x}_k; \boldsymbol{v}_j, \sigma_j) \geq \rho, 1 \leq j \leq B\}$;
        $W_2 = \{C_j | \sigma_j' \leq \tau, C_j \in W_1\}$;
        if $W_2 == \emptyset$
        A new cluster C_B, $B = B + 1$, is created;
        else let $C_a \in W_2$ be the cluster with the largest similarity measure;
        Incorporate $\boldsymbol{x}_k$ into C_a;
    return with the created B clusters;
end Code_Book_SCFNN

4.3.3 Experimental Results

Three characteristics are usually considered for evaluating the effectiveness of an image compression algorithm, i.e., compression ratio, compression speed, and image quality. Compression ratio (CR) for an image, is defined to be

$$\mathrm{CR} = W_I / W_T \tag{4.25}$$

where W_I is the number of bits in the original image and W_T is the number of bits transmitted through the communication channel for the image. Obviously, a larger CR means a less amount of bandwidth requirement and is more efficient in transmission. Image quality concerns the quality of the reconstructed image at the receiving end and is usually indicated by signal-noise ratio (SNR) defined below:

$$\mathrm{SNR} = 10\log_{10} \frac{\frac{1}{N_x N_y} \sum_{i=1}^{N_x} \sum_{j=1}^{N_y} x(i,j)^2}{\mathrm{MSE}}, \tag{4.26}$$

$$\mathrm{MSE} = \frac{1}{N_x N_y} \sum_{i=1}^{N_x} \sum_{j=1}^{N_y} (x(i,j) - \hat{x}(i,j))^2 \tag{4.27}$$

where $x(i,j)$ and $\hat{x}(i,j)$ are original and reconstructed values, respectively, of the pixel located at position (i,j). As usual, we assume that if a reconstructed image has a higher SNR, then it is of higher quality.

To demonstrate the effectiveness of SCFNN, we show the results of three experiments below. In these experiments, each block contains $8 \times 8 = 64$ pixels, i.e., $p = 8$. Firstly, the results obtained from four benchmark images are presented using local code-books. Next, we show the performance of image compression using a global code-book. Finally, we work with another two benchmark images contaminated with the white Gaussian noise.

4.3.3.1 Benchmark Images (Local Code-book)

A local code-book is derived from the blocks of the image to be transmitted, and has to be sent with the code-book indices to the decoder. Therefore, compression with local code-books have better quality of reconstruction, but results in a low compression ratio. We do compression with local code-books on four benchmark images, Elaine, Peppers, Man and Boat, as shown in Figure 4.9. All these images are 256×256 in resolution. MSE, SNR, and CR associated with these images are shown in Table 4.1. Note that 2^b in the first column of these tables indicates the size of the code-book, and the transmission of the associated codebook is considered in the calculation of CR for each image. Figure 4.10(a), Figure 4.10(b) and Figure 4.10(c), show the reconstructed images of Elaine, Man and Boat, respectively.

4.3.3.2 Benchmark Images (Global Code-book)

Next, we show the performance of image compression using a global code-book. A global code-book is defined to be one that is used for encoding/decoding independent of images to be transmitted. With this idea, compression ratio

Fig. 4.9. The original benchmark images: (a) Elaine; (b) Peppers; (c) Man; (d) Boat

Table 4.1. MSE, NSR, and CR for benchmark images using local code-books

Image	MSE	SNR	CR
Elaine(2^8)	72.28	24.62	3.76
Elaine(2^9)	19.26	30.37	1.93
Man(2^8)	211.37	17.28	3.76
Boat(2^8)	125.14	21.86	3.76

Table 4.2. MSE, SNR, and CR for benchmark images using the code-book of Elaine(2^8)

Image	MSE	SNR	CR
Peppers(2^8)	365.45	16.83	64.00
Man(2^8)	676.87	12.22	64.00
Boat(2^8)	379.59	17.04	64.00

can be increased, since each image is transmitted without transmitting its own code-book together. Usually, we test this idea by encoding one image with a code-book that is obtained from another image. We compress Peppers, Man, and Boat based on the code-book obtained from Elaine(2^8) and the results are shown in Table 4.2. Figure 4.10(d), Figure 4.10(e) and Figure 4.10(f) show the reconstructed images of Peppers, Man and Boat, respectively, based on the code-book of Elaine(2^8).

4.3.3.3 Benchmark Images with Noise

We test the effectiveness of SCFNN with another two benchmark images, Lena and Bird, contaminated with the white Gaussian noise, as shown in Figure 4.11. Figure 4.12 show the reconstructed images obtained using local code-books of sizes 256 and 128 of original images, respectively. The values of SNR and MSE for these reconstructed images are, given in Table 4.3.

Table 4.3. MSE, SNR, and CR for benchmark images contaminated with noise

Image	MSE	SNR	CR
Lena(2^8)	104.31	22.23	3.76
Bird(2^7)	67.92	24.25	7.21

Fig. 4.10. Reconstructed images: (a) Elaine(2^9); (b) Man(2^8); (c) Boat (2^8); (d) Peppers based on the Elaine(2^8) code-book; (e) Man based on the Elaine(2^8) code-book; (f) Boat based on the Elaine(2^8) code-book

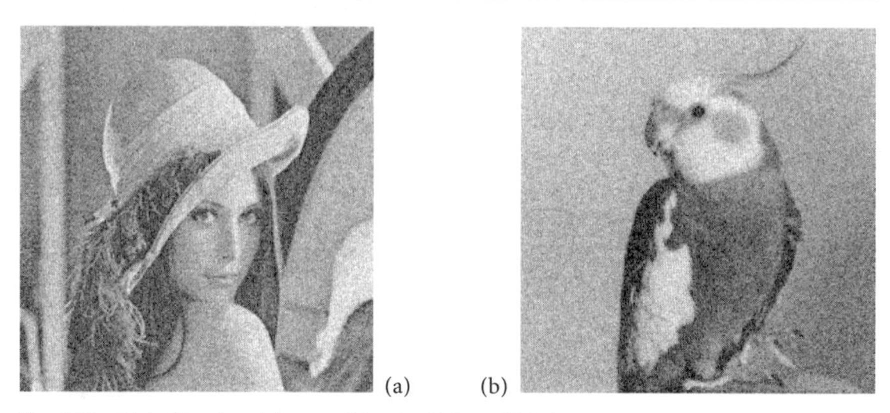

Fig. 4.11. Original benchmark images with noise: (a) Lena; (b) Bird

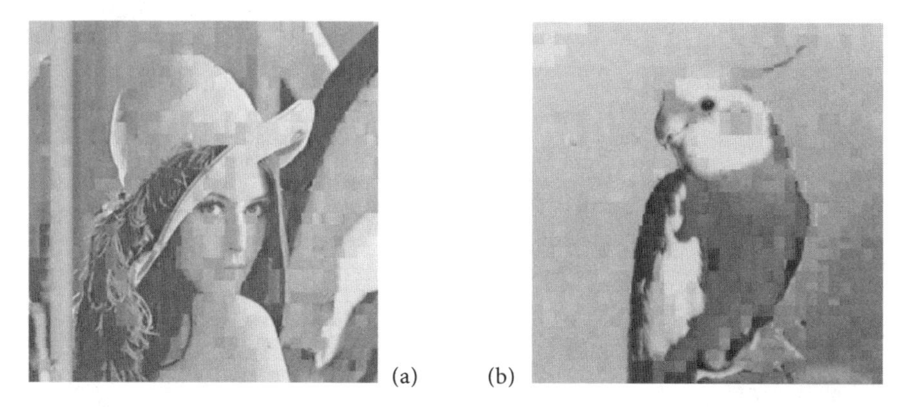

Fig. 4.12. Reconstructed images: (a) Lena(2^8); (b) Bird(2^7)

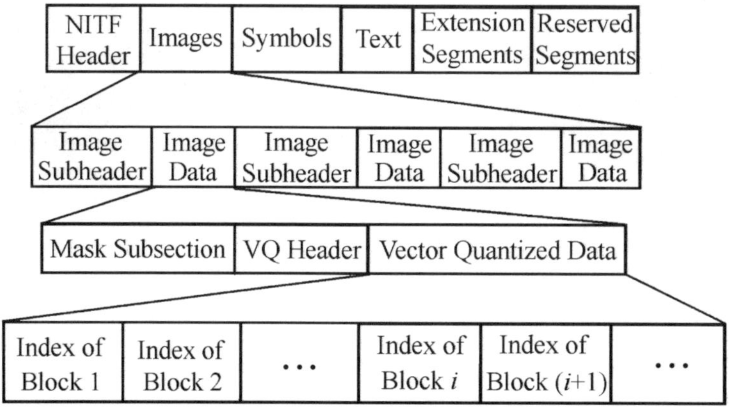

NITF Header	Images	Symbols	Text	Extension Segments	Reserved Segments

Image Subheader	Image Data	Image Subheader	Image Data	Image Subheader	Image Data

Mask Subsection	VQ Header	Vector Quantized Data

Index of Block 1	Index of Block 2	...	Index of Block i	Index of Block (i+1)	...

Fig. 4.13. NITF file Structure with VQ compressed images

4.4 Image Transmission by NITF

We demonstrate the usage of fuzzy neural networks in real communication of images in this section. We adopt NITFF (National Imagery Transmission Format) [1, 7] which is a standard for encoding/decoding VQ compressed images. The standard is introduced first, and then the functions of fuzzy neural networks in the standard are described.

NITF is a multi-component format which is designed to allow up to 999 images and symbols to be combined in a single file. Each component has metadata associated with it. This technique allows for overlays that can readily be removed through an application. NITF is more than just an image file format; it goes beyond supporting the core needed to share imagery between disparate systems. It facilitates the increasing need for greater flexibility in using multiple images with annotation in a composition that relates the images and annotation to one another.

NITF can accept and decompress data that has been compressed using a VQ compression scheme. Images contained in a NITF file can be in either color or gray scale. For simplicity, we only consider gray scale images here.

The components of a NITF file, as shown in Figure 4.13, include:

- NITF File Header. This gives the basic description of the file, e.g., how many subcomponents such as images, symbols, or texts, exist.
- Image Segments. NITF allows each image contained in the file to be compressed. Currently, bi-level, JPEG, JPEG2000 and VQ are the compression methods supported by NITF.
- Symbol Segments.
- Text Segments.
- Remaining segments.

The subheader of each image identifies the image compression method used. If an image is compressed with VQ, then the code-book is placed in the VQ header followed by the compressed image codes. The VQ header also provides information about the organization of the code-book, indicating how many code-words included in the code-book, the size of each code-word, and how the data that makes up the code-words is organized. As a multi-component format, NITF can collect several images in a file and thus is suitable for transmitting a global code-book in one file.

4.4.1 Encoding a VQ Compressed NITF Image

To compress an image with vector quantization, we can use fuzzy neural networks to generate the code-book from the input image and then classify each image block to the nearest code-word, as shown in Figure 4.14. The code-book and the indices of image blocks together, are encoded into a bitstream and become the compressed image data in a NITF file. Two encoding schemes are provided, block-based and row-based.

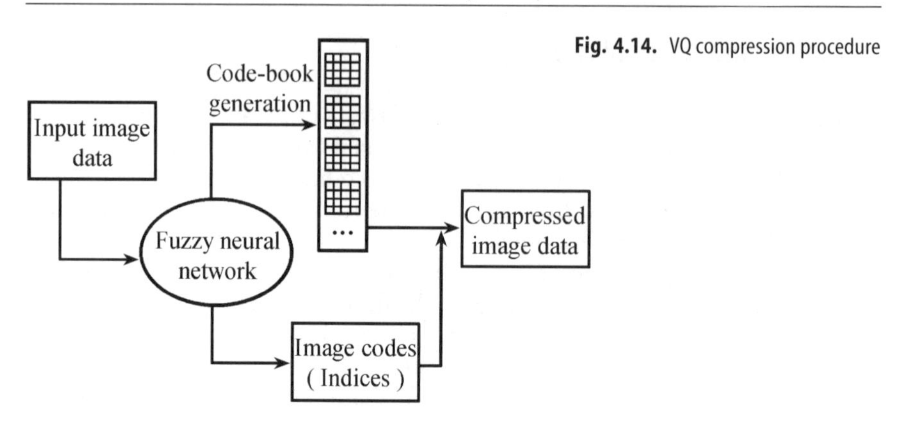

Fig. 4.14. VQ compression procedure

4.4.1.1 Encoding with Block-Based Code-books

For the block-based scheme, a code-book with each code-word of 4×4 pixels in size is created. To illustrate how it works, we do compression on Lena. The image is first partitioned into blocks of 4×4 pixels. Therefore, each block or code-word contains 16 pixels. A code-book of 512 code-words is generated and $64 \times 64 = 4096$ blocks are vector quantized through the network. Following the file header and the image subheader, the 512 code-words are encoded in ascending order, i.e., starting with code-word 1 and ending with code-word 512. Assume that the first code-word is

$$\begin{bmatrix} 153 & 153 & 153 & 153 \\ 155 & 155 & 155 & 155 \\ 155 & 155 & 155 & 155 \\ 154 & 154 & 154 & 154 \end{bmatrix}$$

and the second code-word is

$$\begin{bmatrix} 150 & 151 & 150 & 151 \\ 148 & 148 & 148 & 148 \\ 135 & 132 & 132 & 131 \\ 112 & 112 & 113 & 113 \end{bmatrix}.$$

Then the bitstream of the first code-word, i.e., [153 153 153 ... 154 154 154], is followed by that of the second code-word, i.e., [150 151 150 ... 112 113 113], and so on. The bitstream of the code-book is therefore encoded as [153 153 153 ... 154 154 154 150 151 150 ... 112 113 113 ... ...]. Following the code-book, the indices of the image blocks from left to right and top to bottom are encoded into another bitstream of compressed image data. Suppose we have the following image with each image block replaced by its index:

$$\begin{bmatrix} 127 & 55 & \cdots & 13 \\ 8 & 288 & \cdots & 512 \\ \vdots & \vdots & \vdots & \vdots \\ 64 & 175 & \cdots & 336 \end{bmatrix}$$

The bitstream consists of these indices becomes [127 55 ... 13 8 288 ... ... 512 ... 64 175 ... 336].

4.4.1.2 Encoding with Row-Based Code-books

NITF allows the organization of the VQ code-book to be optimized for the specific use of the VQ data. For the row-based scheme, four code-books with each code-word of 4 pixels in size are created. That is, it stores different rows of 4×4 code-words in different code-books such that the image can be reconstructed line-by-line, instead of block-by-block.

The first code-book is used to group row 1 of all the 4×4 code-words together. For instance, for the previous example, the first code-book has the following form:

$$\begin{bmatrix} 153 & 153 & 153 & 153 \\ 150 & 151 & 150 & 151 \\ \vdots & \vdots & \vdots & \vdots \end{bmatrix}.$$

Row 2, row 3, and row 4 of all the 4×4 code-words are placed in respective code-books as above. Then these row-based code-books are encoded one by one. The bitstream of these code-books thus looks like [153 153 153 153 150 151 150 151 ... 155 155 155 155 148 148 148 148 ... 155 155 155 155 135 132 132 131 ... 154 154 154 154 112 112 113 113 ...]. Note that the quantized image data remains the same as those compressed with the block-based code-book scheme.

4.4.2 Decoding a VQ Compressed NITF Image

The decoding process of a VQ compressed image is shown in Figure 4.15. When a VQ compressed image is received, code-books are read and then the image blocks are reconstructed.

4.4.2.1 Decoding with Block-Based Code-books

For the block-based scheme, we only need to check with one code-book. After the code-book is read, the indices of image blocks are extracted. Through the table lookup operation with the code-book, the code-word indexed by the first vector-quantized image code is used to spatially decompress the 4×4 block at the upper left corner of the image. Decompression continues from left to right and top to bottom, as shown in Figure 4.16, until all the image blocks have been spatially decompressed.

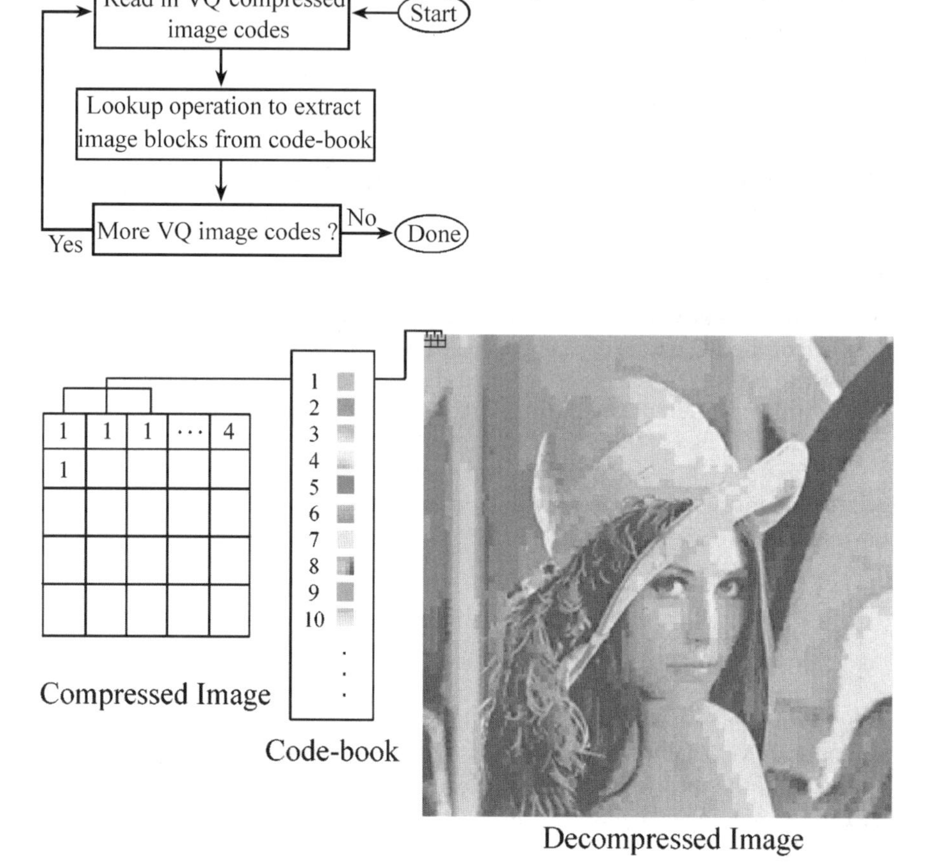

Fig. 4.15. VQ decompression procedure

Fig. 4.16. Spatial decompression with the block-based code-book scheme

4.4.2.2 Decoding with Row-Based Code-books

The process of decoding a VQ compressed NITF image with row-based code-books is very similar to those of the block-based code-book scheme, except that we have to check with four code-books. After the row-based code-books are read, the first 64 indices, i.e., quantized image data, are used to decompress the first row of the image by table-lookup in the first code-book. The second, third and fourth rows of the image are then decompressed by table-lookup in the second, third, and fourth code-books, respectively. The decompression continues from left to right and top to bottom, as shown in Figure 4.17, until all the image rows have been spatially decompressed.

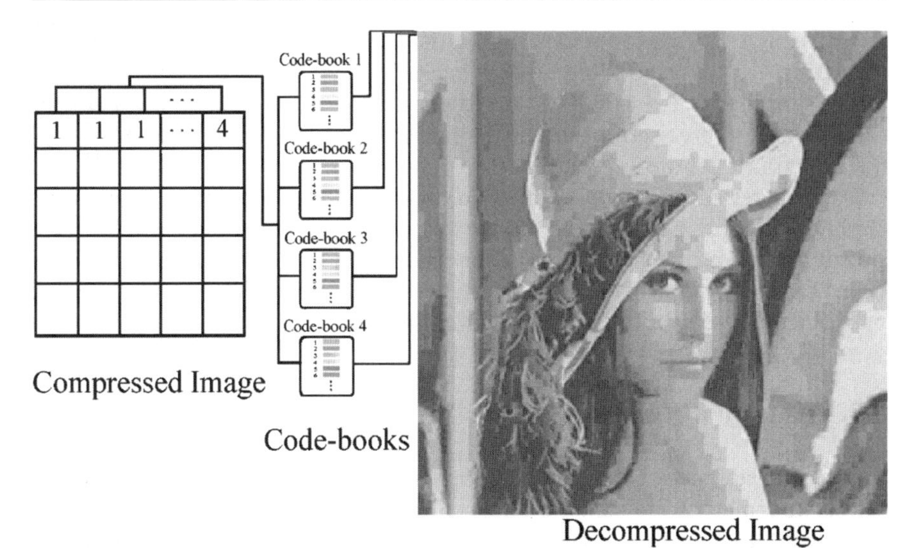

Fig. 4.17. Spatial decompression with row-based code-books

4.5 Neuro-Fuzzy Based Video Compression

A lot of video applications appear on internets and wireless communications, such as video conference, video phone, and distance education, in which facial expression and body gesture are usually the main focus in a video stream. Therefore, the most important topic in object segmentation is the extraction of human objects, including face and body, in image sequences.

Several approaches have been proposed for identifying human objects in a video stream. One approach [8, 16, 19, 26, 27, 37, 46, 54, 58, 63, 70] applies static features or spatial data, such as luminance, chrominance, location, or shape of human objects, to determine the foreground or background regions in a video frame. The advantage of this approach is that only one video frame is required for segmentation. However, finding unique features to be used for identifying human objects is not easy, leading to a high segmentation error. Another approach [12, 23, 29, 36, 39, 40, 44, 45, 65) uses motion information, or temporal data, to detect human objects. Human objects are assumed to have the most significant motion in a video sequence and are extracted by comparing two or more frames in a stream. However, while motion is not obvious or there are other objects having more significant motion, the detected result will be wrong.

We combine spatial and temporal information [12, 65] and employ the SCFNN algorithm to overcome the above difficulties. The basic idea is that the base frame of a video stream is divided into segments and then each segment is categorized as foreground or background, based on a combination of multiple criteria. Firstly, SCFNN is used to group similar pixels in the frame into clusters. Connected segments contained in the clusters are combined, and each segment is checked if it is part of a human face using the values of chrominance and luminance. By referring to the position of the face region and related motion information, the corresponding body is located. Then, the obtained SCFNN network is further tuned by a SVD-based hybrid learning algorithm, which can then be used to precisely locate the human object in the base frame and the remaining frames of the video stream.

We have tested the proposed method on different color video sequences, including standard benchmarks Akiyo and Silent. The obtained results have shown that the method can improve the accuracy of the identification of human objects in video sequences. Also, the method can work well even when the human object presents no significant motion in a sequence.

4.5.1 System Overview

The system adopted for segmenting human objects, combines temporal and spatial information, and consists of three main steps: clustering, detection, and refinement, as shown in Figure 4.18. For simplicity, we assume that only one person appears in the video stream we are interested in. In the clustering stage, the SCFNN algorithm is used to group similar pixels in the base frame of a given video stream, into fuzzy clusters. The number of clusters is determined automatically by the algorithm. Connected segments in the clusters are then combined to form larger segments. In the detection stage, each segment is checked when it is part of a human face using chrominance values and the variance of luminance values. By referring to the position of the face region and related motion information, the corresponding body is found. Then the base frame is divided into three regions: foreground, background, and ambiguous. In the last step, a supervised network is constructed from the fuzzy clusters obtained by SCFNN and is trained by a highly efficient SVD-based hybrid learning algorithm using the data points obtained from the foreground and background regions. The trained fuzzy neural network is then used to decide the category of the pixels in the ambiguous region. Finally, the pixels belonging to the foreground region form the desired human object in the base frame. Note that the same neural network can be used for segmentation of the remaining frames in the same video stream.

By this approach, similar pixels are grouped together in the first stage and are processed collectively afterwards. The face region is determined by referring to

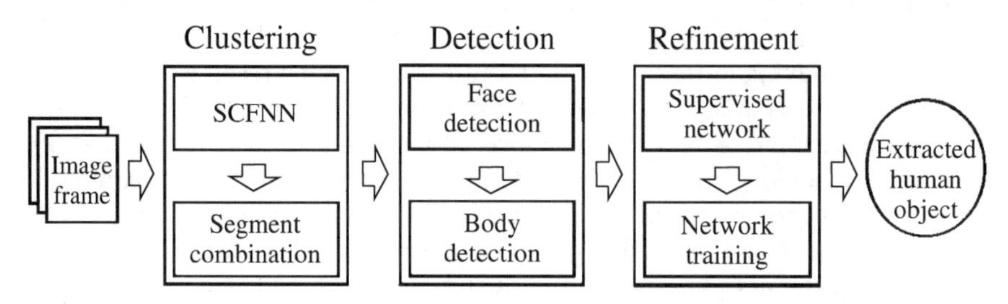

Fig. 4.18. Block diagram of the human-object segmentation system

both chrominance and luminance values. Also, motion information is used to help the determination of the corresponding body. The usage of such information is confined to the neighborhood of a specific area. By using a combination of multiple criteria in determining face and body, the difficulties associated with other methods can be alleviated. Therefore, the human object can be extracted more precisely in a video stream.

4.5.2 Clustering by SCFNN

We use composite signals, chrominance C_r and C_b and luminance Y, as the basis for clustering. Given an image frame of $N_1 \times N_2$ in size, we divide it into $N_1/4 \times N_2/4$ blocks each having 4×4 pixels. Each block is associated with a feature vector $\boldsymbol{x} = [x_1, x_2, x_3]$ where x_1, x_2, and x_3 denote the average C_r, C_b, and Y values, respectively, of all the constituent pixels of the block. The composite signals (C_r, C_b, Y) of a pixel can be denoted by:

$$\begin{bmatrix} C_r \\ C_b \\ Y \end{bmatrix} = \begin{bmatrix} 0.500 & -0.419 & -0.081 \\ -0.169 & -0.331 & 0.500 \\ 0.299 & 0.587 & 0.114 \end{bmatrix} \begin{bmatrix} R \\ G \\ B \end{bmatrix} \tag{4.28}$$

where R is the red component, G is the green component, and B is the blue component of the RGB signal of the pixel [61].

Now, we have N training vectors and $N = N_1/4 \times N_2/4$. Each training vector has 3 dimensions, i.e., $n = 3$. We apply the SCFNN algorithm of Section 4.2.3 to find clusters. Since chrominance and luminance represent different characteristics for each block, we consider chrominance and luminance values separately. Therefore, we need to modify the work related to the two similarity tests in SCFNN. Firstly, we modify the similarity test associated with Eq. (4.20). For a training vector (x_1, x_2, x_3), we calculate the following chrominance similarity measure:

$$d_1(x_1, x_2; C_j) = \prod_{i=1}^{2} g(x_i; v_{ij}, \sigma_{ij}) \tag{4.29}$$

for all $1 \leq j \leq J$. *We say* that $\boldsymbol{x}$ passes the chrominance similarity test on cluster C_j if:

$$d_1(x_1, x_2; C_j) \geq \rho_1 \tag{4.30}$$

where $0 \leq \rho_1 \leq 1$ is a predefined threshold. Secondly, we modify the variance test associated with Eq. (4.21). We calculate the following luminance similarity measure:

$$d_2(x_3; C_j) = g(x_3; v_{3j}, \sigma_{3j}) \tag{4.31}$$

for all cluster C_j on which $\boldsymbol{x}$ has passed the chrominance similarity test. We say that $\boldsymbol{x}$ passes the luminance similarity test on cluster C_j if

$$d_2(x_3; C_j) \geq \rho_2 \tag{4.32}$$

where $0 \le \rho_2 \le 1$ is a predefined threshold. Finally, we choose the cluster with the largest product $d_1(x_1, x_2; C_j)d_2(x_3; C_j)$ to be the one to which $\boldsymbol{x}$ is most similar.

4.5.3 Labeling Segments

Suppose we have J fuzzy clusters, $C_1, C_2, \ldots,$ and C_J, after clustering. A cluster may consist of several parts which are not connected to each other, since we use composite signals, not positions, for clustering. Our desire is to let the pixels in a connected segment be processed collectively. We label each connected segment with a unique name. Let the connected segments be labeled as $S_1, S_2, \ldots,$ and S_L, where L is the total number of such segments. Note that L may be equal to or greater than J. Segments of very small size are considered to be noise, or meaningless parts in the image. Therefore, they are combined to bigger segments in order to reduce their influence on final results. Let $n(S_i)$ denote the size of S_i, i.e., the number of blocks in S_i, and $\kappa = \zeta N_1/N_2/16$ where ζ is a predefined parameter and $0 < \zeta < 1$. We check $n(S_i)$ for each segment S_i. If $n(S_i) \le \kappa$, then S_i is combined into S_j which is the smallest of the segments connected to S_i and $n(S_j) > \kappa$. This process is iterated until all segments are bigger than κ. Let the resulting segments be labeled as $R_1, R_2, \ldots,$ and R_Q, where $Q \le L$ is the total number of such segments. Later on, the image will be processed based on the connected segments, instead of individual pixels or blocks.

Let's apply the above procedure on Figure 4.19(a) which is an image of 360×288 pixels consisting of 90×72 blocks. The image is divided into $J = 24$ clusters containing $L = 931$ connected segments. After the combination process, we have $Q = 120$ segments shown in Figure 4.19(b), in which different segments are represented by different gray values.

(a)

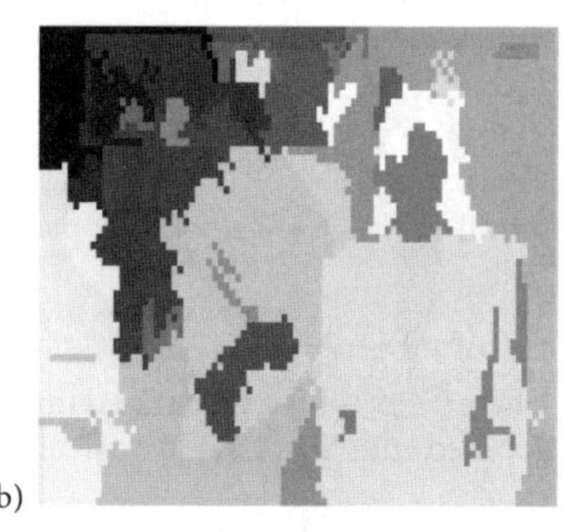

(b)

Fig. 4.19. (a) An example image; (b) Obtained connected segments

4.5.4 Human Object Estimation

We apply chrominance values and the variation of luminance values to estimate the face region. By referring to the position of the face region and motion information, the body region is also estimated.

4.5.4.1 Face Estimation

C_r and C_b values of human skin have been found to occupy only a small region in the C_rC_b space, as the white area shown in Figure 4.20, with approximately $5 \leq C_r \leq 45$ and $-1 \leq C_b \leq -50$ [17]. We use this property for identifying possible face segments. Let $x_1(R_i)$ and $x_2(R_i)$ be the C_r and C_b values of segment R_i. If $(x_1(R_i), x_2(R_i))$ *is* located in the white area of Figure 4.20, then R_i *is* regarded as a possible face segment. Let the possible face segments be denoted as $P_1, P_2, \ldots$, and P_K. Also, a block is called a possible face block if it is contained in a possible face segment.

The Human face is usually a round object with a smooth boundary. The technique of density map [17] can help eliminating branching or annoying parts. Then the image is divided into a set of maximally connected segments such that each segment could not be connected if more elements were added onto it. For convenience, these segments are labeled as $H_1, H_2, \ldots$, and H_G, where G is the total number of such segments. Next, the variation of luminance is used to determine the segment which is most likely to be the human face. According to [17] the, human face usually has the largest standard deviation of luminance in many MPEG-4 applications. We calculate the standard deviation of luminance for each segment H_i, $1 \leq i \leq G$, as follows:

$$\sigma(H_i) = \sqrt{\frac{1}{N_B(H_i)-1} \sum_{B \in H_i} (x_3(B) - \bar{x}_3(H_i))^2} \tag{4.33}$$

where B is any block in H_i, $N_B(H_i)$ *is* the number of blocks in H_i, $x_3(B)$ is the luminance value of B, and $\bar{x}_3(H_i)$ is the mean luminance of H_i defined by

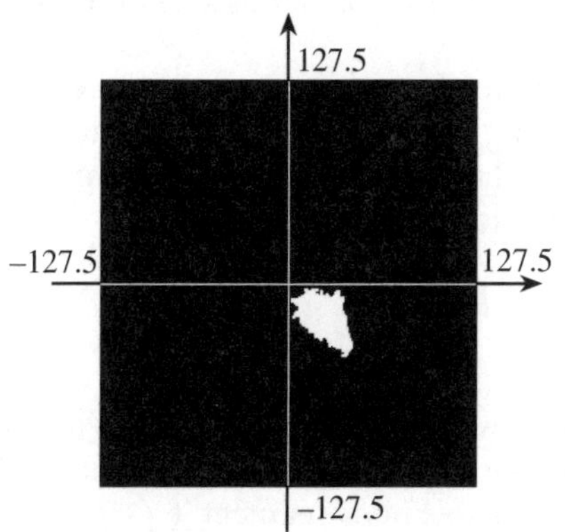

Fig. 4.20. C_r and C_b values of human skin

$$\bar{x}_3(H_i)=\frac{1}{N_B(H_i)}\sum_{B\in H_i} x_3(B)\,. \tag{4.34}$$

Now, the segment H_g, $1 \le g \le G$, with the maximum standard deviation, i.e.,

$$\sigma(H_g) = max(\sigma(H_1), \sigma(H_2), \ldots, \sigma(H_G))\,,$$

is chosen to be the estimation of the face region.

Note that H_g *is* obtained from the calculation on blocks. We'd like to link it to segments and make a possible refinement on it. Consider a possible face segment P_i, $1 \le i \le K$. Let Ψ_i denote the set of blocks which belong to both H_g and P_i, i.e.,

$$\Psi_i = H_g \cap P_i\,. \tag{4.35}$$

We check

$$N_B(\Psi_i) / N_B(P_i) \ge \lambda\,. \tag{4.36}$$

where $N_B(\psi_i)$ and $N_B(P_i)$ are the number of blocks in Ψ_i and P_i, respectively, and λ, $0 < \lambda \le 1$ is a predefined threshold. If Eq. (4.36) holds, then P_i is accepted to be a part of the estimated face. Otherwise, it is not. This process is repeated for all possible face segments. Finally, we have a set of possible face segments that constitute the estimated face, and let these segments be labeled as F_1, F_2, ..., and F_W. For convenience, we use ε_f to denote the estimated face region, i.e., $\varepsilon_f = \{F_i | 1 \le i \le W\}$.

4.5.4.2 Body Estimation

We assume that the body is located directly below the head. A circle below the face region is drawn to detect the corresponding body region. The circle is defined with center being $(c_x, c_y + h)$ and radius being $h/2$, where (c_x, c_y) and h are the center and the height, respectively, of the face region ε_f. Then, the labeled segments, R_i, e.g., referring to Figure 4.19(b), covered partly or totally by the circle region are regarded as possible body segments. A block is called a possible body block if it belongs to a possible body segment. For convenience, we use ε_b to denote the estimated body region, which is the set of all possible body segments.

Based on the possible body segments obtained so far, we can add more segments, if any, to the estimated body by looking into the motion information associated with such segments. Let t represent the index of the current frame. We define the motion index of a segment R_i as follows:

$$V(R_i)=\frac{\sum_{B\in R_i}\sum_{m=0}^{m=k-2}\sum_{j=1}^{j=3}|x_j^{t+m+1}(B)-x_j^{t+m}(B)|}{N_B(R_i)} \tag{4.37}$$

where k is the number of frames in a video sequence to be referenced, $x_j^m(B)$ denotes the x_j value of block B in the mth frame, and $N_B(R_i)$ is the number of blocks in R_i. A segment R_i *is* regarded as a possible body segment if

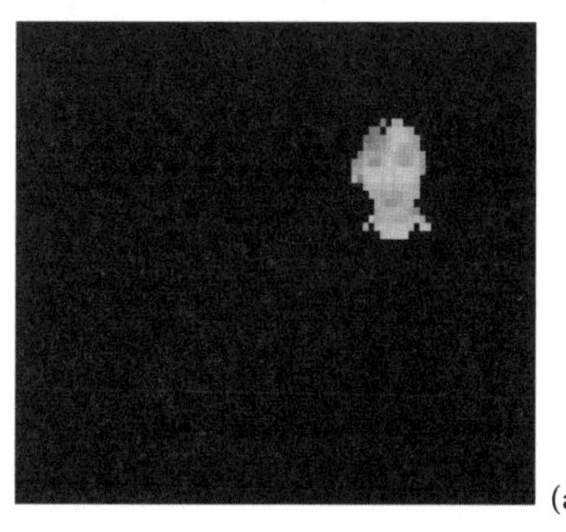
(a)

(b)

Fig. 4.21. (a) The estimated face; (b) the estimated human object

- R_i *is* neither contained in ε_f nor in ε_b;
- R_i *is* connected to a segment in ε_b;
- $V(R_i) \geq \beta$, where β is a user-defined constant.

When such a R_i is found, R_i is added to ε_b. This process is iterated until no R_i satisfies the above conditions.

Finally, we have the estimated human object, ε_u as follows:

$$\varepsilon_u = \varepsilon_f \cup \varepsilon_b \tag{4.38}$$

Figure 4.21(a) shows the estimated face region and Figure 4.21(b) shows the estimated human object for Figure 4.19(a).

4.5.5 Human Object Refinement

As in [19], we divide the base image into foreground, background, and ambiguous regions. A fuzzy neural network is constructed and trained by the data points taken from the foreground and background regions. The blocks in the ambiguous region are then classified by the trained network to the foreground region or the background region. The blocks belonging to the foreground region form the desired human object.

Morphological erosion and dilation are used to find foreground and background regions. Several times of erosion are performed on ε_u and let the resulting image be the foreground region, denoted Z_e. Also, several times of dilation are performed on ε_u and let the resulting image be the background region, denoted Z_d. The blocks belong neither to Z_e, nor to Z_d, constitute the ambiguous region Z_a.

4.5.5.1 Supervised Network Construction

As seen previously, each fuzzy cluster C_j is represented by the product of $g(x_1; v_{1j}; \sigma_{1j})$, $g(x_2; v_{2j}; \sigma_{2j})$, and $g(x_3; v_{3j}; \sigma_{3j})$, representing Gaussian membership functions of C_r, C_b, and Y, respectively, and each $g(x_i; v_{ij}, \sigma_{ij})$, $1 \le i \le 3$, has the center v_{ij} and standard deviation σ_{ij}. C_j can be interpreted as a fuzzy IF-THEN rule of the following form:

$$\text{IF } x_1; \text{ IS } g(x_1; v_{1j}, \sigma_{1j}) \text{ AND } x_2 \text{ IS } g(x_2; v_{2j}, \sigma_{2j}) \text{ AND } x_3 \text{ IS } g(x_3; v_{3j}, \sigma_{3j}) \\ \text{THEN } y \text{ IS } c_j \tag{4.39}$$

where x_1, x_2, x_3, and y are variables for C_r, C_b, Y, and the corresponding output, respectively. The output c_j is set as follows:

$$c_j = \begin{cases} 1 & \text{if } C_j \text{ totally covers a block in } \varepsilon_u, \\ 0 & \text{otherwise} \end{cases} \tag{4.40}$$

for $1 \le j \le J$. A rule with $c_j = 1$ specifies the conditions under which a block belongs to the foreground region. Note that we have J fuzzy rules. These rules form a rough discriminator for classification.

Based on the J rules, a four-layer supervised fuzzy neural network is constructed, as shown in Figure 4.22. The four layers are called the input layer, the fuzzification layer, the inference layer, and the output layer, respectively. Links between layer 1 and layer 2 are weighted by (v_{ij}, σ_{ij}), for all $1 \le i \le 3$, $1 \le j \le J$, links between layer 3 and layer 4 are weighted by c_j, for all $1 \le j \le J$, and the other links are weighted by 1. Note that the first three layers are totally identical to those in a SCFNN network shown in Figure 4.4. Let (x_1, x_2, x_3, y) be an input-output pattern where (x_1, x_2, x_3) is the input vector and y is the corresponding desired output. The operation of the neural network is described as follows.

1. Layer 1. Layer 1 contains three nodes. Node i of this layer produces output $o^{(1)}_i$ by transmitting its input signal x_i directly to layer 2, i.e.,

$$o_i^{(1)} = x_i \tag{4.41}$$

   for all $1 \le i \le 3$.
2. Layer 2. Layer 2 contains J groups and each group contains three nodes. Node (i, j) of this layer produces its output, $o_{ij}^{(2)}$, by computing the value of the corresponding Gaussian function, i.e.,

$$o_{ij}^{(2)} = G_{ij}(o_i^{(1)}) = \exp\left[-\left(\frac{o_i^{(1)} - v_{ij}}{\sigma_{ij}}\right)^2\right] \tag{4.42}$$

   for all $1 \le i \le 3$ and $1 \le j \le J$.
3. Layer 3. Layer 3 contains J nodes. Node j's output, $o_j^{(3)}$, of this layer is the product of all its inputs from layer 2, i.e.,

$$o_j^{(3)} = \prod_{i=1}^{3} o_{ij}^{(2)} \tag{4.43}$$

   for all $1 \le j \le J$.

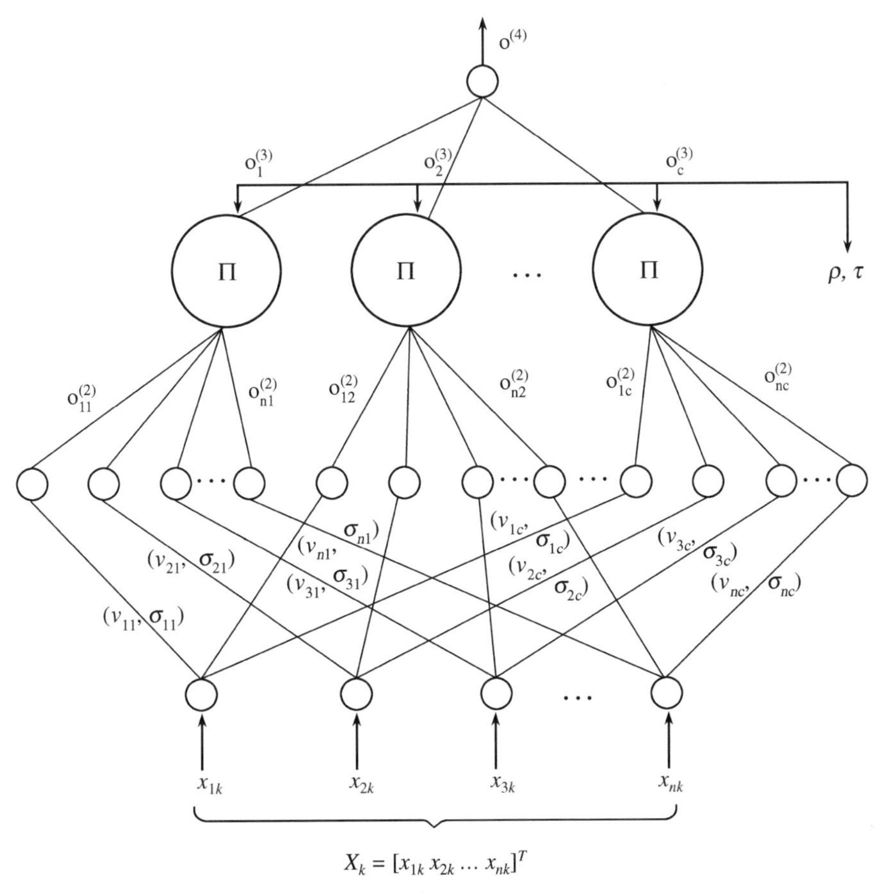

Fig. 4.22. Architecture of the fuzzy neural network

4. Layer 4. Layer 4 contains only one node whose output, $o^{(4)}$, represents the result of the centroid defuzzification, i.e.,

$$o^{(4)} = \frac{\sum_{j=1}^{J} o_j^{(3)} \cdot c_j}{\sum_{j=1}^{J} o_j^{(3)}} \tag{4.44}$$

Note that Layers 1–3 operate identically as the SCFNN network does. Apparently, v_{ij}, σ_{ij}, and c_j are the parameters that can be tuned to improve the precision of the discriminator. The tuning is done by hybrid learning, described below.

4.5.5.2 Hybrid Learning

As mentioned earlier, the training data for the network are taken from the foreground and background regions. Let the set of training data be denoted as

$T = \{(x_{1k}, x_{2k}, x_{3k}, y_k) | 1 \le k \le N_T\}$ where (x_{1k}, x_{2k}, x_{3k}) is the input vector, denoting C_r, C_b, and Y values, respectively, of the training data, and $y_k = 1$ for the data taken from the foreground region and $y_k = 0$ for the data taken from the background region.

The learning algorithm we use is a combination of a recursive SVD-based least squares estimator and the gradient descent method, which was demonstrated to be efficient for the network architecture of Figure 4.22 [48]. In each iteration, the learning of v_{ij}, σ_{ij}, and c_j are treated separately. To optimize c_j, v_{ij} and σ_{ij} stay fixed, and the recursive SVD-based least squares estimator is applied. To refine v_{ij} and σ_{ij} , c_j stays fixed and the batch gradient descent method is used. The process is iterated until the desired approximation precision is achieved.

Let $k.o^{(4)}$ and $k.o_j^{(3)}$ denote the actual output of layer 4 and the actual output of node j in layer 3, respectively, for the kth training pattern. By Eq. (4.44), we have:

$$k.o^{(4)} = a_{k1}c_1 + a_{k2}c_2 + \ldots a_{kJ}c_J \tag{4.45}$$

where

$$a_{k1} = \frac{k.o_j^{(3)}}{\sum_{j=1}^{J} k.o_j^{(3)}} \tag{4.46}$$

for $1 \le j \le J$. Apparently, we would like $|y_k - k.o^{(4)}|$ to be as small as possible for the k_{th} training pattern. For all N_T training patterns, we have N_T equations in the form of Eq. (4.45). Clearly, we would like

$$J(X) = ||\boldsymbol{B} - \boldsymbol{AX}|| \tag{4.47}$$

to be as small as possible, where $\boldsymbol{B}$, $\boldsymbol{A}$, and $\boldsymbol{X}$ are matrices of $N_T \times 1$, $NT \times J$, and $J \times 1$, respectively, and

$$\boldsymbol{B} = \begin{bmatrix} y_1 \\ y_2 \\ \vdots \\ y_{N_T} \end{bmatrix}, \quad \boldsymbol{A} = \begin{bmatrix} a_{11} & a_{12} & \cdots & a_{1J} \\ a_{21} & a_{22} & \cdots & a_{2J} \\ \vdots & \vdots & \vdots & \vdots \\ a_{N_T 1} & a_{N_T 2} & \cdots & a_{N_T J} \end{bmatrix}, \quad \boldsymbol{X} = \begin{bmatrix} c_1 \\ c_2 \\ \vdots \\ c_J \end{bmatrix} \tag{4.48}$$

As mentioned earlier, we treat v_{ij} and σ_{ij} as fixed, so $\boldsymbol{X}$ is the only variable vector in Eq. (4.47). The optimal $\boldsymbol{X}$ which minimizes Eq. (4.47) can be found by a recursive estimator based on the technique of singular value decomposition (SVD) [24, 25]. The method considers training patterns one by one, starting with the first pattern until the last pattern, therefore demanding less time and space requirements [48].

On the other hand, parameters v_{ij} and σ_{ij}, $1 \le i \le 3$ and $1 \le j \le J$, are refined by the gradient descent method, treating c_j, $1 \le j \le J$, as fixed. The error function used is defined as follows:

$$E = \frac{1}{2N_T} \sum_{k=1}^{N_T} (y_k - k.o^{(4)})^2 \; . \tag{4.49}$$

We adopt the batch backpropagation mode in order to work properly with the recursive SVD-based estimator. The learning rules for v_{ij} and σ_{ij} are [48]:

$$v_{ij}^{new} = v_{ij}^{old} - \eta_1 \left(\frac{\partial E}{\partial v_{ij}} \right) \tag{4.50}$$

$$\frac{\partial E}{\partial v_{ij}} = \frac{2}{N_T} \sum_{k=1}^{N_T} \left\{ [k.o^{(4)} - y_k] \frac{[c_j - k.o^{(4)}][x_{ik} - v_{ij}]k.o_j^{(3)}}{\sigma_{ij}^2 \sum_{r=1}^{J} k.o_r^{(3)}} \right\}. \tag{4.51}$$

$$\sigma_{ij}^{new} = \sigma_{ij}^{old} - \eta_2 \left(\frac{\partial E}{\partial \sigma_{ij}} \right) \tag{4.52}$$

$$\frac{\partial E}{\partial \sigma_{ij}} = \frac{2}{N_T} \sum_{k=1}^{N_T} \left\{ [k.o^{(4)} - y_k] \frac{[c_j - k.o^{(4)}][x_{ik} - v_{ij}]^2 k.o_j^{(3)}}{\sigma_{ij}^3 \sum_{r=1}^{J} k.o_r^{(3)}} \right\}. \tag{4.53}$$

where η_1 and η_2 are learning rates.

4.5.5.3 Final Human Object

After training is completed, the trained network is used to classify the blocks in the ambiguous region Z_a. Each block of the ambiguous region Z_a is fed to the trained neural network. If the corresponding network output of a block is greater than or equal to a threshold ϕ, the block is categorized as foreground. Note that ϕ is a predefined parameter and $0 < \phi < 1$. Finally, the maximally connected segment [31] in the foreground region for the desired human object in Figure 4.23(a) shows the three regions Z_e, Z_d, and Z_a. For the example the image of Figure 4.19(a), and the final human object obtained is shown in Figure 4.23(b).

As in [19], the trained network is used for finding human objects of the other frames in the same video stream, without the necessity of reconstruction or re-training. Each block of such a frame is fed to the network. A block is categorized as foreground if the corresponding network output is greater than or equal to a

(a)

(b)

Fig. 4.23. (a) Foreground, background, and ambiguous regions; (b) refined human object

threshold ϕ. The largest maximally connected segment taken to be the desired human object of the underlying frame.

4.5.6 Experimental Results

We show segmentation results on two benchmark video streams, Akiyo (368×240) and Silent (360×288). The first frame of each stream is selected as the base frame. Note that the numbers shown in the parentheses indicate the resolution of each frame of the underlying stream. For example, each frame of Akiyo contains 368×240 pixels. Figures 4.24 and 4.25 show the human objects extracted from the base frames of these video streams. There are two sub-figures in each figure. The first sub-figure shows the original base frame image, the second sub-figure shows the human object extracted. To evaluate the extraction accuracy quantitatively, we use the error index E_I defined as the ratio of the number of mismatched pixels and the number of total pixels in an image, i.e.,

$$E_I = \frac{N_p^m}{N_1 N_2} \tag{4.54}$$

where N_p^m is the number of mismatched pixels and $N_1 \times N_2$ is the total number of pixels in an image. A mismatched pixel is a non-human pixel mistaken for a human pixel or a human pixel mistaken for a non-human pixel. Obviously, the smaller the E_I is, the more accurate the extraction. The error indices associated with Figures 4.24 and 4.25 are given in the first two columns in Table 4.4.

Next, we show the generalization capabilities of the approach. We extract the human object from the 50th image frames of these video streams. As men-

(a)

(b)

Fig. 4.24. The base frame of Akiyo: (a) original image; (b) extracted human object

Table 4.4. Error indices for object extraction

Akiyo, 1st	Silent, 1st	Akiyo, 50th	Silent, 50th
0.0245	0.0111	0.0236	0.0191

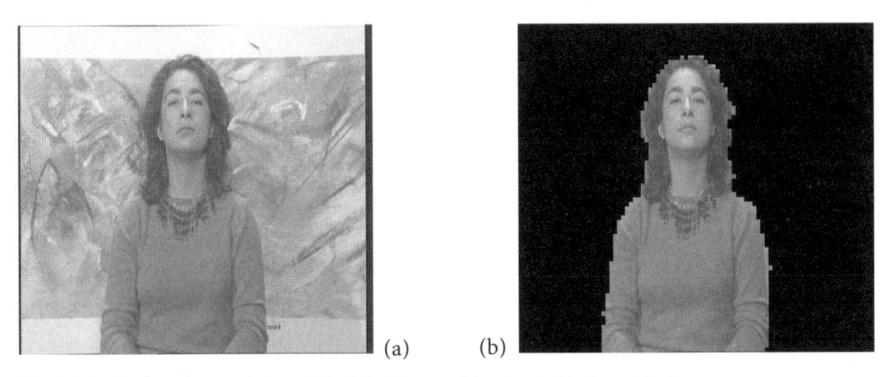

Fig. 4.25. The base frame of silent: (a) original image; (b) extracted human object

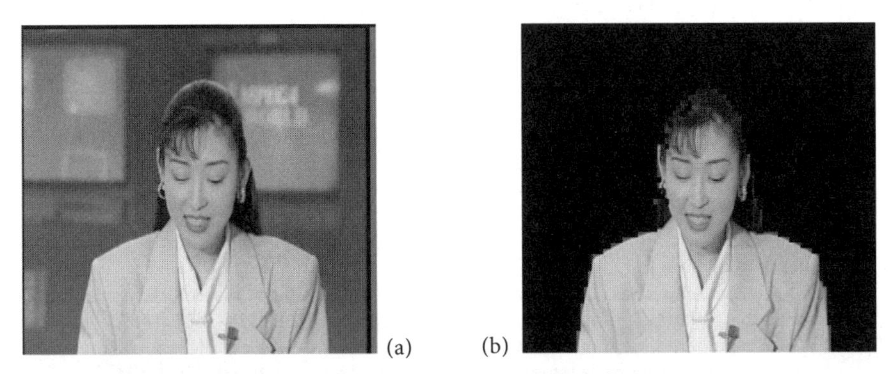

Fig. 4.26. The 50th frame of Akiyo: (a) original image; (b) extracted human object

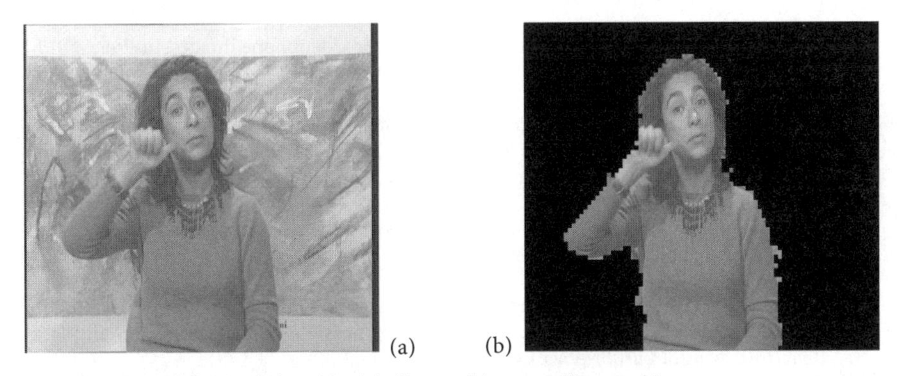

Fig. 4.27. The 50th frame of Silent: (a) original image; (b) extracted human object

tioned before, we may use the trained network of the base frame to extract human objects in the other frames of the same stream, without reconstruction and retraining. The results are shown in Figures 4.26 and 4.27. The error indices associated with Figures 4.26 and 4.27 are given in the last two columns in Table 4.4. Applications in Video transmission can be found in [4, 5, 56].

References

[1] *Basic Image Interchange Format (BHF) Annex B*, 1998. ISO/IEC 12087-5.

[2] *Information technology - Generic coding of moving pictures and associated audio information: Video*, 2000. ISO/IEC 13818-2.

[3] *Information technology - Multimedia content description interface - Part 5: Multimedia description schemes*, 2002. ISO/IEC 15938-5.

[4] *MPEG-4 video verification model version 18.0*, 2001. ISO/IEC JTC1/SC29/WG11 N3908.

[5] *Overview of the MPEG-4 Standard*, 2002. ISO/IEC JTC1/SC29/WG11 N4668.

[6] Special session on robust video. In *Proc. of the IEEE International Conference on Image Processing*, Kobe, Japan, 1999.

[7] *Vector Quantization Decompression for the National Imagery Transmission Format Standard*, 1994. MIL-STD-188-199.

[8] A. M. Alattar and S. A. Rajala. Facial features localization in front view head and shoulders images. In *Proc. of IEEE International Conference on Acoustic, Speech, and Signal Processing*, pages 3557-3560, Arizona, USA, March 1999.

[9] R. E. J. Bezdek and W. Full. FCM: The fuzzy c-means clustering algorithm. *Computers and Geoscience*, 10(2-3): 191-203, 1984.

[10] J. C. Bezdek and N. R. Pal. Two soft relatives of learning vector quantization. *Neural Networks*, 8(5): 729-743, 1995.

[11] S. M. Bohte, H. L. Poutre and J. N. Kok. Unsupervised clustering with spiking neurons by sparse temporal coding and multilayer RBF networks. *IEEE Trans. Neural Networks*, 13(2): 426-435, 2002.

[12] C. Bregler. Learning and recognizing human dynamics in video sequences. In *Proc. of Conference on Computer Vision and Pattern Recognition*, pages 568-574, San Juan, Puerto Rico, June 1997.

[13] G. A. Carpenter and S. Grossberg. ART 2: Self-organization of stable category recognition codes for analog input patterns. *Applied Opties*, 26(23): 4919-4930, 1987.

[14] G. A. Carpenter, S. Grossberg, and D. B. Rosen. ART 2-A: An adaptive resonance algorithm for rapid category learning and recognition. *Neural Networks*, 4(4): 493-504, 1991.

[15] G. A. Carpenter, S. Grossberg, and D. B. Rosen. Fuzzy ART: Fast stable learning and categorization of analog pattern by an adaptive resonance system. *Neural Networks*, 4(6): 759-771, 1991.

[16] D. Chai and K. N. Ngan. Automatic face location for videophone images. In *Proc. of IEEE TENCON on Digital Signal Processing Applications*, pages 137-140, Perth, Western Australia, November 1996.

[17] D. Chai and K. N. Ngan. Face segmentation using skin-color map in videophone applications. *IEEE Trans. Circuits Syst. Video Technol.*, 9(4): 551-564, 1999.

[18] M. R. Civanlar and A. M. Teklap. Real-time video over the Internet. *Signal Processing: Image Communication*, 15(2): 1-5, 1999.

[19] N. Doulamis, A. Doulamis, and S. Kollias. Improving the performance of MPEG compatible encoding at low bit rates using adaptive neural networks. *Real-Time Imaging*, 6(5): 327-345, 2000.

[20] R. O. Duda and P. E. Hart. *Pattern Classification and Scene Analysis*. John Wiley, 1973.

[21] T. Eltoft and R. J. P. deFigueiredo. A new neural network for cluster-detection-and-labeling. *IEEE Trans. Neural Networks*, 9(5): 1021-1035, 1998.

[22] P. Fleury, S. Bhattacharjee, L. Piron, T. Ebrahimi, and M. Kunt. MPEG-4 video verification model: A solution for interactive multimedia applications. *J. Electronic Imaging*, 7(3): 502-515, 1998.

[23] D. M. Gavrila. The visual analysis of human movement: A survey. *Computer Vision and Image Understanding*, 73(1): 82-98, 1999.

[24] G. H. Golub and C. Reinsch. Singular value decomposition and least squares solutions. *Numerische Mathematik*, 14: 403-420, 1970.

[25] G. H. Golub and C. F. Van Loan. *Matrix Computations*. The Johns Hopkins University Press, London, 1991.

[26] V. Govindaraju, D. B. Sher, R. K. Srihari, and S. N. Srihari. Locating human faces in newspaper photographs. In *Proc. of IEEE Computer Society* Conference on *Computer* Vision *and Pattern Recognition,* pages 549–554, San Diego, CA, USA, June 1989.

[27] V. Govindaraju, S. N. Srihari, and D. B. Sher. A computational model for face location. In *Proc.* of *3rd International Conference on Computer* Vision, pages 718–721, Osaka, Japan, December 1990.

[28] S. Grossberg. Adaptive pattern classification and universal recoding: 1. parallel development and coding of neural feature detectors. *Biological Cybernetics,* 23: 121–134, 1976.

[29] C. Gu and M. C. Lee. Semiautomatic segmentation and tracking of semantic video objects. *IEEE Trans. Circuits Syst. Video Technol.,* 8(5): 572–584, 1998.

[30] M. Hagan, H. Demuth, and M. Beale. *Neural Network Design.* Thomson Learning, 1996.

[31] R. M. Haralick and L. G. Shapiro. *Computer and Robot Vision, Vol. 1.* Addison-Wesley, Reading, MA, 1992.

[32] C. M. Huang and R. W. Harris. A comparison of several vector quantization codebook generation approaches. *IEEE Trans. Image Processing,* 2(1): 108–112, 1993.

[33] C. F. Juang. A TSK-type recurrent fuzzy network for dynamic systems processing by neural network and genetic algorithms. *IEEE Trans. Fuzzy Syst.,* 10(2): 155–170, 2002.

[34] C. F. Juang and C. T. Lin. An on-line self-constructing neural fuzzy inference network and its applications. *IEEE Trans. Fuzzy Syst.,* 6(1): 12–32, 1998.

(35] C. F. Juang and C. T. Lin. A recurrent self-organizing neural fuzzy inference network. *IEEE Trans. Neural Networks,* 10(4): 828–845, 1999.

[36] S. H. Kim and H.G. Kim. Face detection using multi-modal information. In *Proc. of 4th IEEE International Conference on Automatic Face and Gesture Recognition,* pages 14–19, Grenoble, France, March 2000.

[37] S. H. Kim, N. K. Kim, S. C. Ahn, and H. G. Kim. Object oriented face detection using range and color information. In *Proc. of 3th IEEE International Conference on Automatic Face and Gesture Recognition,* pages 76–81, Nara, Japan, April 1998.

[38] T. Kohonen. *Self-Organization and Associative Memory.* Springer-Verlag, 3rd edition, 1989.

[39] 1. Kompatsiaris and M. G. Strintzis. Spatiotemporal segmentation and tracking of objects for visualization of videoconference image sequences. *IEEE Trans. Circuits Syst. Video Technol.,* 10(4): 1388–1402, 2000.

[40] C. M. Kuo, C. H. Hsieh, and Y. R. Huang. A new temporal-spatial image sequence segmentation for object-oriented video coding. In *Proc. of IEEE Asia Pacific Conference on Multimedia Technology and Applications,* pages 117–127, 1-Shou University, Taiwan, ROC, Dec. 2000.

[41] S. J. Lee and C. S. Ouyang. A neuro-fuzzy system modeling with selfconstructing rule generation and hybrid SVD-based learning. *to appear in IEEE Trans. Fuzzy Syst.*

[42] Y. Lin, G. A. Cunningham III, and S. V. Coggeshall. Using fuzzy partitions to create fuzzy systems from input-output data and set the initial weights in a fuzzy neural network. *IEEE Trans. Fuzzy Syst.,* 5(4): 614–621, 1997.

[43] J. Mao and A. K. Jain. A self-organizing network for hyper-ellipsoidal clustering (HEC). *IEEE Trans. Neural Networks,* 7(1): 16–29, 1996.

[44] S. J. Mckenna, S. Jabri, Z. Duric, and H. Wechsler. Tracking interactive people. In *Proc. of 4th International Conference on Automatic Face and Gesture Recognition,* pages 348–353, Grenoble, France, March 2000.

[45] T. B. Moeslund and E. Granum. A survey of computer vision-based human motion capture. *Computer Vision and Image Understanding,* 81(3): 231–268, 2001.

[46] F. Moscheni, S. Bhattacharjee, and M. Kunt. Spatio-temporal segmentation based on region merging. *IEEE Trans. Pattern Analysis and Machine Intelligence,* 20(9): 897–915, 1998.

[47] C. S. Ouyang and S. J. Lee. Knowledge acquisition from input-output data by fuzzy-neural systems. In *Proc. of 1998 IEEE International Conference on Systems, Man, and Cybernetics,* pages 1928–1933, San Diego, California, USA, 1998.

[48] C. S. Ouyang and S. J. Lee. A hybrid learning algorithm for fuzzy neural networks. In *Proc. of 8th Int. Conf. Neural Information Processing (ICONIP2001)*, pages 311–316, Shanghai, China, Nov. 2001.
[49] K. Panchapakesan. *Image processing through vector quantization.* PhD thesis, The University of Arizona, Department of Electrical & Computer Engineering, 2000.
[50] G. Patane and M. Russo. The enhanced LBG algorithm. *Neural Networks,* 14(9): 1219–1237, 2001.
[51] G. Patane and M. Russo. Fully automatic clustering system. *IEEE Trans. Neural Networks,* 13(6): 1285–1298, 2002.
[52] F. Pereira. New trends on image and video coding. In *Proc. of the 10th Portuguese Conference on Pattern Recognition,* Lisboa, Portugal, 1998.
[53] I. Rojas, H. Pomares, J. Ortega, and A. Prieto. Self-organized fuzzy system generation from training examples. *IEEE Trans. Fuzzy Syst.,* 8(1): 23–36, 2000.
[54] D. Saxe and R. Foulds. Toward robust skin identification in video images. In *Proc. of 2nd International Conference on Automatic Face and Gesture Recognition,* pages 379–384, Killington, VT., 1996.
[55] K. Sayood. *Introduction to Data Compression.* Morgan Kaufmann, 2nd edition, 2000.
[56] T. Sikora. The MPEG-4 video standard verification model. *IEEE Trans. Circuits Syst. Video Technol.,* 7(1): 19–31, 1997.
[57] P. K. Simpson. Fuzzy min-max neural networks – Part 2: Clustering. *IEEE Trans. Fuzzy Syst.,* 1(1): 32–45, 1993.
[58] K. Sobottka and I. Pitas. Face localization and facial feature extraction based on shape and color information. In *Proc. of IEEE International Conference on Image Processing,* pages 483–486, Lausanne, Switzerland, September 1996.
[59] E. C. K. Tsao, J. C. Bezdek, and N. R. Pal. Fuzzy Kohonen clustering networks. *Pattern Recognition,* 27(5): 757–764, 1994.
[60] N. Vlajic and H. Card. Vector quantization of images using modified adaptive resonance algorithm for hierarchical clustering. *IEEE Trans. Neural Networks,* 12(5): 1147–1162, 2001.
[61] H. Wang and S. F. Chang. A highly efficient system for automatic face region detection in MPEG video. *IEEE Trans. Circuits Syst. Video Technol.,* 7(4): 615–628, 1997.
[62] C. H. Wei and C. S. Fahn. The multi-synapse neural network and its application to fuzzy clustering. *IEEE Trans. Neural Networks,* 13(3): 600–618, 2002.
[63] C. S. Won. A block-based MAP segmentation for image compressions. *IEEE Trans. Circuits Syst. Video Technol.,* 8(5): 592–601, 1998.
[64] C. C. Wong and C. C. Chen. A hybrid clustering and gradient descent approach for fuzzy modeling. *IEEE Trans. Syst., Man, Cybern. B,* 29(6): 686–693, 1999.
[65] C. R. Wren, A. Azarbayejani, T. Darrell, and A. P. Pentland. Pfinder: Real-time tracking of the human body. *IEEE Trans. Pattern Analysis and Machine Intelligence,* 91(7): 780–785, 1997.
[66] S. Wu, M. J. Er, and Y. Gao. A fast approach for automatic generation of fuzzy rules by generalized dynamic fuzzy neural networks. *IEEE Trans. Fuzzy Syst.,* 9(4): 578–594, 2001.
[67] S. Wu and M. J. Er. Dynamic fuzzy neural networks – A novel approach to function approximation. *IEEE Trans. Syst., Man, Cybern. B,* 30(2): 358–364, 2000.
[68] J. Yen and R. Langari. *Fuzzy logie: Intelligence, Control and Information.* Prentice Hall, 1999.
[69] L. A. Zadeh. Fuzzy sets. *Inform. and Contr.,* 8(3): 338–353, 1965.
[70] M. Zobel, A. Gebhard, D. Paulus, J. Denzler, and H. Niemann. Robust facial feature localization by coupled features. In *Proc. of 4th IEEE International Conference on Automatic Face and Gesture Recognition,* pages 2–7, Grenoble, France, March 2000.
[71] L. Zong and N. G. Bourbakis. Digital video and digital TV: A comparison and the future directions. *Real-Time Imaging,* 7(6): 545–556, 2001.

5 A Neuro-Fuzzy System for Source Location and Tracking in Wireless Communications

Ana Perez-Neira, Joan Bas and Miguel A. Lagunas

5.1 Introduction

Traditionally, high resolution Direction of Arrival (DOA) estimation has been associated with algorithms rather than with processing schemes or architectures. This chapter deals with processing schemes of low complexity that help to attain the performance of higher computational complex algorithms.

Motivated by a previous work on feasible implementations of the Estimate and Maximize algorithm [1], in [2] the authors solve the multiple source location and tracking problem as decoupled single source tracking problems. In this way, it is possible to reduce the complexity for either radar or mobile communication applications. The tracking processor consists of parallel processors, each one of them devoted to a single source. At each parallel branch the spatial power density is measured after the constrained scanning (system P in Fig. 5.1), the so-called notch periodogram. If no other user interferes in the periodogram, its maximum corresponds to the desired user location. However, the resolution of the constrained scanning beam is limited by twofold: the array resolution and the degradation of the tracking for close users. It is the component of resolution that may potentially create a problem. The switch block detects a resolution threshold and smoothly commutes to the proposed fuzzy DOA estimator (System F), which obtains the estimates for the two closely located users (θ_i, θ_j).

The concept of global tracker includes not only the DOA detection scheme, but also the parameter filtering. In this way, it is possible to cope with eventual

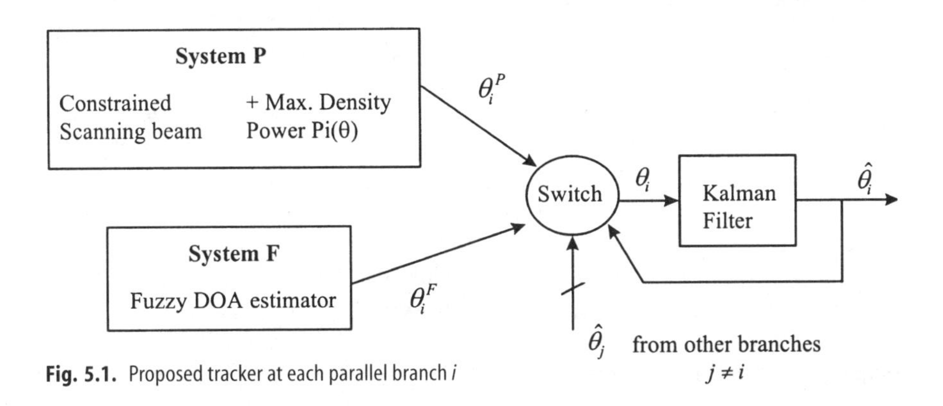

Fig. 5.1. Proposed tracker at each parallel branch *i*

fadings of bounded time duration, as it may occur in crossing radial trajectories of two targets. Therefore, additional Kalman filters for each source are needed. For a detailed explanation on the design of System P and its interaction with the Kalman trackers we refer to [1]. Next section presents the signal model in which the neuro-fuzzy DOA estimator is based.

5.2 Problem Statement

5.2.1 Signal Model

A digital wireless communication system based on adaptive arrays for the location of NS mobile users is addressed. The communication system of each user is formed by an array of Q identical and omni-directional radio receivers. The NS users operate simultaneously in the same frequency and time slot. Letting $a_i(\theta_k)$ be the gain and phase signature of a source coming from direction θ_k when it impinges on the i_{th} sensor, then the signal output of the i_{th} sensor at time n results in: (Fig. 5.2)

$$x_i(n) = \sum_{k=1}^{NS} a_i(\theta_k) s_k(n) + v_i(n) \tag{5.1}$$

where $v_i(n)$ is the noise generated at the i_{th} sensor (usually thermal noise) and $s_k(n)$ is a scaled and phased-delayed version of the signal emitted by user k.

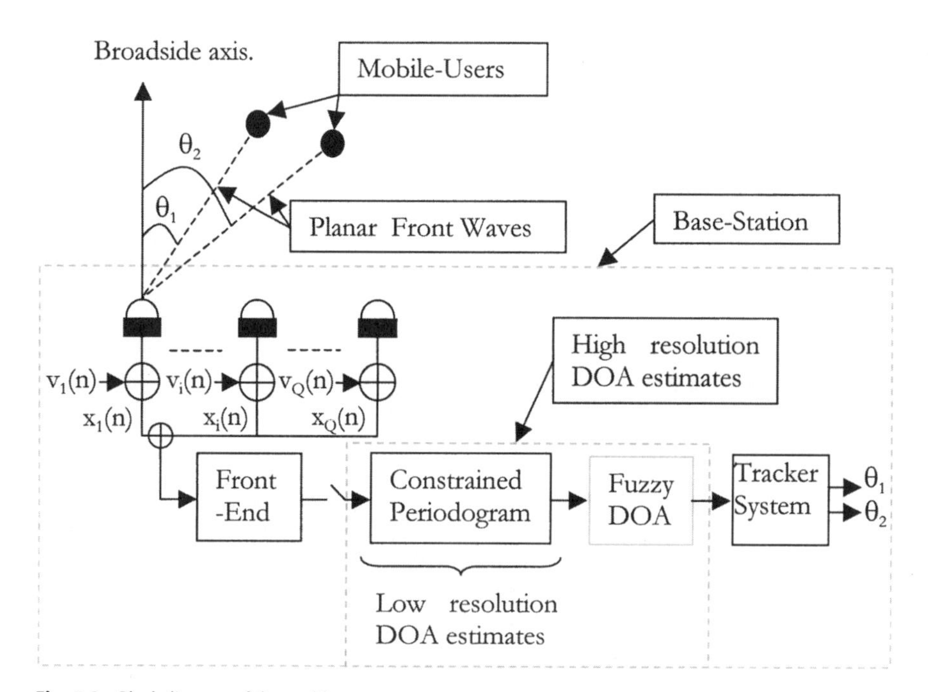

Fig. 5.2. Block diagram of the problem statement

Then, for the common case of resorting to a linear uniform array, the spatial signature $a_i(\theta_k)$ can be formulated as:

$$a_i(\theta_k) = \exp(-j2\pi di \sin\theta_k) \tag{5.2}$$

d being the space between two consecutive sensors in wavelengths (Fig. 5.3). It has been assumed that the antenna array consists of Q sensors of known characteristics (i.e. calibrated array). In addition, the data received $x_i(n)$ from the Q sensors is collected in the so-called snapshot vector $\boldsymbol{x}(n)$. The problem of interest is to estimate and track the NS sources' angular positions θ_k from the data $\boldsymbol{x}(n)$. In the signal model, it is accepted that the distances between the mobiles and the base station are large. This fact is relevant since it implies important consequences for the signal model. Firstly, the local scatters placed near the mobile users have a negligible radius in comparison with the distance between the base station and the receiver. Secondly, the fading of the antenna elements is assumed to be fully correlated. Thirdly, the angle of arrival is estimated from a small number of snapshots (i.e. less than 10 snapshots). Thus, the relative array velocity does not change over the observation interval and is assumed to be equal to zero.

The DOA estimation problem can be analysed from a spectral estimation viewpoint. In this way from a finite-length record of second-order stationary random process, an estimate of its power spectral density (PSD) is obtained. The main limitation on the quality of most PSD estimates is due to the quite small number of data samples usually available for processing. Most commonly, the amount of data available is limited by the fact that the signal under study can be considered second-order stationary only over short observation intervals. This is the case of the mobile communication problem addressed in this work.

There are two main approaches to the PSD estimation problem: the nonparametric and the parametric [3]. The nonparametric approach does not make any assumption on the functional form of the PSD. This is in contrast to the parametric approach. The parametric approach can consequently only be used when there is enough information about the studied signal in order to formulate a model. Otherwise, the nonparametric approach should be used. This work is focused on nonparametric techniques. However, the resolution in these techniques is constrained by the sample length. For this reason, a system is pro-

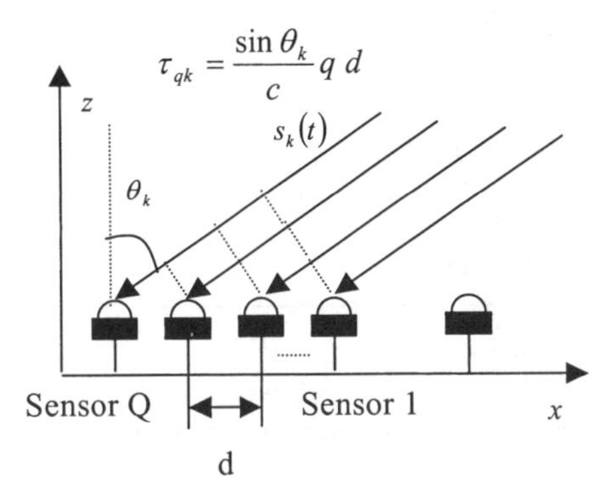

Fig. 5.3. Uniform linear antenna array consisting of Q sensors

posed to improve the resolution that they offer. As the final goal is also to obtain a low complexity system, the design departs from the periodogram, which can be implemented by resorting to the Fast Fourier Transform (FFT) algorithm.

5.2.2 The Periodogram as a Motivational Tool for a Neuro-Fuzzy System

The conventional periodogram for DOA estimation is described by Eq. (5.3), which, for a niform linear array, is the spatial Discrete Fourier Transform. Note that the window length Q is equal to the number of sensors in the array and the spatial extent of an aperture determines the resolution with which two plane waves can be separated. The larger the extent, the more directional the aperture is on any specific direction. Recalling the spatial filtering interpretation of the aperture smoothing function, the main lobe could be considered as a filter's pass band and the side lobes as a stop band. Hence, the main lobe's width defines the aperture's ability to discriminate the propagating waves, i.e. the aperture's resolution. For instance, if the Rayleigh resolution criterion is considered, the resolution can be approximated by $1/Q$. Figure 5.4 depicts the periodogram for a single source impinging from the broadside to an array of 9 sensors. Figure 5.5 depicts the periodogram when another source of equal power impinges 4°

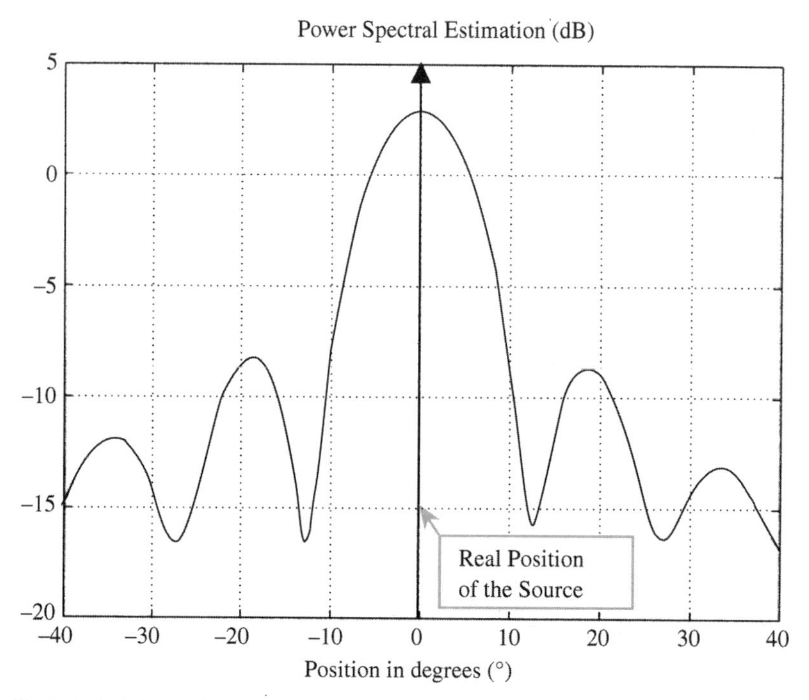

Fig. 5.4. Periodogram for a single source placed in the broadside position. The number of array sensors is 9 and SNR = 5 dB

apart. Note that the periodogram presents just one peak and is not able to resolve both sources. Nevertheless, when Figs. 5.4 and 5.5 are compared, an expert could affirm that inside the main lobe of

$$P(\theta)=\frac{1}{N}\sum_{i=1}^{N}|\boldsymbol{a}^{H}(\theta)\boldsymbol{x}(n)|^{2} \tag{5.3}$$

Figure 5.5, there is more than one source. Therefore, it is possible to conclude that by observing the periodogram bandwidth (width of its main lobe), an expert is able to "improve" the resolution of the periodogram $P(\theta)$. This fact motivates the use of fuzzy logic to incorporate expert knowledge to improve the estimates offered by the periodogram. The "expert" knowledge is used to initialise the rules of the fuzzy system (raw estimation). Afterwards, the rules are fine tuned by a neural network learning algorithm. The resulting neuro-fuzzy system will model the non-linear mapping from the feature space of the periodogram, i.e. beamwidth, to the space of DOA.

5.2.3 Fuzzy Logic for Model-Free Function Approximation

There is a large amount of literature now on different kinds of Fuzzy Logic Systems (FLS) that are universal approximators, being very well summarized in [4].

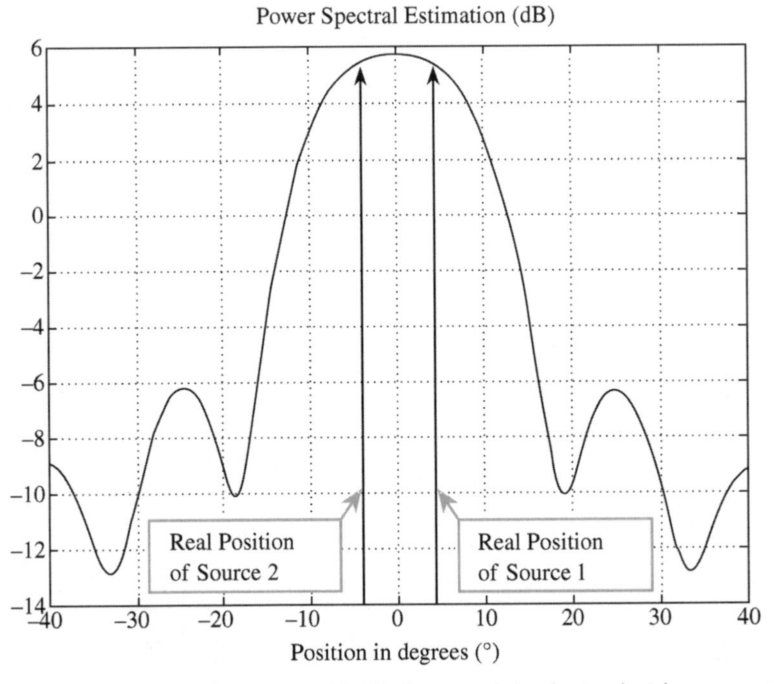

Fig. 5.5. Periodogram of two sources with 5dB of power and placed at 4 and −4 degrees respectively

Among them, this work resorts to the so-called additive FLS, which presents the following formulation:

$$y = \sum_{i=1}^{M} c_i \, \phi_i(z) \tag{5.4}$$

Where z is a vector containing the input variables and y is the output variable (i.e. a Multiple input Single output system has been considered). The so-called Fuzzy Basis Functions (FBF) are $\phi_i(z)$ and, to date, are the only basis functions that can include linguistic information. The relationship between FBF and other basis functions have been extensively studied by Kim and Mendel [5].

The additive FLS of (5.4) has been proven to be able to approximate any continuous function to any degree of accuracy for arbitrary fuzzy sets whenever it uses the centroid defuzzification and product implication [6]. That means that if z is a compact (closed and bounded) subset of R^n and f is continuous, then for any small $\varepsilon > 0$,

$$\sup_{z \in Z} |f(z) - y(z)| < \varepsilon \tag{5.5}$$

Figures 5.6 uses the concept of "patches" to manifest graphically that the fuzzy system approximates a function f by covering it with IF-THEN rule patches and averaging patches that overlap. If a finite number of rule patches are given, it is possible to uniformly approximate any continuous function on a compact domain.

As it was referred to previously, the main goal of this chapter is to perform a non-linear mapping from the feature space of the periodogram, i.e. beamwidth to the space of DOA, specifically the distance (in degrees) between the two sources that are hidden in the main beam of the periodogram. Figure 5.7 depicts an example of this non-linear mapping for two different angular positions of the main beam. The following section, describes the architecture of the proposed fuzzy-neural network, in more detail.

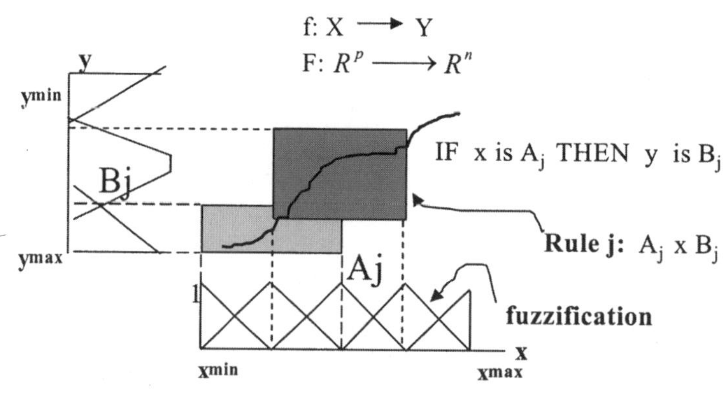

A(.): input fuzzy set
B(.): output fuzzy set

Fig. 5.6. The fuzzy system approximates a function by covering the graph with rule patches

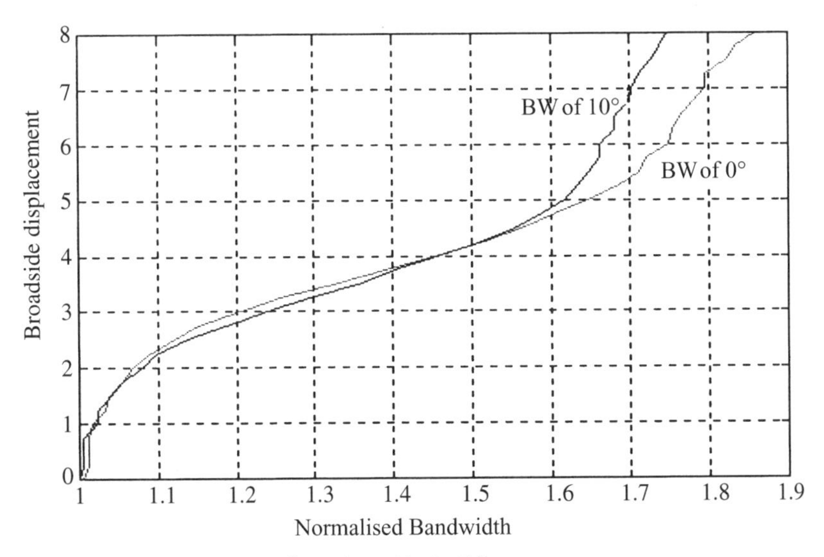

Fig. 5.7. Non-linear mapping to be performed by the FLS

5.3 The Architecture of the Fuzzy-Neural Network

Figure 5.8 pictures the architecture of the fuzzy logic system under the neural viewpoint. This graph emphasizes that along the way, we could choose the fuzzy or the neural angle of view, depending on our needs at the time. If the system is viewed as a fuzzy model, it gives insight into the real system, and it provides a method of simplifying the neural network structure. The feed-forward neural network is the underlying architecture that supports the fuzzy system. In particular, backpropagation neural network training can be applied to help the fuzzy system to match desired input-output pairs [7]. The architecture in Fig. 5.8 graphically describes the three layers that comprise a FLS: fuzzification, inference and defuzzification. There are two inputs, $P_1 + P_2$ neurons in the fuzzification layer and M neurons in the inference layer. Hence, once P_1, P_2, and M are determined, the structure of the network becomes known. Next, the mission of each layer is analysed.

5.3.1 Fuzzification

Every neuron in the fuzzification layer represents a fuzzy membership function for one of the input variables. The fuzzifier or fuzzification layer is a mapping from an input space to input fuzzy sets in a certain input universe of discourse. The most widely used fuzzifier is the singleton fuzzifier because of its simplicity and lower computational requirements. For this reason, the singleton fuzzifier is the fuzzification method that has been chosen in this work. A singleton fuzzi-

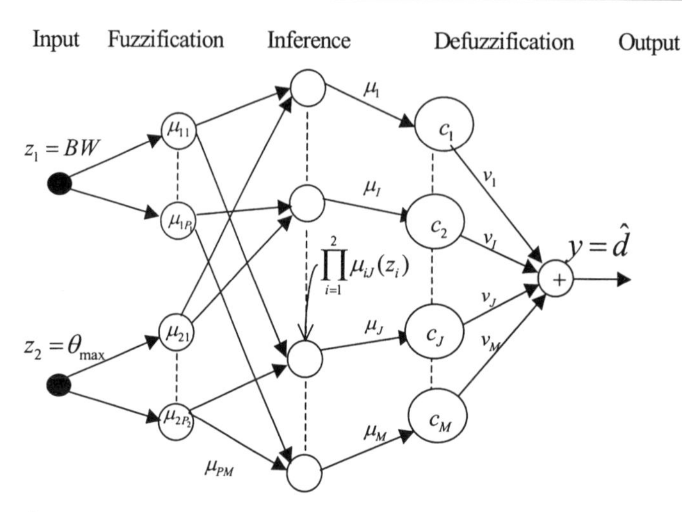

Fig. 5.8. The architecture of the fuzzy neural network

fier maps a specific value $z_i(n)$ at time n to the fuzzy set $A_{ji}(z_i(n))$ $(ji = 1 \ldots P_i)$ with degree $\mu_{ji}(z_i(n))$. Note that in case of low Signal to Noise Ratios (SNR), or when a high input uncertainty exists, non-singleton fuzzy sets can be used instead of singleton ones [8].

As outlined in a previous section, one of the inputs to the fuzzy system is the width of the main lobe of the periodogram (BW). From this input, the FLS computes the distance between the two sources hidden in the main lobe. The question now is which of the various existing definitions for beamwidth should be taken. Because of its computational load, the root mean square and the mean beamwidth definitions are disregarded. On the other hand, the 3 dB beamwidth appears as a suitable one. Nevertheless, in order to make the measurement sensitive to the presence of more than one source for different SNR and variations in the position of the sources, the 10 dB beamwidth formulated in (5.6) is considered instead:

$$10 \log_{10} P\,(\pm BW_{10}) = 10 \log_{10}(P\,(\theta_{Max})) - 10 \qquad (5.6)$$

In order to make this measurement independent from the number of sensors, the input $z_1 = BW$ to the fuzzy system will be normalized by the BW_{10} of periodogram $P(\theta)$, when a single source impinges from the broadside. Additionally, to increase the sensitivity of the BW to the different positions of the two sources within the main lobe, the exponential of BW is used. Figure 5.9 depicts the mentioned increase of sensitivity. The spatial frequency, which for the uniform linear array in Eq. (5.2) is defined as $u_k = d \sin \theta_k$, depends (in a non-linear way) on the DOA θ_k and the width of the main lobe. Note that the beamwidth of the periodogram changes according to the position of its maximum (Fig. 5.6). For this reason, the maximum of the periodogram should be another input to the fuzzy logic system $(z_2 = \theta_{max})$. From these two inputs, the FLS estimates the half distance between

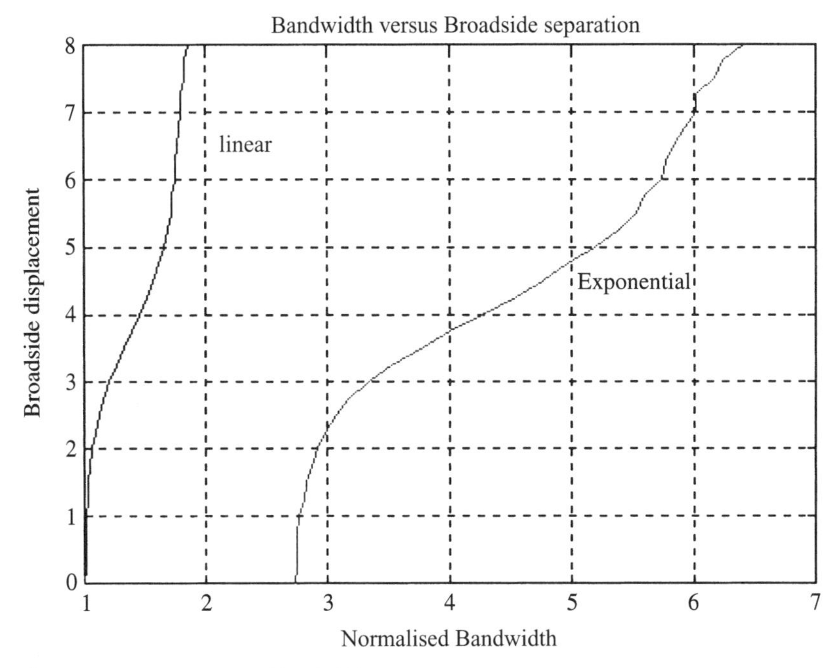

Fig. 5.9. Linear vs. exponential bandwidth

them, denoted as d, and the position of the two sources hidden in the main lobe. To carry out the source localization from the FLS output "d" we resort to:

$$\hat{\theta}_1 = \theta_{\max} - d \tag{5.7a}$$

$$\hat{\theta}_2 = \theta_{\max} + d \tag{5.7b}$$

The ranges of BW, $\theta_{\max}$ and d are the following: $1 \leq BW \leq 2$, $-25° \leq \theta_{\max} \leq 25°$, $0 \leq d \leq \delta_P$, where δ_P refers to the resolution of the periodogram. Therefore, the goal of the FLS is to resolve sources closer than δ_P in the field of view of ($\pm 25°$).

Once the range of each input and output variable has been determined, it is possible to characterize them by a set of fuzzy sets. Figure 5.10 depicts the membership functions that describe both variables when Gaussian functions are used. Although Gaussian shapes are more appropriate to train, the triangular shapes are more suitable in order to simplify computation. In regards to the number and position of the fuzzy sets, in order to reduce computation, the whole field of view or range of $\theta_{\max}$ has been measured in $P_2 = 5$ values $\{\pm 20°, \pm 10°, 0\}$ (Fig. 5.10a). Note that equally spaced membership functions are possible only if the mean values (centres) can be chosen by the designer (e.g. for $\theta_{\max}$.). Nonetheless, if these values are trained to the data that are associated, then unequally spaced functions are obtained in general.

The fuzzy sets B_i that describe the output y are designed to be symmetric and of equal area. In this way, if the centroid defuzzification method is used, only the individual centroid of each fuzzy set B_i matter in the approximation is carried out by the FLS [9]. Initially, equally spaced centroids are set, as is presented in Sect. 5.4.

5.3.2 Inference

In a FLS, the behaviour of a fuzzy system is determined by a set of M IF-THEN rules $R_{m,}$ that associates output fuzzy sets with input fuzzy sets. Our system is a two-input/single-output system where the m_{th} fuzzy rule is R_m:

$$R_m\text{: IF } z_1 \text{ is } A_{1m} \text{ AND } z_2 \text{ is } A_{2m} \text{ THEN } y \text{ is } B_m \tag{5.8}$$

The next section deals in more detail with the extraction of these rules from numerical data. These associations are gathered in a bank of parallel fuzzy associative memory (FAM) rules. In addition, the bank of FAM rules can be easily constructed, processed and modified in software, or in digital VLSI circuitry.

The inference in Fig. 5.8 is in charge of matching the preconditions of rules in the fuzzy system rule base with input state linguistic terms, to perform implications. Each input $z_i(n)$ is matched in parallel against all the IF-part A_{im} sets in (5.8) for each rule m. The inference "fires" or "activates" the rules to some degree μ_m. The procedure of inference used to develop the fuzzy DOA estimator is the so-called correlation-product inference, which obtains the weight as it is formulated in (5.9).

$$\mu_m = \prod_{i=1}^{2} \mu_{im}(z_i(n)) \qquad m=1\ldots M \qquad Rules \tag{5.9}$$

Then each fired IF-part set A (Fig. 5.10) scales its THEN-part set B and B shrinks down to this height.

5.3.3 Defuzzification

The third step or layer is the defuzzification, where the system computes the output y as the centroid or centre of gravity of this final output set. For symmetric output fuzzy sets B_m, Eq. (5.10) formulates the centroid defuzzification as follows:

$$y = \frac{\sum\limits_{m=1}^{M} \mu_m(z)\, c_m}{\sum\limits_{m=1}^{M} \mu_m(z)} \tag{5.10}$$

where c_m represents the centroid of the output fuzzy set B_m. An appealing feature of the centroid defuzzification is that all centroid additive fuzzy systems try to approximate the function $f(.)$ with a random conditional mean $E\{Y/Z=z\}$. Substituting (5.9) in (5.10), the mapping performed by the FLS between z and y

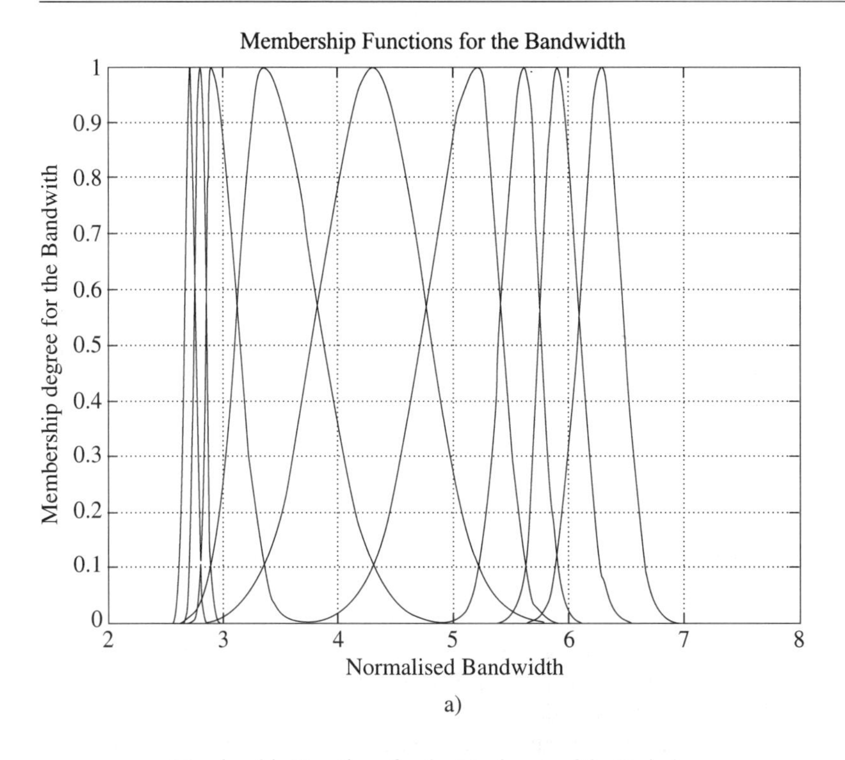

a)

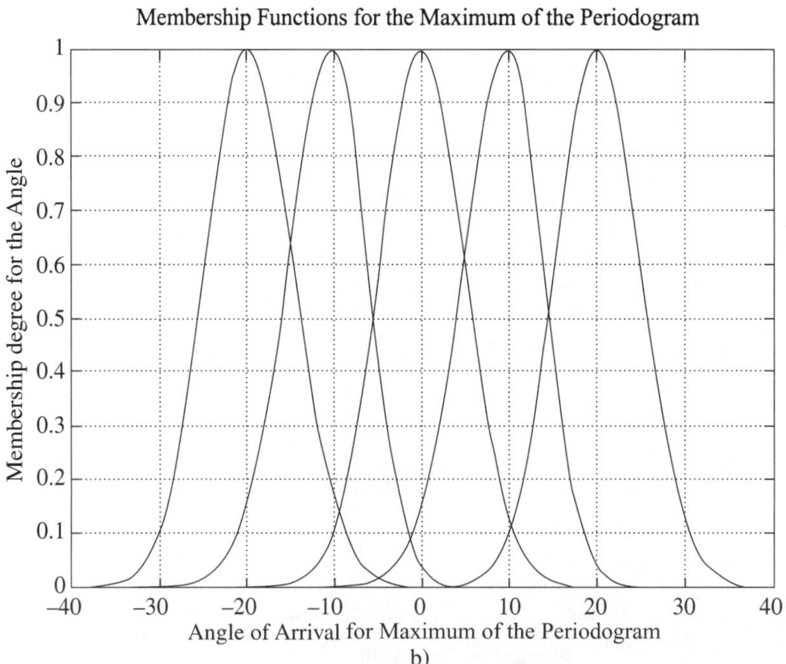

b)

Fig. 5.10. Fuzzy Sets for a) input z_1 and b) z_2

is obtained in (5.11).

$$y = \sum_{m=1}^{M} c_m \frac{\prod_{i=1}^{2} \mu_{im}(z_i)}{\sum_{m=1}^{M} \mu_m(z)} = \sum_{m=1}^{M} c_m \, \phi_m(z) = \sum_{m=1}^{M} v_m \tag{5.11}$$

where the FBF that were introduced in (5.11) are defined as:

$$\phi_m(z) = \frac{\prod_{i=1}^{2} \mu_{im}(z_i)}{\sum_{m=1}^{M} \mu_m(z)} \tag{5.12}$$

Note that the FBF are determined only by the IF parts of the rules (antecedents). Figure 5.11 points out a very interesting property of the FBF: the $\phi_j(z_i)$ whose centres are inside the interval {2.5, 7} look like Gaussian functions, whereas the $\phi_j(z_i)$ whose centres are on the boundaries of the interval look like sigmoid functions. It is known in the neural network literature that Gaussian radial basis functions are good at characterizing local properties, whilst neural networks with sigmoid non linearities are good at characterizing global properties. The FBFs seem to combine the advantages of both the Gaussian radial basis functions and the sigmoidal neural networks. Other differences of the designed neuro-fuzzy system from a neural network based on Radial Basis Functions are

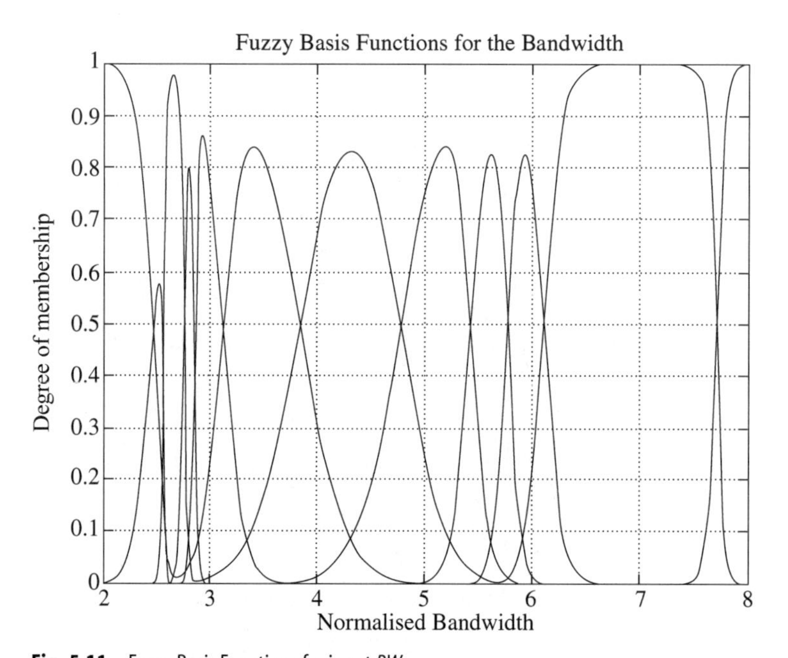

Fig. 5.11. Fuzzy Basis Functions for input *BW*

twofold. Firstly, the neural network is represented by a fuzzy system and insight into the real system is gained, and secondly, our input fuzzy membership functions can be defined by functions other than Gaussian (e.g. triangular or trapezoidal).

In order to specify the FBF expansion, it is necessary to: i) determine the rule base and ii) use non-linear optimisation techniques. Specifically, the rule base and features of the antecedent fuzzy sets A (i.e. mean and width) are tuned by using the data itself (coarse tuning) and the backpropagation algorithm (fine tuning).

5.4 Design of the Rule Base

5.4.1 Initialization

The selection of the fuzzy rules has a substantial effect on the performance of the FLS. The rules only translate the behaviour of the main lobe, when two sources are closer than the array resolution (δ_P). As this resolution depends on the maximum of the periodogram θ_{max} (i.e. there is more resolution in the broadside than in the end fire), a different set of rules is developed for each θ_{max}. As mentioned previously, in order to reduce computation, the whole field of view or range of θ_{max} has been quantized in $P_2 = 5$ values.

The elaboration of the Fuzzy Associative Memories, FAM [6] for each quantized maximum position θ_{max} is a set of N training pairs that have been collected previously. Each pair relates a BW with the distance "d" between the present sources. In these training pairs, the output d, has been quantized in N different states $d_i = i°$ $(i = 0 \ldots N-1)$, the output membership functions B_i are centred at these quantized values. Thus, the number of fuzzy sets to describe BW is $P_1 = N$. For each θ_{max}, each distance d_i maps to the corresponding measured normalised bandwidth BW_i. These bandwidth measures are the support values to design the unequally spaced fuzzy set $A_i(z_2)$ in Fig. 5.10b (note that $d_1 = 1.5$ instead of $d_1 = 1$ in order for $A_1(z_2)$ and $A_2(z_2)$ not to overlap). Table 5.1

Table 5.1. Rules from training pairs for an array of 9 antennas

distance: d \ θ_{max}	z_2: Bandwidth (exponential)	
	0°	10°
0	2.7121	2.7456
1.5	2.8027	2.8577
2	2.8949	2.9447
3	3.3527	3.6693
4	4.3119	4.2631
5	5.2292	5.0531
6	5.6226	5.2593
7	5.913	5.4739
8	6.2931	5.7546

shows the $M = P_2xP_1$ rules that are obtained for the case of a 9 antenna array (note that just 3 fuzzy sets are considered to describe θ_{max}. Therefore, the field of view could not exceed [–15° 15°]. The ratio between the powers of both sources is considered less or equal to 7 dB. This assumption is valid in communication systems with power control, where power groups can be allocated in time-frequency space, such that their dissimilar power does not negatively affect the SDMA.

To obtain each of the $\{BW_i, \theta_{max\,i}, d_i\}$ relations of Table 5.1, over 50 realizations of $P(\theta)$ are averaged, each one is computed with 10 snapshots, and over the values of $SNR_{dB} = \{5, 6, 7, 8, 9\}$.

The number of rules in the FLS, depends on the design method used to construct it. If no tuning of the FLS parameters is done, then there could be as many as $M = N$ rules. If tuning is used, and we abide by the commonly used design principle that there must be fewer tuned design parameters than training pairs, then we must use less than N. For the specific problem of DOA estimation, the number of output fuzzy sets has been set to a multiple of the Rayleigh resolution (i.e. the number of sensors Q). This choice, proves to give good results for $P_1 = Q$ or $2Q$. In addition, this choice keeps the computational burden (i.e. the number of rules) within acceptable limits. Nevertheless, other options are also possible, as those summarized in [8, 10].

The FAM table presented in Table 5.1 associates the membership function of each training pair. Just by vector/matrix multiplications, each input fires in parallel the M rules. Note that these rules are obtained from numerical data: the fuzzy sets that appear in the antecedents and consequents of the rules are established by the data. To obtain Table 5.1, it is necessary to know the position of the sources, calculate the periodogram and measure its main lobe bandwidth and its maximum. Note that the bandwidth of the main lobe of the periodogram depends on its maximum. Thus, for the same displacement, different normalised bandwidths are obtained for different maximums. This is the motivating factor to use the maximum of the periodogram as an input to the fuzzy system.

The system surface is the relationship between the input variables and the output ones. Certain points on the control surface correspond to the exact rules, where the conjunction of the antecedents is unity for a specific rule and none of the other rules fire. Between these points, the fuzzy logic system of Eq. (5.11) approximates the output. If the data were noiseless or deterministic, the FLS would perform an interpolation.

At this point, the number of neurons in the input layer (inputs), the number of neurons in the inference layer (M rules), and the number of neurons in the fuzzification layer ($P_1 + P_2$), are known. Hence, the structure of the neural network is established, as are the initial values of the parameters that control the network.

The mapping in Table 5.1 is only exact after computing the mean of the different BW for each $d_i = i°$, and when the periodogram is estimated with a large number of snapshots. Nonetheless, as our work focuses on DOA estimation for tracking, less than 100 (e.g. 10) snapshots are considered. For this reason, to improve the function approximation, an LMS algorithm is used to train the rules by learning the estimator parameters (i.e. width of the input of the input membership functions and position of the output membership functions). The rules are learned from various realizations of the periodogram. For each realization, the rules and input/output fuzzy sets are adaptively designed in order to minimize the cost function (5.13).

$$\Psi = E\{|d_{act} - d(z)|^2\} \tag{5.13}$$

where d_{act} is the current distance between the sources and the measured maximum of the periodogram. By means of initializing with the proposed FAM in Table 5.1, the convergence of the system speeds up considerably. Next section is devoted to the application of the neural network training algorithm, which can be carried out off-line.

5.4.2 Training of the Neuro-Fuzzy System

Given a collection of N input-output numerical data training pairs $(z^{(1)}: y^{(1)}), \ldots, (z^{(N)}: y^{(N)})$, tuning is essentially equivalent to determine a system that provides an optimal fit to the input-output pairs, with respect to a cost function. In neural systems, the brain behaviour is translated into a mathematical model. This model is composed by the neurons and the interconnections between neurons. The neurons represent the interface between two consecutive layers whereas their interconnections are identified by a weight according to their importance. As with the biological neurons, the mathematical neurons can also learn. This learning can be differentiated in two methodologies: *Supervised* and *Non Supervised*. The Supervised learning is guided by the desired value. Therefore, it is possible to calculate the error measure accurately, to compensate and tune the neuron parameters appropriately. The error measure is defined as the difference between the current network output (y_k) and the reference signal ($d_{act\,k}$) as (5.14) shows:

$$e_k = y_k - d_{act\,k} \tag{5.14}$$

On the other hand, in the *Non Supervised* technique, a reliable reference signal is not used, except only an estimate of it. Therefore, this fact implies that *Non Supervised* techniques will always present a lower performance than *Supervised* systems. Figure 5.12 depicts a block scheme about *Supervised* and *Non Supervised* structures.

Finally, this work is focused on the *Supervised* learning method using the *back-propagation* algorithm as an adaptive technique.

5.4.3 Back-Propagation Algorithm

The parameters to optimise in the proposed fuzzy system are: the width and the media of the input and output membership functions. The parameters are

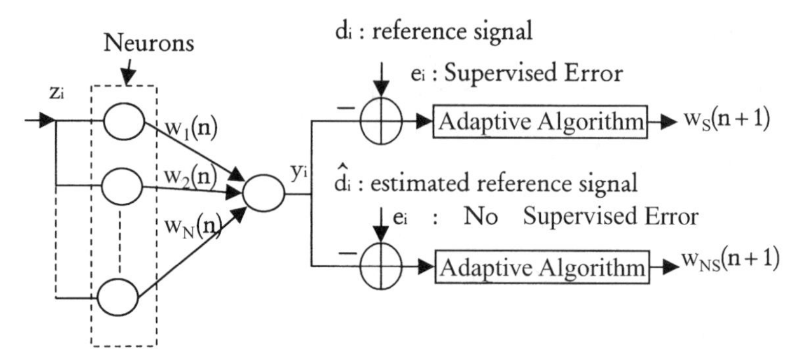

Fig. 5.12. Supervised and No Supervised structures

tuned in order to minimize a cost function at each iteration k

$$J_k = \frac{1}{2} | y_k - d_{act\,k} |^2 \tag{5.15}$$

where y follows (5.11). This expression can be grouped in order to make the notation easier in the following way:

$$y = \frac{a}{b} \tag{5.16}$$

where $a = \sum_{m=1}^{M} c_m \, \mu_m(z)$ and $b = \sum_{m=1}^{M} \mu_m(z)$. As it is used, the product inference $\mu_m(z) = \prod_{i=1}^{2} \mu_{im}(z_i)$, where $\mu_{im}(z_i)$ is the membership degree for the ith input and the mth IF-THEN rule. This work derives the backpropagation algorithm for the two most common cases: Gaussian and triangular membership functions.

The Gaussian membership function $\mu_{im}(z_i)$ belongs to the set of P_i membership functions that describes the ith input (Fig. 5.10) and has been defined as follows:

$$\mu_i^j(z) = \exp\left(-\frac{(z_j - \bar{z}_i^j)^2}{2(\sigma_i^j)^2} \right) \tag{5.17}$$

and the triangular membership function is enunciated as:

$$\mu_i^j(z) = \begin{cases} 1 - \dfrac{|z_j - \bar{z}_i^j|}{\sigma_i^j} & |z_j - \bar{z}_i^j| \le \sigma_i^j \\ 0 & |z_j - \bar{z}_i^j| \ge \sigma_i^j \end{cases} \tag{5.18}$$

where $\bar{z}_i^j$ and σ_i^j represent the media and width of the input membership function, respectively. Figure 5.13 depicts a graphical scheme about the adaptation of these parameters in a neuro-fuzzy system.

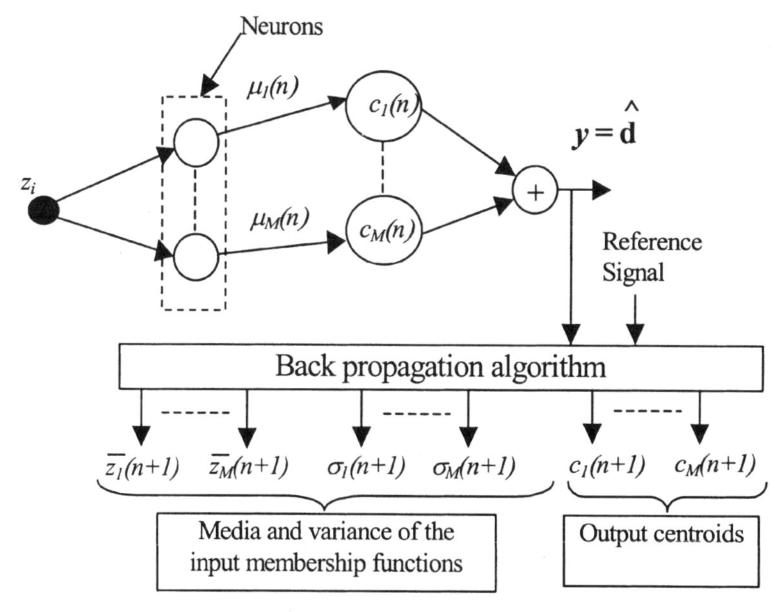

Fig. 5.13. Adaptive scheme of the neuro-fuzzy system parameters

In the backpropagation algorithm, the learning rule is formulated as:

$$\xi(k+1) = \xi(k) - \alpha \frac{\partial J_k}{\partial \xi_k} \tag{5.19}$$

where ξ is the parameter to adapt, α is the step size, $\partial J_k / \partial \xi_k$ is the gradient of the instantaneous error power with respects to the parameter that is being adapted.

5.4.3.1 Adaptation of the Output Centroids

The back-propagation law for the optimisation of output centroids c_m, $m = 1 \dots M$ is:

$$\boldsymbol{c}(k+1) = \boldsymbol{c}(k) - \alpha \frac{\partial J_k}{\partial \boldsymbol{c}} \tag{5.20}$$

where the M centroids have been collected in vector $\boldsymbol{c}$ and k stands for the iteration number. The gradient differentiation in (5.20) can be developed easily, if the chain rule is applied to the last equation. Then, the gradient of error power can be expressed as follows:

$$\frac{\partial J_k}{\partial \boldsymbol{c}} = \frac{\partial J_k}{\partial y_k} \cdot \frac{\partial y_k}{\partial a} \cdot \frac{\partial a}{\partial \boldsymbol{c}} = (y_k - d_k) \frac{1}{b} \boldsymbol{\mu}(z) \tag{5.21}$$

where $\boldsymbol{\mu}(\boldsymbol{z})$ is the vector of the antecedent memberships degrees for input $\boldsymbol{z}$ and for all rules: $\mu(z) = [\mu_1(z) \cdots \mu_m(z)]^T$. Then, substituting (5.21) in (5.20) results in:

$$c(k+1)=c(k)-\alpha(y_k-d_k)\frac{1}{b}\mu(z) \tag{5.22}$$

5.4.3.2 Adaptation of the Input Membership Function Mean

To train the input membership function, mean $\overline{z}_{im}$ is considered constant with respect to the rule index m and trained for each input j ($j = 1, 2$) $\overline{z}_i^j$ $i = 1 \dots P_i$. For each mean the training rule is:

$$\overline{z}_i^j(k+1)=\overline{z}_i^j(k)-\alpha\frac{\partial J_k}{\partial \overline{z}_i^j} \tag{5.23}$$

The gradient differentiation in (5.23) can be easily developed by applying the chain rule to it, as follows:

$$\frac{\partial J_k}{\partial \overline{z}_i^j}=\frac{\partial J_k}{\partial y_k}\cdot\frac{\partial y_k}{\partial \mu_i^j}\cdot\frac{\partial \mu_i^j}{\partial \overline{z}_i^j}=(y_k-d_k)\frac{1}{b}\left(\frac{\partial a}{\partial \mu_i^j}-y_k\frac{\partial b}{\partial \mu_i^j}\right)\frac{\partial \mu_i^j}{\partial \overline{z}_i^j} \tag{5.24}$$

The last term is the gradient of the input membership degree, with respects to their means. Therefore, this gradient will be different according to the type of membership function used. Next, the expression of the optimum input membership media for the Gaussian and triangular ones, is developed.

For instance, as a result of the definition of the Gaussian membership function given in (5.17), the training Eq. (5.23) results in (5.25):

$$\overline{z}_i^j(k+1)=\overline{z}_i^j(k)-\alpha(y_k-d_k)\frac{1}{b}\left(\frac{\partial a}{\partial \mu_i^j}-y_k\frac{\partial b}{\partial \mu_i^j}\right)\mu_i^j\frac{(z_j-\overline{z}_i^j)}{(\sigma_i^j)^2} \tag{5.25}$$

5.4.3.3 Adaptation of the Input Membership Function Width

The back-propagation law for the optimisation of the width of each of the input membership function σ_i^j, is:

$$\sigma_i^j(k+1)=\sigma_i^j(k)-\alpha\frac{\partial J_k}{\partial \sigma_i^j} \tag{5.26}$$

The gradient differentiation in (5.26) can be developed easily by applying the chain rule as follows:

$$\frac{\partial J_k}{\partial \sigma_i^j}=\frac{\partial J_k}{\partial y_k}\cdot\frac{\partial y_k}{\partial \mu_i^j}\cdot\frac{\partial \mu_i^j}{\partial \sigma_i^j}=(y_k-d_k)\frac{1}{b}\left(\frac{\partial a}{\partial \mu_i^j}-y_k\frac{\partial b}{\partial \mu_i^j}\right)\frac{\partial \mu_i^j}{\partial \sigma_i^j} \tag{5.27}$$

Note that the last expression depends on the gradient of the input membership degree with respects to its width. Therefore, this gradient will be different for the membership function that is used. Next, the expression of the optimum input membership width for the Gaussian and triangular membership functions, is developed.

For instance, according to the definition of the Gaussian membership function formulated in (5.17), the training Eq. (5.26) results in:

$$\sigma_i^j(k+1) = \sigma_i^j(k) - \alpha(y_k - d_k)\frac{1}{b}\left(\frac{\partial a}{\partial \mu_i^j} - y_k \frac{\partial b}{\partial \mu_i^j}\right)\mu_i^j \frac{2(z_j - \bar{z}_i^j)^2}{(\sigma_i^j)^3} \tag{5.28}$$

As an example, Fig. 5.14 shows the learning curve for one of the means in the case of Gaussian fuzzy sets. Figure 5.15 depicts the learning curve for the case of triangular fuzzy sets.

5.4.4 Steps to Follow in the Design of the Rule Base

In the supervised training that has been developed, five steps have to be distinguished (Fig. 5.16):

Step 1: Initialise the backpropagation algorithm according to the FAM rule table, shown in Table 1.

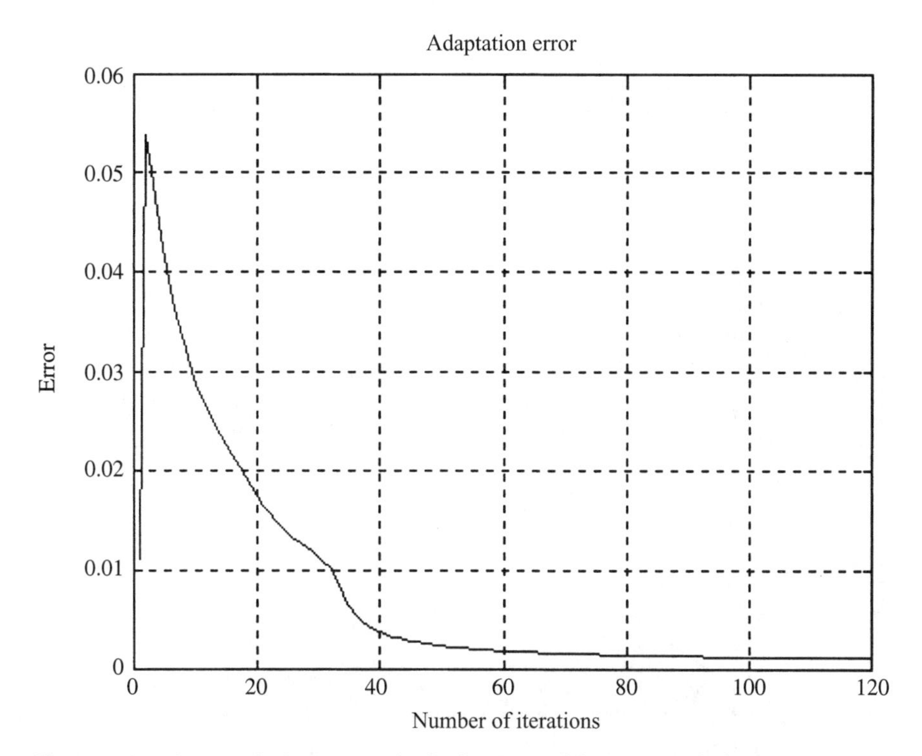

Fig. 5.14. Learning curve for Gaussian membership functions and displacement of 5 degrees

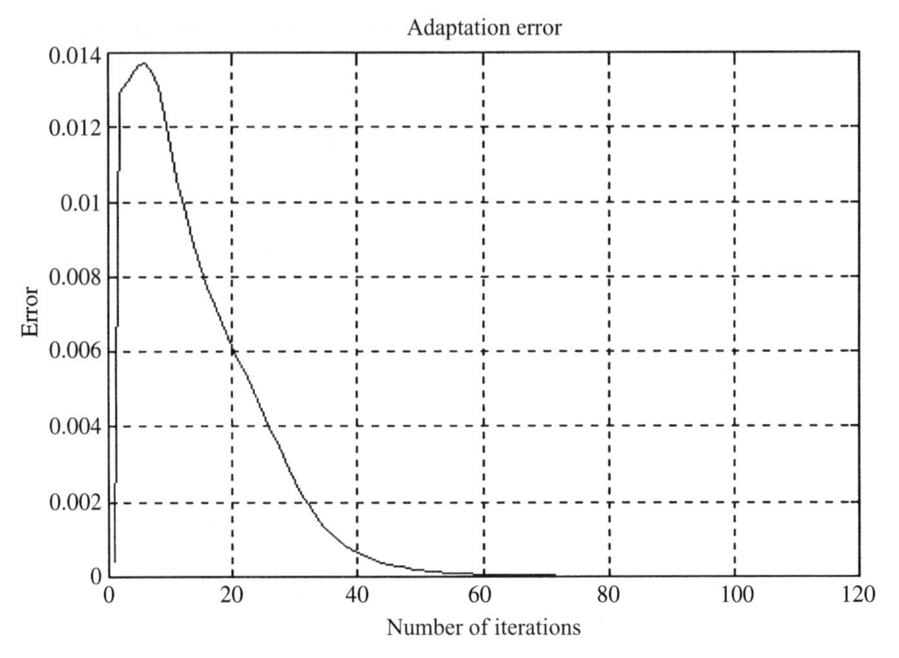

Fig. 5.15. Learning curve for triangular membership functions and displacement 5 degrees

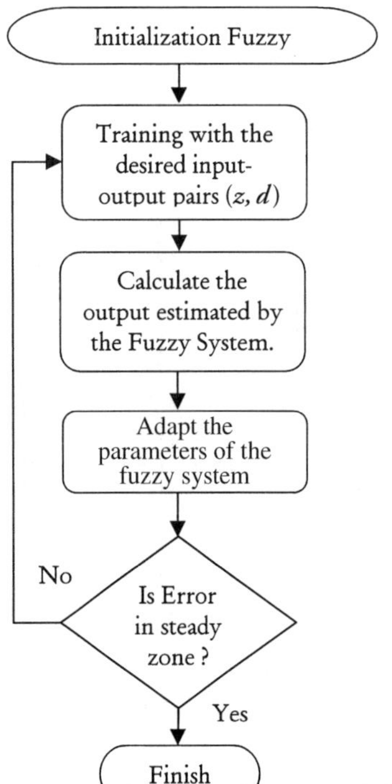

Fig. 5.16. The five steps in the design of a Fuzzy System

Step 2: Present the desired input-output pairs (z, d) to the first layer of the fuzzy system. The input data $\boldsymbol{z}$ is a vector formed by the maximum of the periodogram $P(\theta)$ and the bandwidth of its main lobe. The periodogram is computed by taking 10 snapshots in order to make the algorithm suitable for mobile sceneries. Although this number of snapshots is low, this fact introduces variance in the estimation. Note that the periodogram is unbiased but not consistent. Consequence, its variance does not tend to zero as the number of snapshots increases.

Step 3: Compute the output of the fuzzy system using (5.11).

Step 4: Adapt the parameters using the obtained equations. The backpropagation algorithm starts from the output layer and works backward to the first layer.

Step5: Take another measurement $(z^{(i)}: y^{(i)})$ and iterate steps 2 to 5, until the error curve has reached its steady zone.

5.5 Simulations

The simulations presented in this section have been carried out for two coherent sources that impinge on a 9 sensor array. Coherent sources have been considered. The neuro-fuzzy system consists of 15 rules for each θ_{max} (i.e. the initial output centroids are set at $d_i = [0, 1.5, 2, 2.5, 3, 3.5, 4, 4.5, 5, 5.5, 6, 6.5, 7, 7.5, 8]$). The results measure the performance of the neuro-fuzzy system after the supervised training with 120 input-output samples or iterations. As a benchmark, the normalized Maximum Likelihood or Capon estimator has been considered, which stands as the best nonparametric DOA estimator. Two sets of simulations have been carried out: i) with Gaussian fuzzy sets and ii) with triangular fuzzy sets.

5.5.1 Gaussian Fuzzy Sets

Figure 5.17 compares the NML/Capon method (see 5.8) with the fuzzy estimate when the two sources present a SNR, equal to 5 dB each. To compute the mean square error, 30 Monte Carlo simulations have been run. Figure 5.18 depicts the comparison when one source presents 8 dB and the other one 5 dB of SNR.

This plot reveals the better performance of the neuro-fuzzy estimator. In this simulation, the system has been trained with SNR between 5 and 9 dB.

Figure 5.19 plots the membership functions for the input $z_1 = BW$. Training the fuzzy system in lower SNR, the width of the membership functions are increased, as Fig. 5.20 depicts. Figure 5.21 plots the performance of the neuro-fuzzy system after training. Two cases have been considered, one when the two sources present 0 dB of SNR and the other when their SNR is equal to two dB. Next, is a presentation of the results when triangular fuzzy sets are used.

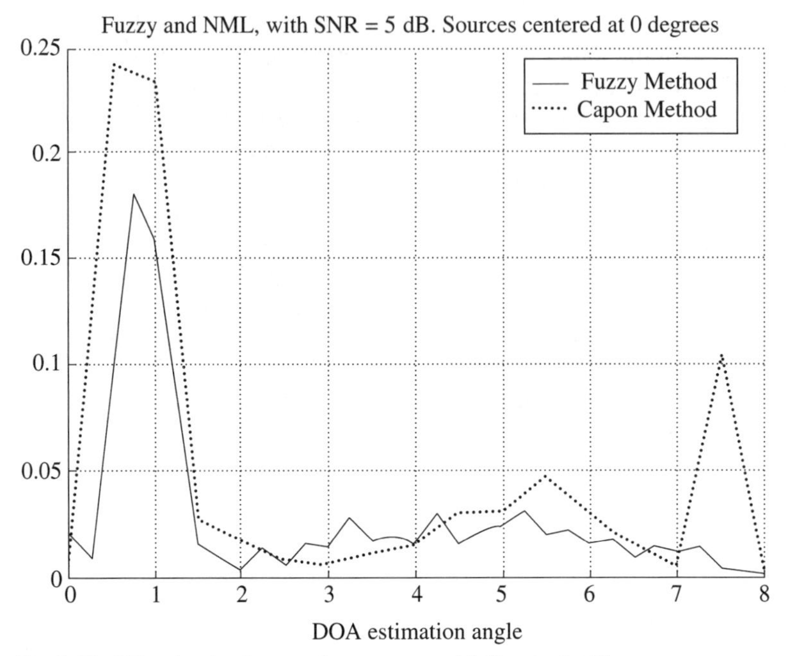

Fig. 5.17. DOA estimation for two coherent sources of 5 dB centred at 0°

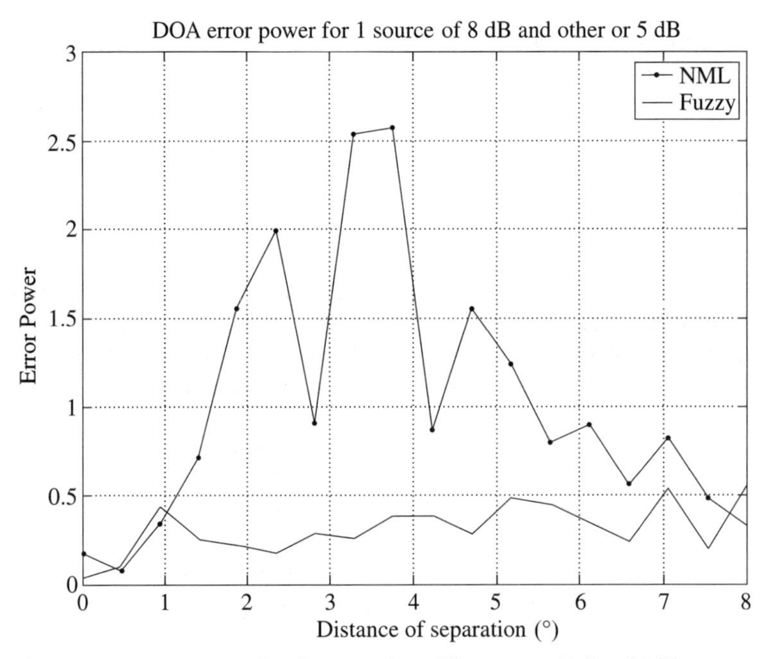

Fig. 5.18. DOA error power when the sources have different power (8 dB and 5 dB)

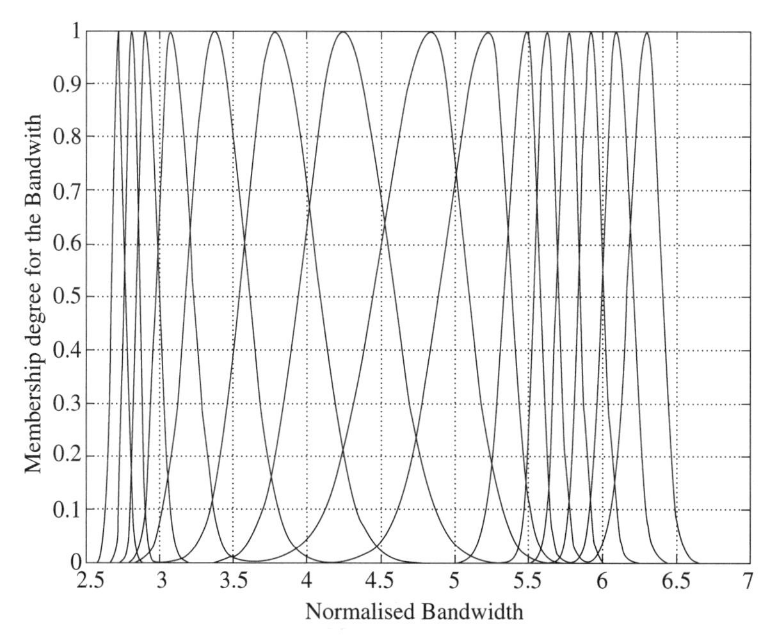

Fig. 5.19. Gaussian membership functions trained with data that belong to the range of SNR 5–9 dB

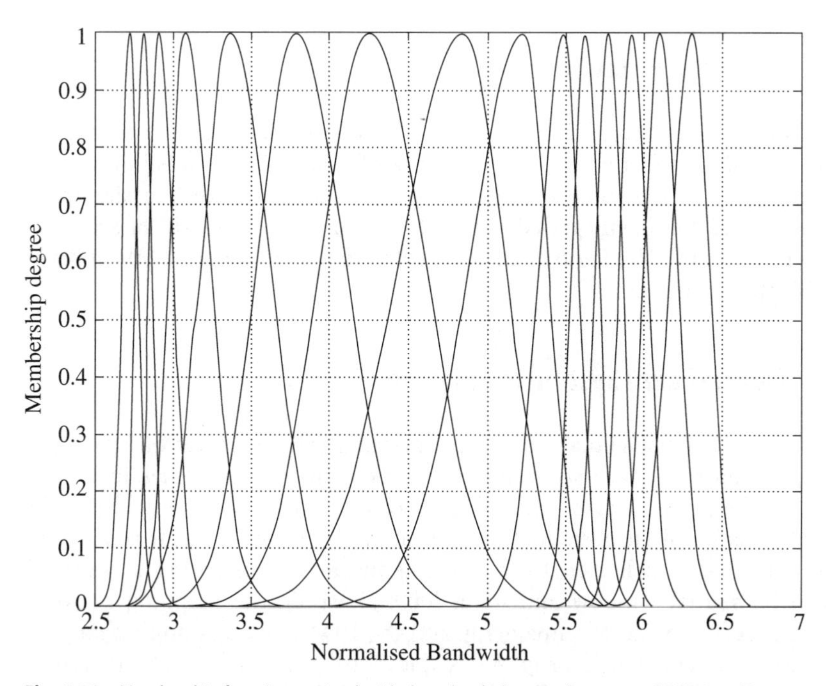

Fig. 5.20. Membership functions trained with data that belong to the range of SNR 0–4 dB

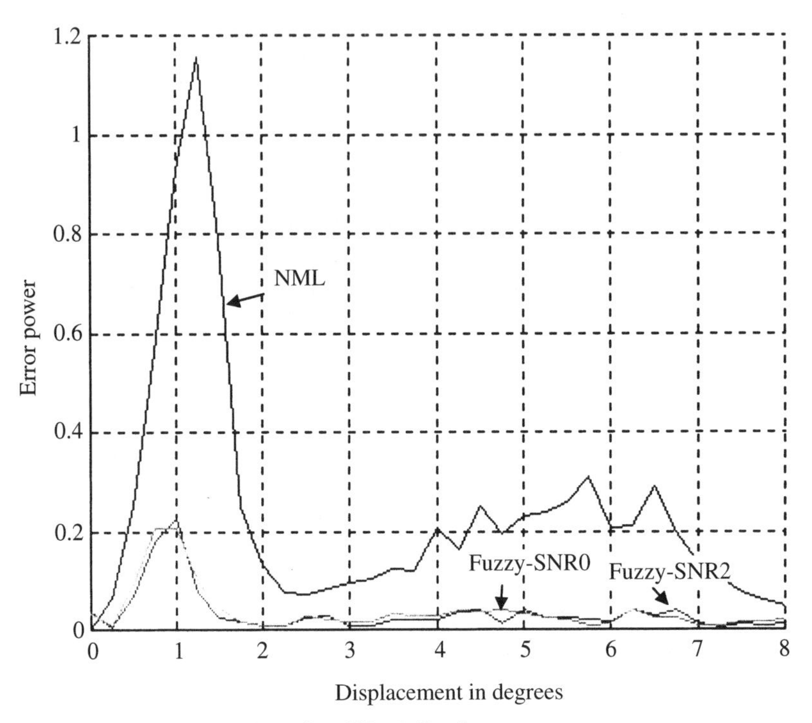

Fig. 5.21. Error power to SNR = 0 dB vs SNR = 2 dB vs Capon

5.5.2 Triangular Fuzzy Sets

If low computational complexity is desired, triangular fuzzy sets can be used. Figure 5.22 illustrates that the performance of triangular membership functions does not degrade much, with respect to that obtained with Gaussian fuzzy sets. Figure 5.22 has been carried out for two sources, one of 5 dB of SNR and the other of 7 dB of SNR. Figure 5.23 depicts the membership functions of input $z_2 = BW$ after training.

5.6 Neuro-Fuzzy System Evaluation

This chapter has presented a neuro-fuzzy system for high resolution DOA estimation, in non-stationary sceneries. The system further processes in a non-linear way the periodogram, extending its resolution beyond the width of its main lobe. Classical DOA estimation techniques carry out a linear processing of the data that is collected from the antenna array presenting a trade-off between its spatial window and main lobe. However, the mapping between the data collected by an antenna array and the DOA system is non-linear. The neuro-fuzzy system that is proposed, exploits this fact and breaks the commented trade-off.

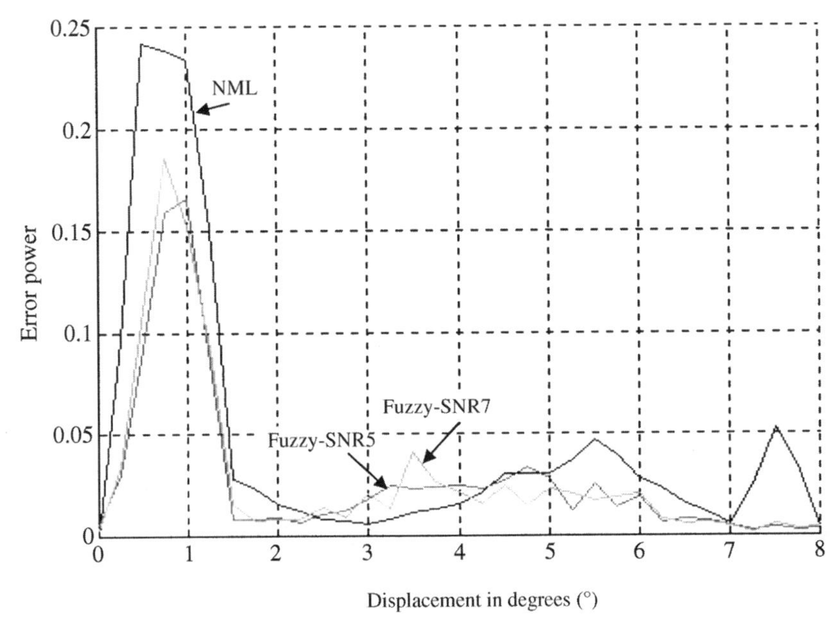

Fig. 5.22. Error power to SNR = 5 dB vs SNR = 7 dB vs Capon

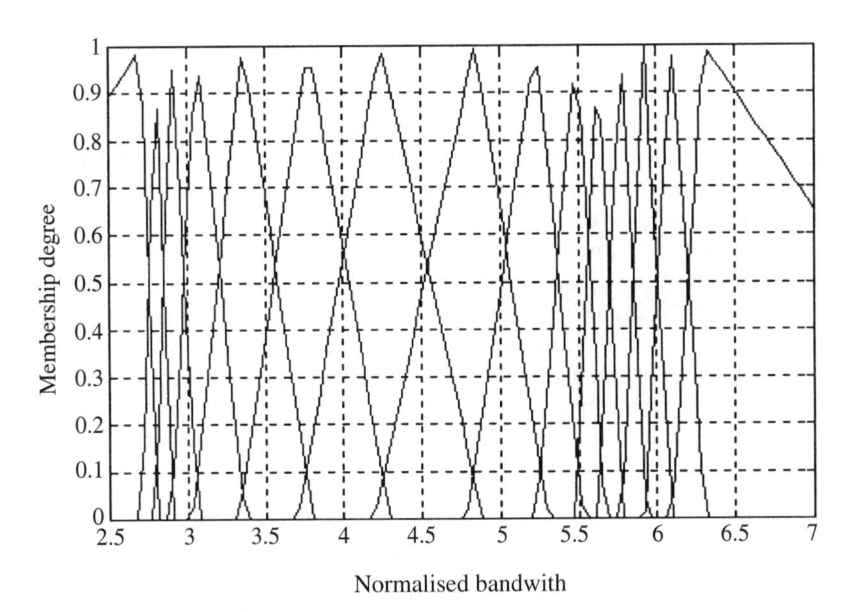

Fig. 5.23. Triangular membership functions trained with data that belong to the range of SNR 5–9 dB

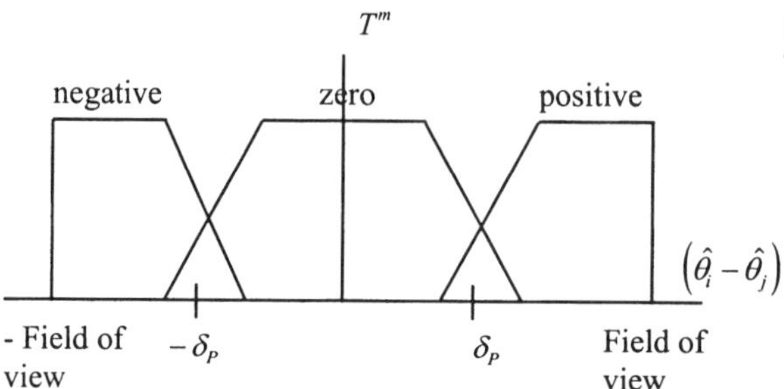

Fig. 5.24. Membership functions for the fuzzy switch

According to the simulations, it has been demonstrated that the proposed system outperforms the high-resolution Normalized Capon method (NML). Nevertheless, the computational burden is lower than that of the NML. After an off-line training stage, the resulting rules or FAM can be stored in a look-up table; thus, in contrast to the NML, no matrix inversion is needed. In case that on-line training is desired, the system could also be implemented in a general purpose DSP. For that application, triangular fuzzy sets can then be used without degrading the performance of the system. In [11] there is a detailed description on how an additive fuzzy logic system with triangular fuzzy sets can be implemented on a DSP.

It is remarked that the proposed DOA estimator acts as a sub-system that helps the periodogram. Traditionally, high resolution Direction of Arrival (DOA) estimation has been associated with algorithms, rather than with a processing scheme or architecture. The main aim of this chapter has been to demonstrate that processing architectures of low complexity can help to attain the performance of computational complex algorithm. That is, offering systems suitable for a communication front-end where the impact of using antenna arrays is desired to be as low as possible. The proposed architecture was introduced in Fig. 5.1 and consists of: i) notch periodogram for scanning the DOA of each user, ii) neuro-fuzzy system that enhances the resolution of the periodogram for close or colliding users, iii) switch that commutes between the periodogram and the neuro-fuzzy system and, iv) Kalman tracker.

The switch in Fig. 5.1 commutes between the proposed fuzzy DOA estimator and the constrained scanning can be simply designed as a hard-decision threshold detector: "if the DOA distance between close sources is less than the array resolution, the DOA estimates are then produced by the fuzzy system." However, fuzzy systems are also highly suitable to fuse data from different systems, producing smooth transitions even in noisy sceneries. For this reason, a fuzzy system to build a fuzzy switch, has been proposed. The rule for commuting between systems P (Periodogram) and F (Fuzzy) of Fig. 5.1, is:

$$R^m\text{: If } (\hat{\theta}_i - \hat{\theta}_j) \text{ is } T^m \text{ THEN } \theta_i \text{ is } \theta^m_{ik} \qquad j, i = 1 \ldots NS, m = 1 \ldots M \text{ rules}; k = 1, 2$$

where $(\hat{\theta}_i - \hat{\theta}_j)$ measures the distance between the angle of user i and that of user j (produced at branch i and at each of the other (NS-1) branches, respectively), T^m are the membership functions defined in Fig. 5.24. Finally θ^m_{ik} is the DOA es-

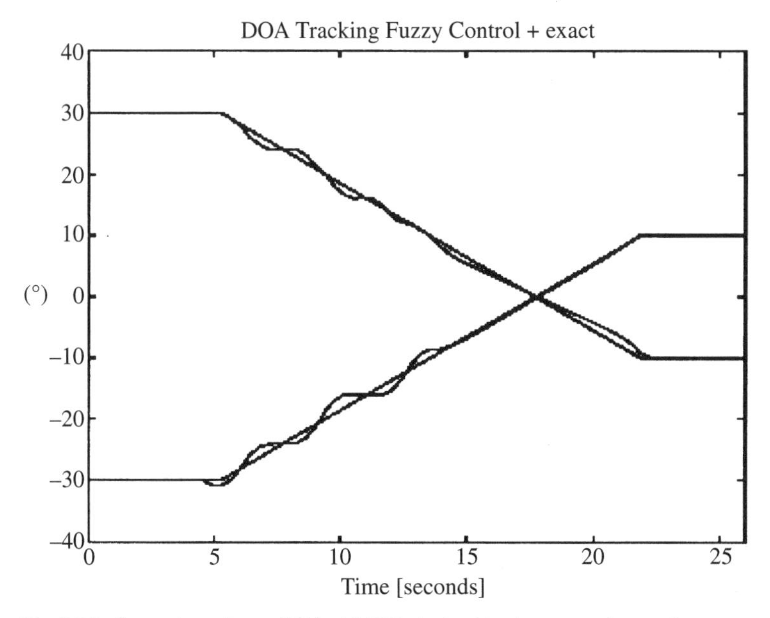

Fig. 5.25. Comparison of actual DOA with DOA obtained by the proposed neuro-fuzzy system

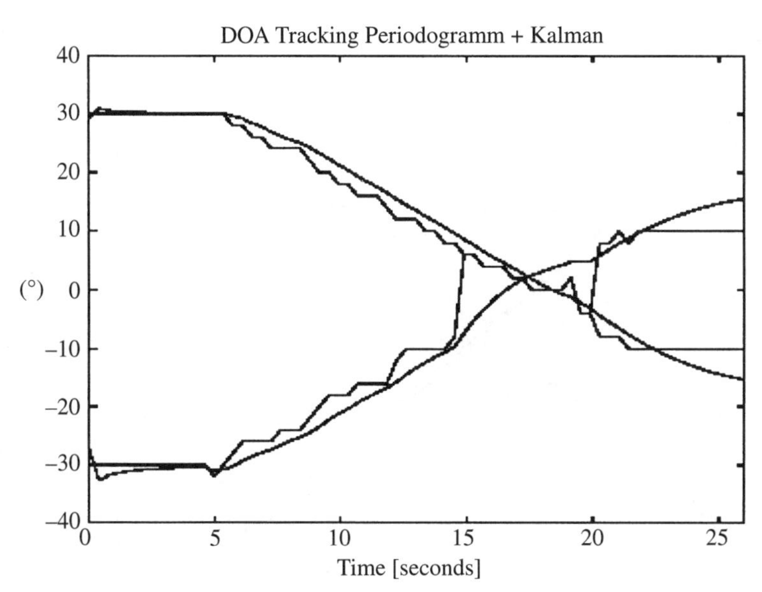

Fig. 5.26. Comparison of DOA estimation from periodogram and from periodogram enhanced with a Kalman tracker

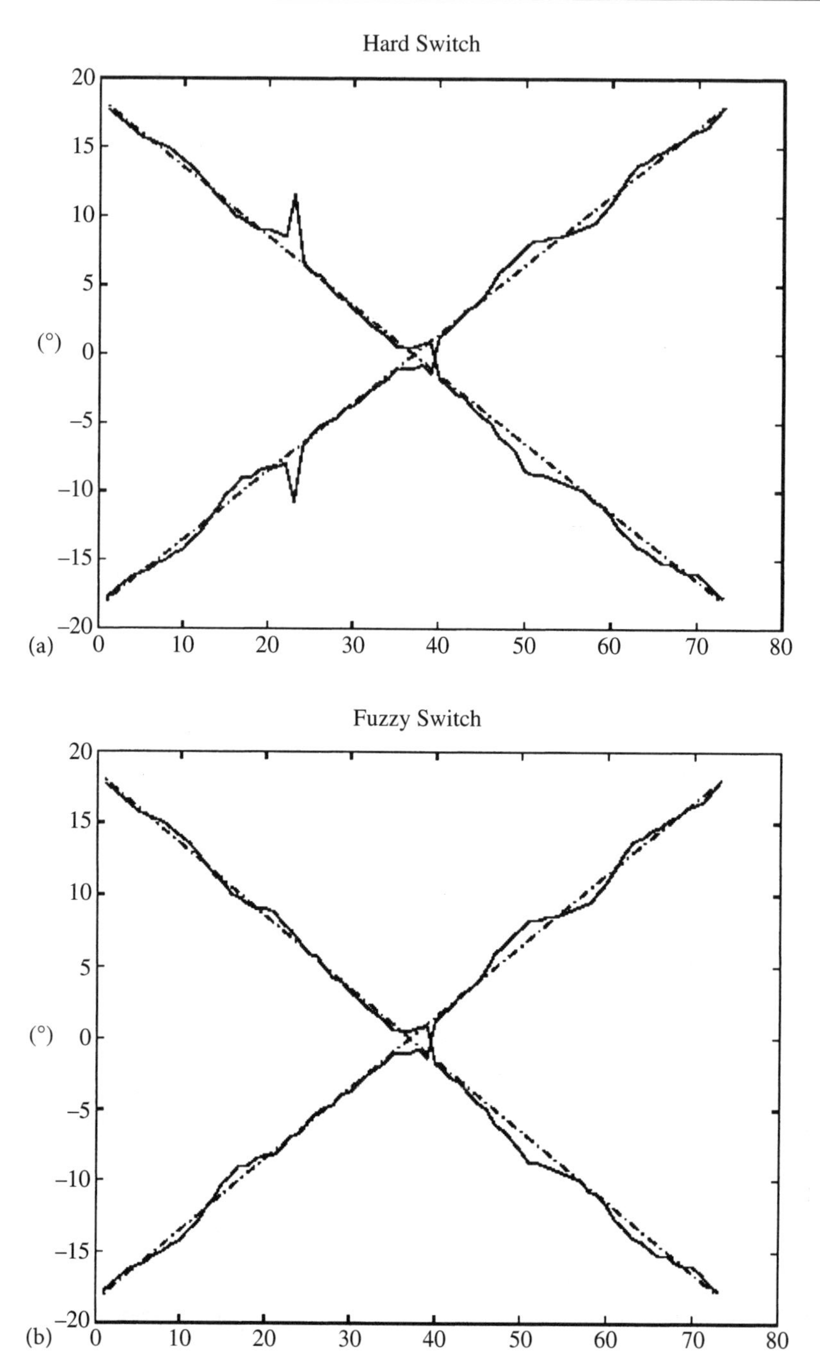

Fig. 5.27. Comparison of the performance of a) hard switch and b) fuzzy switch. The hard switch results in spikes in the DOA estimation

timate produced by either the scanning beam ($k=1$ and then $\theta_{i1}^{m} = \theta^{P}$) or by the fuzzy estimator ($k = 2$ and then $\theta_{i2}^{m} = \theta^{F}$). The output of the fuzzy switch is then computed as:

$$\theta_i = f(\hat{\theta}_i - \hat{\theta}_j) = \frac{\sum_{m=1}^{M} T^m (\hat{\theta}_i - \hat{\theta}_j)\, \theta_{ik}^{m}}{\sum_{m=1}^{M} T^m (\hat{\theta}_i - \hat{\theta}_j)} \tag{5.29}$$

Figure 5.25 shows the results of the tracker system proposed in Fig. 5.1. (i.e. parallel Kalman trackers with the fuzzy logic detection and estimation sub-systems) after using trial data [12]. The results are compared against the actual sources' position. The scenery consists of 2 users of 30 dB each , moving towards each other with constant velocity.

Figure 5.26 compares the performance of the periodogram estimates with the performance of the system proposed in [1] that uses periodogram and Kalman trackers, but does not introduce the fuzzy system. Finally, Fig. 5.27 shows the behaviour of the DOA detector when a hard switch is used instead of the fuzzy one.

Glossary

DOA Direction of Arrival
FAM Fuzzy Associative Memory
FBF Fuzzy Basis Function
FLS Fuzzy Logic System
NML Normalized Maximum Likelihood or Capon estimator
PSD Power Spectral Density
SNR Signal to Noise Ratio

References

[1] Pérez-Neira A, Lagunas M.A. (1996) High performance DOA trackers derived from parallel low resolution detectors, 8th IEEE Signal processing workshop on Statistical Signal Array Processing, pp. 558–561, Corfu (Greece).
[2] Pérez-Neira A, Lagunas M.A., Kirlin R.L. (1997) Cross-Coupled DOA trackers, IEEE Trans. on Signal Processing, vol. 45, no. 10.
[3] Johnson D.H. (1993) Array Signal Processing, Prentice Hall PTR.
[4] Kreinovich, Mouzouris G.C, Nguyen H.T (1998) Fuzzy rule based modeling as a universal approximation tool, Fuzzy Systems, Modeling and Control, pp. 135–195, Kluwer Ac. Publ., Boston.
[5] Kim H.M, Mendel J.M., (1995) Fuzzy Basis Functions: Comparisons with other Basis Functions, IEEE Transactions on Fuzzy systems, vol. 3, nº 2.
[6] Kosko B. (1997) Fuzzy Engineering.
[7] Shigeo A. (1997) Neural Networks and Fuzzy Systems, Kluwer Academic Publishers.
[8] Mendel J (2001) Rule-based fuzzy logic systems: introduction and new directions, Prentice-Hall PTR.

[9] Klir G.J., Yuan B. (1995) Fuzzy sets and Fuzzy Logic Theory and Applications, Prentice Hall.
[10] Wang L. (1994) Fuzzy Systems and Control. Design and Stability Analysis, Prentice Hall PTR.
[11] Bas J., Pérez-Neira A., (2001) A DSP-based Fuzzy System for Real-Time Interference Cancellation in DS-SS Systems, EUSFLAT, Leicester (United Kingdom).
[12] European Project TSUNAMI (1995) Report on the analysis of the field trial results, Doc. Nº: R2108/UPC/WP3.6/DR/P/035/b1.
[13] Wang L.X. (1994) Adaptive Fuzzy Systems and Control, Prentice Hall, Engleewood Cliffs, New Jersey.
[14] Nelles O. (2001) Nonlinear System Identification, Springer-Verlag.

6 Fuzzy-Neural Applications in Handoff

6.1 Introduction

Handoff is a process of transferring a mobile station from one base station or channel, to another. The channel change due to handoff occurs through a change in a time slot, frequency band, codeword, or a combination of these, for time division multiple access (TDMA), frequency division multiple access (FDMA), code division multiple access (CDMA), or a hybrid scheme, respectively. The handoff process determines the spectral efficiency (i.e., the maximum number of calls that can be served in a given area) and the quality perceived by users. Efficient handoff algorithms cost-effectively preserve and enhance the capacity and Quality of Service (QoS) of communication systems.

Many of the existing handoff algorithms do not exploit the advantages of multi-criteria handoff, which can give better performance than single-criterion algorithms, due to the flexible and complementary nature of handoff criteria. The existing algorithms do not exploit knowledge on the sensitivities of handoff parameters to different characteristics of a cellular environment. The adaptation and learning capabilities of artificial intelligence (AI) tools have not been fully utilized. The existing algorithms fail to consider the behavior of other handoff algorithms in a given cellular environment and to provide a systematic procedure for the adaptation of handoff parameters to the dynamic cellular environment. Recent novel approaches for the design of high performance handoff algorithms that exploit attractive features of several existing algorithms, provide adaptation to the dynamic cellular environment and allow systematic tradeoff among different system characteristics. Handoff can also be used as a vehicle for controlling Quality of Service for Wireless Systems.

In subchapter 6.2, it is shown how a high performance handoff algorithm can be designed using a neurofuzzy system. Two basic approaches, the neural encoding based approach and the pattern recognition based approach, are discussed here. Various cellular systems such as a microcellular system, an overlay system, and a system with soft handoff support, are also considered. This subchapter also shows that high performance handoff algorithms with low complexity can be designed using neurofuzzy systems.

Subchapter 6.3 aims at controlling QoS of code division multiple access mobile communication systems, by proposing two new practical methods applied on the conventional Soft Handoff (SHO) based on simple step control (SSC) method and neurofuzzy inference system with gradient descent method. The output parameters selected to be controlled by adapting SHO thresholds, are

Trunk-Resource Efficiency (TRE), the forward-link bit energy to noise power spectral density ratio (E_b/N_o) and outage probability (P_{out}). TRE is concerned with call blocking, while E_b/N_o are selected to be one, two or three of the control plant inputs, according to the method applied. QoS-controlling SHO (TRE-controlling SHO and E_b/N_o-controlling SHO) can improve the system performance. However, the method to be selected for use, depends on the environment and the requirement of the system operator and/or users.

6.2 Application of a Neuro-Fuzzy System to Handoffs in Cellular Communications

Nishith D. Tripathi, Jeffrey H. Reed and Hugh F. VanLandingham

6.2.1 Introduction

Handoff is an important component of a cellular communication system. Handoff is a process in which the communication link between the mobile station and the cellular infrastructure is changed through a change in the base station or a radio channel. Fundamentals of the process of handoff are described here. Efficient and high-performance handoff algorithms can cost-effectively enhance the capacity and the quality of service of a cellular system.

6.2.1.1 Overview of a Cellular System

In a cellular communication system, a geographical area is divided into small regions called cells (Fig. 6.2.1). The cell size is determined by the capacity and coverage considerations. A group of cells compose a cluster, and the frequency spectrum is reused by each cluster.

The number of cells per cluster depends on the multiple access technology. For example, there is one cell per cluster for a system based on the CDMA technology, while there are typically four or seven cells per cluster for a cellular system based on the TDMA technology. Figure 6.2.1 depicts one cell per cluster, and each cell consists of three sectors. The process of dividing a cell into sectors, is called sectorization. In sectorization, instead of omni-directional antennas that cover the entire cell, directional antennas, that illuminate specific portions of a cell are utilized. For example, Fig. 6.2.1 illustrates 120° sectorization in which three directional antennas together serve the whole cell. Each cell is served by a base station (BS), and the user (or mobile) terminal is referred to as the mobile station (MS). The radio link from the BS to the MS is called the forward link or downlink, while the link from the MS to the BS is called the reverse link or the uplink. To reduce interference on both forward and reverse links, sectorization is frequently implemented in a cell system.

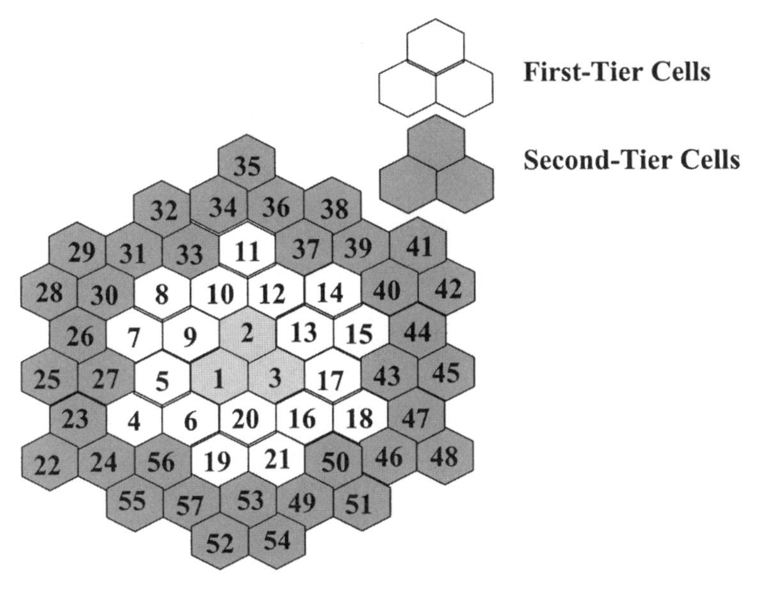

Fig. 6.2.1. A Tri-sectored Cellular System

6.2.1.2 Handoff Fundamentals

Handoff is a process that transfers the service to the MS from one BS or channel to another. The channel change in handoff may be through a time slot, frequency band, code, or combination of these for TDMA, FDMA, CDMA or a hybrid scheme, respectively [1]. Figure 6.2.2 shows a simple handoff scenario. Consider an MS moving from Sector C of Cell 1 (served by BS 1) to Sector A of Cell 2 (served by BS 2). When the MS is close to BS 1, it is connected to BS 1. As the MS moves away from BS 1 and toward BS 2, the strength of the signal emitted from BS 1 and received at the MS decreases, leading to potentially poor signal quality. To maintain the quality of the communication link and to prevent a call drop, a handoff is made at an appropriate time instant in the handoff region. The MS now communicates with BS2 in Cell 2. The radio propagation environment may cause signal finding, making multiple handoffs necessary during the travel shown in Fig. 6.2.2.

There are several types of handoff such as intracell, intercell, hard, and soft. When a handoff is made within the currently serving cell (e.g., through a change in frequency), it is called an intracell handoff. The intracell handoff can reduce interference, because interference on different frequency channels could be different. A handoff between the cells is referred to as an intercell handoff. Handoff may be hard or soft. During hard handoff (HHO), the connection to the old BS is broken before a connection to the candidate BS is made. Hence, the HHO is "break before make" type handoff. The HHO occurs when handoff is made between separate radio systems, different frequency assignments or different air interface characteristics or technologies. The SHO is a "make before break," in which the connection to the old BS is not broken until a connection to

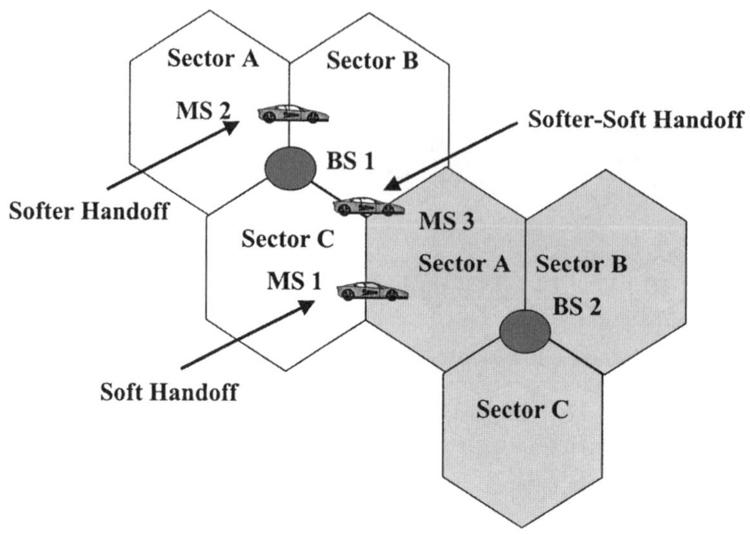

Fig. 6.2.2. Handoff Scenarios in a Cellular System

the new BS is made. Usually, more than one sectors communicate simultaneously with the MS. There are different types of SHO. When sectors of the same BS are involved in communication with the MS, the handoff is called softer handoff. When one sector from each BS is involved, the handoff is called soft handoff. In Fig. 6.2.2, the MS 1 is in soft handoff with BS 1 and BS 2, and MS 2 is in softer handoff. When multiple sectors of one BS and one or more sectors of another BS communicate with the MS, the resulting SHO is called softer-soft handoff. For example, MS 3 is in softer-soft handoff. Soft handoff is more frequently encountered in a CDMA system, than hard handoff.

Handoff is necessary because of the factors related to radio link, network management or service options. Examples of radio link related handoff factors include RSS, SIR and system related constraints (e.g., signaling delay). Low RSS, low SIR, and long propagation delay, are indicators of potentially poor signal quality, and handoff is made to maintain sufficiently large RSS and SIR and short propagation delay. Examples of the factors related to network management are traffic congestion in a cell and existence of a shorter communication path. Unavailability of a service in a cell, causes a handoff that is related to service options.

6.2.1.3 Desirable Features of Handoff

An algorithm that makes a handoff decision should consider several desirable features of handoff [2–9]. A brief summary of such features is provided below:

Fast Handoff. A call drop can be prevented by making the handoff fast enough, so that the MS does not experience significant signal degradation.

Reliable Handoff. Handoff should be reliable, and the quality of service after handoff should be good.

Successful Handoff. Handoff should be successful, meaning that a free channel should be available after handoff.

Preservation of Cell Borders. Handoff should preserve planned cell borders, to avoid congestion and high interference levels.

Minimum Handoff Frequency. The number of handoffs should be minimized to reduce processing load and reduce the likelihood of call drop, due to the unavailability of a free resources in the new cell.

Traffic Balancing. Balancing of traffic among adjacent cells can utilize resources efficiently. For example, overall new call breaking probability will be low, when traffic balancing is employed.

6.2.1.4 Handoff Complexities

Several factors complicate the process of handoff. These handoff complexities are briefly touched upon below:

Cellular Structures and Topographical Features. There are different types of cellular system deployment scenarios. There could be big cells in rural or suburban areas (called macrocells), small cells in urban areas (called microcells) or an overlay system with a given area, both containing macrocells and microcells. The propagation environment is quite different in macrocells or microcells. For example, large-scale fading (referred to as shadow fading), has larger variations in urban microcells than in rural macrocells. Furthermore, urban microcells special cases such as street corner effect exits, are characterized by a sudden drop (such as 10 dB) in signal strength over a short distance (such as 10–15 meters).

Traffic. Traffic distribution is a function of time and space [10]. The handoff process should work uniformly well in different traffic scenarios. Examples of approaches to deal with traffic nonuniformities include traffic balancing in adjacent cells, use of different cell sizes, nonuniform channel allocation and dynamic channel allocation.

System Constraints. Several systems have constraints over common characteristics, such as transmit power and propagation delay. The handoff process should consider such system constraints.

Mobility. The quality of a communication link is influenced by the degree of mobility. A high speed MS moving away from a serving BS, experiences signal degradation faster than a low speed MS. Hence, mobility plays an important role in the handoff process.

6.2.2 Handoff Algorithms

The process of handoff is one of the critical design components of a cell system responsible for the capacity. The overall handoff procedure consists of two distinct phases [11], *the initiation phase* (in which the decision about handoff is made) and *the execution phase* (in which a new resource is assigned to the MS or the handoff request fails). Handoff algorithms usually carry out the first phase. An efficient handoff algorithm cost-effectively enhances the overall performance of a cellular system.

6.2.2.1 Handoff Criteria

The variables used as inputs or triggers to the handoff algorithms are known as handoff criteria. The handoff criteria discussed below, include RSS, SIR, distance, transmit power, traffic, cell and handoff statistics, and velocity.

RSS. This orientation is simple, direct, and widely used. The RSS is an indication of the distance between the MS and the BS and could also indicate the quality perceived by users if the interference does not vary significantly over time. However, if the interference changes, RSS does not necessarily indicate good quality, as interference could also be considerable for such communication link. Fluctuations in RSS are due to the dynamics of propagation environment. They can cause frequent handoffs.

SIR. This parameter is common to system capacity and dropped call rate. A high SIR implies good voice quality, shorter frequency reuse distance, and lower dropped call rate. The SIR may fluctuate due to the prop environment, leading to frequent handoffs. Furthermore, a poor SIR may be temporarily experienced close to the serving BS and inter cell handoff is undesirable in such cases. An intra cell handoff may suffice.

Distance. This criterion is useful in preserving planned cell boundaries. The distance between the BS and the MS can be estimated from measurements such as the signal strength [12] and signal propagation delays [13]. When the system employs a mechanical for the determination of the MS location, distance can become readily available for use as a handoff criterion. In a micro cellular system, the precision of the distance measurement decreases, due to shorter distances.

Transmit Power. The use of this criterion can reduce the transmit power requirements for both the fluid and reverse links, reducing the interference levels and increasing the battery life of the MS.

Traffic. The use of traffic intensities in adjacent cells, can help balance traffic.

Call and Handoff Statistics. Examples of useful statistics are total time spent by a call in a given cell, arrival time of a call, and elapsed time since last handoff [4, 14].

Velocity. Velocity can be used to provide a handoff algorithm that adapts to the mobility. Velocity-adaptive algorithms can provide similar performance at different speeds. One way to estimate mobile velocity, is through the estimate of the Doppler frequency. (predicted using signal envelope measurements) [15].

6.2.2.2 Conventional Handoff Algorithms

Handoff algorithms are distinguished from one another in two ways, handoff criteria *(conventional handoff algorithms)* and processing of handoff criteria

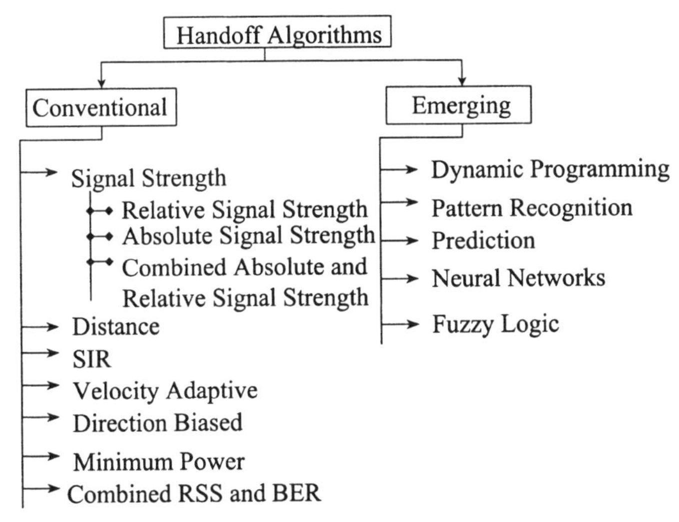

Fig. 6.2.3. An Overview of Conventional and Emerging Handoff Algorithms

(emerging handoff algorithms). Conventional handoff algorithms compare measured criteria with thresholds (i.e., parameters of the algorithms) to make a handoff decision. Emerging handoff algorithms employ new techniques or tools, such as neural networks and fuzzy logic to make a handoff decision and typically involve processing of multiple criteria, simultaneously. Figure 6.2.3 outlines the conventional and emerging handoff algorithms.

The conventional handoff algorithms discussed below, include: *signal strength algorithms, distance based algorithms, SIR based algorithm, velocity adaptive algorithm, direction-biased algorithm, minimum power algorithm, and combined RSS and BER algorithm.*

Signal Strength Based Algorithms. There are three basic types of signal strength algorithms, *relative signal strength algorithms, absolute signal strength algorithms and combined absolute and relative signal strength algorithms.* In the relative signal strength algorithm [9], the BS that receives the strongest signal from the MS, is connected to the MS. This algorithm is advantageous in that the MS communicates with the BS, corresponding to least path loss. Such algorithm may lead to frequent handoffs, as shadow fading causes significant changes in signal strength. A variation of this basic algorithm incorporates hysteresis. Such algorithm makes handoff if the RSS from a candidate BS exceeds the RSS from the current BS by the amount of hysteresis. The North American Personal Access Communication Systems (PACS) standard, implements a handoff algorithm with hysteresis and dwell timer [3]. Hysteresis reduces the number of unnecessary handoffs but it could increase call drops because some necessary handoffs may also be avoided [8]. Note that hysteresis inherently introduces a delay in handoff. It is possible to balance the number of handoffs and handoff delay through appropriate settings of hysteresis and averaging of signal strength measurements. Such design is considered in [11]. Both rectangular and exponential averaging are evaluated. The averaging mechanism is a function of the mobile speed and shadow fading. A scheme that can estimate the standard deviation of shadow fading using the squared deviations of the RSS at the MS, is proposed in [16]. The standard deviation of shadow fading is an important propagation parameter and optimal handoff parameters are sensitive to this parameter. A longer averaging time and smaller hysteresis, provides robustness. However, to detect sudden changes in signal strength (such as those encountered in the street corner effect), less averaging and larger hysteresis are required. To address such competing needs, shadow fading standard deviation is estimated in [15]. An adaptive averaging window is desirable because fading fluctuates widely in the case of a very short averaging interval and handoff delay becomes very long in the case of a very long averaging interval. Since shadow fading is a function of the distance traveled, it is indirectly influenced by velocity. Reference [15] achieves shadow fading dependent averaging interval by estimating the maximum Doppler frequency. The effects of the signal averaging time and hysteresis on the settings of handoff parameters are analyzed in [17] and [18]. Shadow fading plays an important role in the performance of a signal strength based algorithm. The microcellular environment presents more complexities to the signal strength based algorithm. A large hysteresis can help avoid the ping-pong effect for the LOS handoff (i.e., a handoff between two LOS BSs), but will delay an NLOS handoff (i.e., a handoff between a LOS BS and an NLOS BS), potentially causing higher call drops in the case of NLOS handoffs [3]. A small hysteresis will cause frequent handoffs.

Absolute Signal Strength Algorithms. This algorithm requests handoff when the RSS drops below a threshold. Example threshold values are –100 dBm for a noise-limited system and –95 dBm for an interference-limited system [14]. An adaptive threshold that is a function of the path loss slope L of the RSS and the level crossing rate (LCR) of the RSS, can provide good performance. For example, higher slope L or LCR indicates faster movement of the MS away from the BS, and, threshold can be adjusted to make a faster handoff. On the other hand, lower L or LCR is indicative of slower MS movement, and, threshold can be varied for a slower handoff. Such adaptive threshold helps reduce the number of unnecessary handoffs, while completing the necessary handoffs successfully. A two-level algorithm is an extension of the basic single-threshold (or single-level) algorithm [14]. Two handoff thresholds, L_1 and L_2, are used, with $L_1 > L_2$. When the RSS drops below L_1, a handoff request is made. The handoff request is accepted only if a candidate BS provides stronger RSS (e.g., Situation 1 in Fig. 6.2.4). Such handoff requests are made periodically so that the possibilities of the MS being in a signal strength hole in the current cell or the candidate cell being busy can be analyzed effectively. If the RSS drops below L_2, handoff is made to the candidate BS C depicted by Situation 2 in Fig. 6.2.4.

Combined Absolute and Relative Signal Strength Algorithms. The combined absolute and relative strength algorithm makes a handoff, if the RSS of the current BS drops below an absolute signal strength and the RSS of the candidate BS exceeds the RSS of the current BS by some amount (e.g., h dB). The absolute signal strength threshold prevents handoff, when the current BS is sufficient for

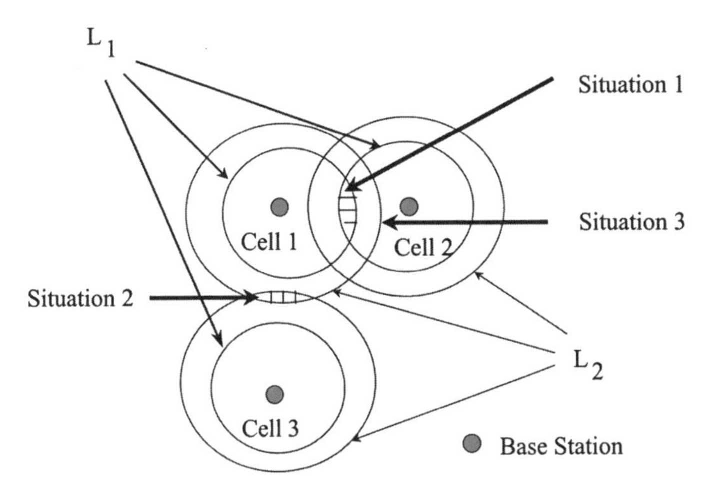

Fig. 6.2.4. A Two-Level Handoff Algorithm

the good signal quality. The parameters of the algorithm can be tuned to strike a balance between the area of the handoff region and the number of handoffs.

Distance Based Algorithm. This algorithm uses distance as the handoff criterion and connects to the nearest BS. Planned cell boundaries can be preserved by such algorithm. German cellular system C450, uses distance as a handoff criterion [19]. For macrocells, the accuracy of distance estimation is good, and this algorithm can perform well. However, in microcells, inaccuracies in distance measurements limit the usefulness of the algorithm based solely on distance. For example, the MS receives stronger signal from the distant but LOS BS, while it receives weaker signal from the closer but NLOS BS. In such case, the distance based algorithm results in poor signal quality, while the signal strength based algorithm provides good signal quality.

SIR Based Algorithms. SIR is one of the measures that quantify signal quality. For example, the cell edge SIRs of 18 dB for AMPS and 12 dB for GSM, are considered to be sufficient for good signal quality. In the case of CDMA systems (such as IS-95 and IS-2000), BSs with strong pilot (E_c/I_0) contribute to a good communication link. According to the SIR based algorithm, when the SIR for the current BS falls below a threshold, and another BS can provide a signal with sufficient SIR, a handoff is made. The algorithm can also include hysteresis in addition to the absolute threshold. The advantage of the algorithm is that it usually makes only necessary handoffs. For example, a low SIR may be due to high interference power or low signal power. It is desirable to make a handoff in either case. The drawback of the algorithm is that handoffs may not occur near nominal cell boundaries, and high transmit power may be required to maintain the communication link between the BS and the MS, that are far apart from each other. The measurement of SIR is often difficult during the call. To avoid such a problem, interference power is measured before the call is connected, and combined signal and interference power are measured during the call. Note that the algorithm can use uplink SIR instead of downlink SIR as a handoff criterion. For example, Reference [20] describes an uplink SIR based algorithm, for a power-controlled system. Each user attempts to meet a target SIR. When the uplink SIR for a user falls below a threshold (usually less than the target SIR), a handoff is made.

Velocity Adaptive Algorithms. It is desirable to make handoff request quickly for fast moving vehicles, because of faster degradation in signal quality. In general, short temporal averaging windows for signal strength measurements are useful. However, if the length of the averaging window is kept constant, optimal handoff performance is achieved only at one speed. To obtain good handoff performance at different speeds, the length of the averaging window can be made a function of the speed [21]. The velocity adaptive algorithm is an alternative to the umbrella cell approach that overlays existing small cells, and that serves high speed users. The velocity adaptive algorithm represents savings in the infrastructure costs. The level crossing rate of the RSS can be used to estimate velocity [22]. The threshold level is set as the average value of the Rayleigh distribution of the RSS. Such velocity estimation method, is based on the idea that the velocity is proportional to the Doppler frequency and that the level crossing rate and Doppler frequency are related to each other. Reference [21] analyzes velocity adaptive handoff algorithms for microcellular systems. Three methods considered to estimate velocity are *the level crossing rate method, the zero-crossing rate method*, and *the covariance approximation method.* A spatial averaging distance of 20 λ to 40 λ (where λ is the wavelength of the signal), is found to be adequate for microcells. The temporal averaging window can be adapted to velocity, by keeping the sampling period constant and varying the number of samples per averaging window or vice versa.

Direction Biased Algorithms. Direction biasing is a method of exploiting the knowledge of the direction of the MS motion. A direction-biased algorithm encourages handoffs to the BSs toward which the MS is moving, and discourages handoffs to the BSs from which the MS is receding. Direction-biasing, reduces the number of handoffs and leads to improvement in overall handoff process [23].

Minimum Power Algorithm. A minimum power algorithm attempts to minimize the uplink transmit power by choosing an appropriate combination of a BS and a channel [24]. Such algorithm leads to less interference in the system but increases the number of unnecessary handoffs. A timer or a similar mechanism can help reduce the number of handoffs. Reference [6] uses a power budget criterion to assign a BS with the least path loss to the MS. This criterion reduces the overall transmit power and CCI.

RSS and BER Algorithm. A handoff algorithm that uses RSS and BER as handoff criteria is discussed in [25]. Two separate thresholds are designed for RSS and BER. The target BS quality, if not available, can be excluded from the comparison process (e.g., in GSM). A hysteresis threshold is also used with RSS. The effects of two RSS parameters and one BER parameter on handoff probability are investigated through simulations. Measured data such as RSS and BER are incorporated onto the software simulator for combined RSS and BER algorithm in [25].

6.2.2.3 Emerging Handoff Algorithms

The emerging algorithms discussed below, include: *the dynamic programming based algorithm, pattern recognition algorithm, prediction algorithm, neural network based algorithm* and *fuzzy logic based algorithm.*

Dynamic Programming Based Algorithms. Dynamic programming provides a systematic way to optimize the handoff process and usually requires estimates of some parameters because of model dependencies. The complexity of dynamic programming restricts its application to simple handoff scenarios. For example, [26] considers handoff as a cost/reward optimization. The handoff is viewed as a switching cost, while the reward is a function of characteristics such as signal strength, SIR, channel fading, and shadowing. Dynamic programming creates optimal handoff policies. Another application of dynamic programming is considered in [Kel5a] to achieve a balance between the number of handoffs and the number of service failures, which is the event by which the RSS drops below a threshold required for desired QoS. An optimal handoff solution is derived through dynamic programming, which assumes a priori knowledge of the MS travel. A sub-optimal solution called locally optimal or greedy algorithm is also found. Another dynamic programming based handoff strategy is evaluated in [28], using a Markov formulation. The optimization function involves a cost for switching (i.e., handoff) and a reward for QoS improvement. The solution is expressed as the optimal value of the hysteresis.

Prediction Based Algorithm. Prediction based algorithms typically predict values of handoff criteria such as RSS to make faster and more efficient handoffs. For example, [29] compares prediction algorithms with relative signal strength and combined absolute and relative signal strength algorithms. A good tradeoff between the number of handoffs and the overall signal quality is achieved by an adaptive prediction algorithm. The algorithm predicts shadow fading correlation properties, to reduce the number of unnecessary handoffs. The future RSS is predicted based on past RSS measurements through an adaptive FIR filter. The filter coefficients are adapted to reduce prediction error. The current and predicted RSS values are used to calculate handoff priority that eventually affects the overall handoff decision.

Neural and Fuzzy Handoff Algorithms. Recently, handoff algorithms that utilize neural networks and fuzzy logic have appeared in literature. For example, [1] resorts to a binary hypothesis test (performed by a neural network) to initiate a signal strength based handoff decision. Reference [30] uses a neural network based methodology for a multi-criteria handoff decision. The fuzzy logic algorithm of [31] is shown to reduce the number of handoffs, because the overlap region between the adjacent cells that influences handoff decisions can be modeled by inherent fuzziness of fuzzy logic. The performance of a fuzzy logic classifier in a microcellular environment is evaluated in [32]. The signal strength measurements are used as inputs to the classifier. A fuzzy handoff algorithm that includes signal strength, distance, and traffic is outlined in [33]. A fuzzy logic based handoff procedure that systematically includes multiple handoff criteria to reduce the number of handoffs without excessive cell overlapping, is analyzed in [7].

6.2.3 Analysis of Handoff Algorithms

Two major aspects of the analysis of the performance of a handoff algorithm are *performance metrics* and *performance evaluation mechanisms*. The performance metrics (also called performance measures) quantify the performance, while the evaluation mechanisms provide a way of collecting statistics of the performance metrics.

6.2.3.1 Performance Metrics

Widely used performance metrics, include: [3, 4, 6, 23].

Call blocking probability, is the probability that a new call (i.e., call origination) is blocked from accessing the network.

Handoff blocking probability, is the probability that a handoff is blocked.

Handoff rate, is the number of handoffs per unit time.

Handoff probability, is the probability that a handoff is made during the duration of the call.

Call drop probability, is the probability that a call ends prematurely.

Handoff delay, is the time difference between the initiation of a handoff request and the actual execution of the handoff request.

Interference probability, is the probability that the signal-to-interference ratio is less than a threshold [9].

Assignment probability, is the probability that the MS is connected to a particular BS [9].

Some performance metrics depend on channel allocation strategies in addition to a handoff strategy. Furthermore, a performance metric may need to be inferred, if it cannot be measured directly. For example, the CDF of the number of calls in a cell can be used to infer call and handoff blocking probabilities. The CDFs of RSS and SIR serve as indicators of the call drop probability.

6.2.3.2 Performance Evaluation Approaches

There are three basic approaches for evaluation of the performance of a handoff algorithm. These approaches are: *the analytical approach, simulation approach*, and *emulation approach.*

The analytical approach, provides a preliminary idea about the performance of a handoff algorithm for simplified scenarios. This approach is valid only for a set of constraints (e.g., assumed RSS profiles). Modeling of real-world scenarios makes this approach complex and often mathematically intractable. The analytical approaches of [34] and [17] can help determine the averaging interval and hysteresis value to balance the number of unnecessary handoffs and delay in handoff in a simplified handoff scenario, where the MS travels from one BS to another. Reference [35] develops an analytical model for analysis of a handoff algorithm, that uses absolute and relative signal strength measurements. A soft handoff algorithm for a CDMA system is analyzed in [36] via an analytical approach.

The emulator approach, utilizes a software simulator to process measured variables such as RSS and BER. A handoff algorithm with a set of parameters is implemented in the software simulator, and the performance of the algorithm is evaluated for different settings of the algorithm parameters. The advantage of this method is that better insight into the behavior of the algorithm is gained, due to the use of actual (as opposed to simulated) propagation measurements. The disadvantage is that periodic measurement efforts are required, and it may not be feasible to compare different handoff algorithms, due to different evaluation platforms. Reference [25] uses measured data (obtained in an urban environment in Southern England) in handoff simulation.

The simulation approach, is the most widely used handoff evaluation approach because of its flexibility to model complex real-world scenarios. A common test-bed can be created to evaluate multiple handoff algorithms on the same platform. The scope of an analytical approach is limited to simple handoff scenarios, while full-fledged field-testing is costly and time consuming. Hence, the simulation approach can be viewed as a compromise between the two approaches. The basic components of a simulation model are *cell model, mobility model, traffic model*, and *propagation model*.

Cell Model – Different deployment scenarios require different cell models. For example, a macrocellular system may need a different cell model from a microcellular system. A 49-cell toroidal system is often used to mimic a macrocellular system (Fig. 6.2.2), while half-square, full-square, or rectangular plan can be sued to mimic a microcellular system in a Manhattan-type (or urban) environment. The cell model considered here consists of a group of four cells shown in Fig. 6.2.5.

The cell radius is 10 km, and, the maximum overlap between Cell A and Cell C is 2 km. The EIRP (Effective Isotropic Radiated Power) from the BS is

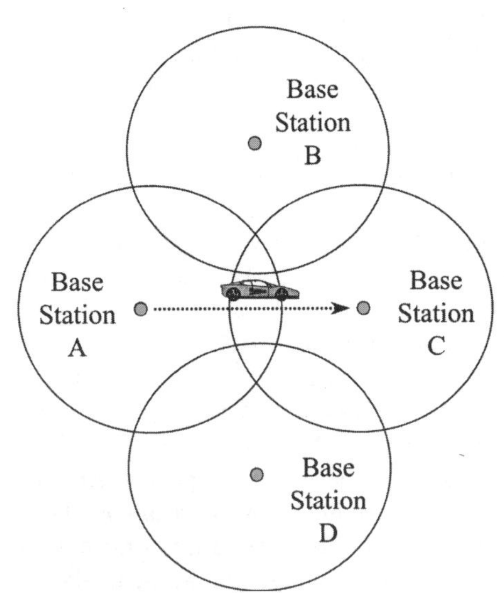

Fig. 6.2.5. A Four-Cell Neighborhood

100 Watts. In typical CDMA systems, the BS maximum transmit power is in the range of 12 to 20 Watts. The MS travels from BS A to BS C. This journey is repeated five hundred times.

Mobility Model – For the simulation results reported here, the MS velocity is kept constant throughput the journey (e.g., 45 mph, 65 mph, and 85 mph). In general, in a multi-cell environment, a truncated Gaussian distribution is often used to characterize different mobile speeds. The MS velocity is assumed to be constant throughout the call duration.

Traffic Model – The traffic (i.e., the number of calls) is assumed to be uniform in all the four cells. In a macrocellular system, mobile traffic can be in a random direction. However, when a specific cell plan is considered for an urban microcellular system, the mobile travel is allowed only along the streets. For several simultaneous calls, the new call arrival process is modeled as a Poisson distributed, with a certain mean arrival rate. Note that if the inter-call arrival time is exponentially distributed, the cell arrival process follows Poisson distribution. The call duration is exponentially distributed.

Propagation Model – A propagation model usually consists of a path loss model, a large-scale fading (e.g., shadow fading) model, and a small-scale fading model. For the purpose of handoff evaluation, small-scale fading (such as multipath fading modeled as Rayleigh fading) can be neglected, because of the averaging of measurements.

Hata's model is used as a path loss model [37]. Path-loss is given by:

$$L_{50(urban)} = 69.55 + 26.16 \log(f_c) - 13.82 \log(h_{te}) - a_{h_{re}} + (44.9 - 6.55 \log(h_{te})) \log(d) \quad (6.2.1)$$

where f_c is the carrier frequency (in MHz), h_{te} is the effective BS (or transmitter) antenna height (in m), h_{re} is the effective MS (or receiver) antenna height (in m), d is the transmitter-receiver (T-R) separation distance in km, and $a_{h_{re}}$ is the correction factor for different sizes of the coverage area. In simulations, the following values are used: $f_c = 900$ MHz, $h_{te} = 30$ m, and $h_{re} = 3$ m. For a medium size city,

$$a_{h_{re}} = (1.1 \log(f_c) - 0.7)\ h_{re} - (1.56 \log(f_c) - 0.8)\,. \quad (6.2.2)$$

In general, the path loss models of Hata and Okumura can be used for macrocells while different models are used for line-of-sight (LOS) and non-line-of-sight (NLOS) propagation in microcells.

Shadow fading is assumed to follow log-normal distribution [38]. The shadow fading is correlated. For example, an exponential autocorrelation model is proposed in [39] for shadow fading, which is validated through measurements. The correlation at 100 m distance is about 0.82 for a large cell in a suburban environment. A correlated shadow fading process can be generated from uncorrelated Gaussian samples. Assume that a sequence n, is an uncorrelated process with samples that have 0-mean and σ_2 standard deviation. Another sequence x is a 0-mean, σ_1 standard deviation, and exponential autocorrelation function given by:

$$\rho(d)=\sigma_1^2 \exp\left(\frac{-d}{d_0}\right) \tag{6.2.3}$$

where d is the distance between two samples, d_0 is a parameter that can be tuned to specify correlation at a given distance. For example, $d_0 = 500$ m for the autocorrelation value of 0.82 at a distance of 100 m. In other words, $e^{\left(\frac{d}{d_0}\right)} = e^{\left(\frac{-100}{500}\right)} = 0.82$. $x(D + d_s) = \alpha x(D) + n(D)$ where d_s is the sample distance, and

$$\alpha=\exp\left(\frac{-d}{d_0}\right) \text{ and} \tag{6.2.4}$$

$$\sigma_2^2=\sigma_1^2(1-\alpha^2) \tag{6.2.5}$$

The simulation model described above is a modified version of the model described in [17]. The number of calls in four cells is changed every seventy-five simulation steps, which represents 15% of the total simulation time. To model interference, a set of co-channel BSs, located at a distance of $D=\sqrt{N}\,R$ from the center of the cell under consideration, is simulated. Here, R is the cell radius, and N is the cluster size (i.e., the number of cells in a cluster). The actual number of interfering BSs is selected uniformly between 0 and 6 every seventy-five simulation steps. Only the first tier of interference is considered.

6.2.4 Neural Encoding Based Neuro-Fuzzy System

A high-performance handoff algorithm can be designed using multiple criteria as inputs and processing the inputs such that appropriate tradeoffs among various system features. A fuzzy logic system (FLS) is a good candidate for such processing. Known sensitivities of the parameters of the conventional handoff algorithms can be used to design an FLS, which adapts the parameters of the algorithm, to enhance performance in a given dynamic cellular environment. The drawbacks of an FLS are large storage and computational requirements. To circumvent these requirements while preserving the potential for high performance, neural encoding of the FLS is proposed [40, 41]. The working of the FLS is learned by a low-complexity neural network. The trained neural network replaces the FLS in an adaptive FLS-based algorithm. The neural network paradigms such as a multi-layer perceptron (MLP) and a radial basis function network (RBFN), are proven to be universal approximators. The specific properties of these paradigms such as input-output mapping capability and compact data presentation capability, allow the design of an adaptive algorithm that retains the high performance of the FLS based algorithm and that has lower storage and computational requirements.

Figure 6.2.6 illustrates the development of a neurofuzzy configuration by exploiting characteristics of a fuzzy logic system and a neural network system.

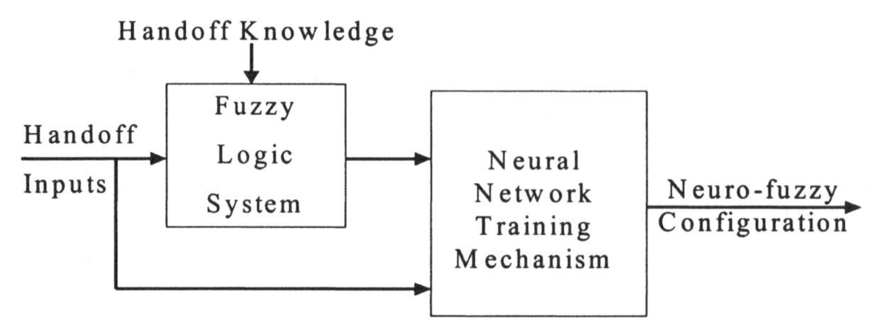

Fig. 6.2.6. Development of a Neurofuzzy Configuration

The first step toward the development of the neurofuzzy system proposed in this chapter, is to create a fuzzy logic system (FLS) using the knowledge of the system. Basically, the knowledge of the expected behavior of the system and the desired controlling influences of handoff actions is used to determine rules of the FLS. Since a number of conditions and circumstances must be considered for a comprehensive high-performance handoff algorithm, the resultant FLS can be quite complex. In practice, computational requirements place significant constraints on the structure and framework of the mechanism that can be implemented. If some methodology with simpler implementation can capture the knowledge contained in the fuzzy logic rule base, it would facilitate the development of a high-performance handoff algorithm. Several paradigms of neural networks may provide the required simplified framework. Since the MLP and the RBFN have been proven to be capable of representing the complex relationships in a compact form, the neural network training mechanism utilizes the inputs and outputs of the designed FLS as the constituents of a training set. This training mechanism creates the neurofuzzy modules that have the form of a neural network, with the knowledge of a fuzzy logic system. Such neurofuzzy system is used as an integral component of an adaptive handoff algorithm.

Following are the stages involved in the development of a neurofuzzy system for a high-performance handoff algorithm, and include: *The determination of a conventional handoff algorithm and the FLS that adapts the parameters of such algorithm, The training of the MLP or the RBFN to mimic the FLS,* and *The performance evaluation of the neurofuzzy system based algorithm.* These stages are discussed next.

6.2.4.1 Determination of the Conventional Algorithm and the FLS

Figures 6.2.7 and 6.2.8, illustrate the first stage.

The input to the conventional algorithm are pre-processed signal strength and *SIR* measurements. This pre-processing is a function of the *MS* velocity. The parameters of the conventional handoff algorithm are $RSS_{threshold}$, $RSS_{hysteresis}$, and $SIR_{threshold}$. Two parameters, $RSS_{threshold}$ and $RSS_{hysteresis}$, are adapted by the *FLS*. The flowchart of the conventional algorithm is shown in Fig. 6.2.6. If RSS_c or SIR_c is sufficiently strong, the current communication link

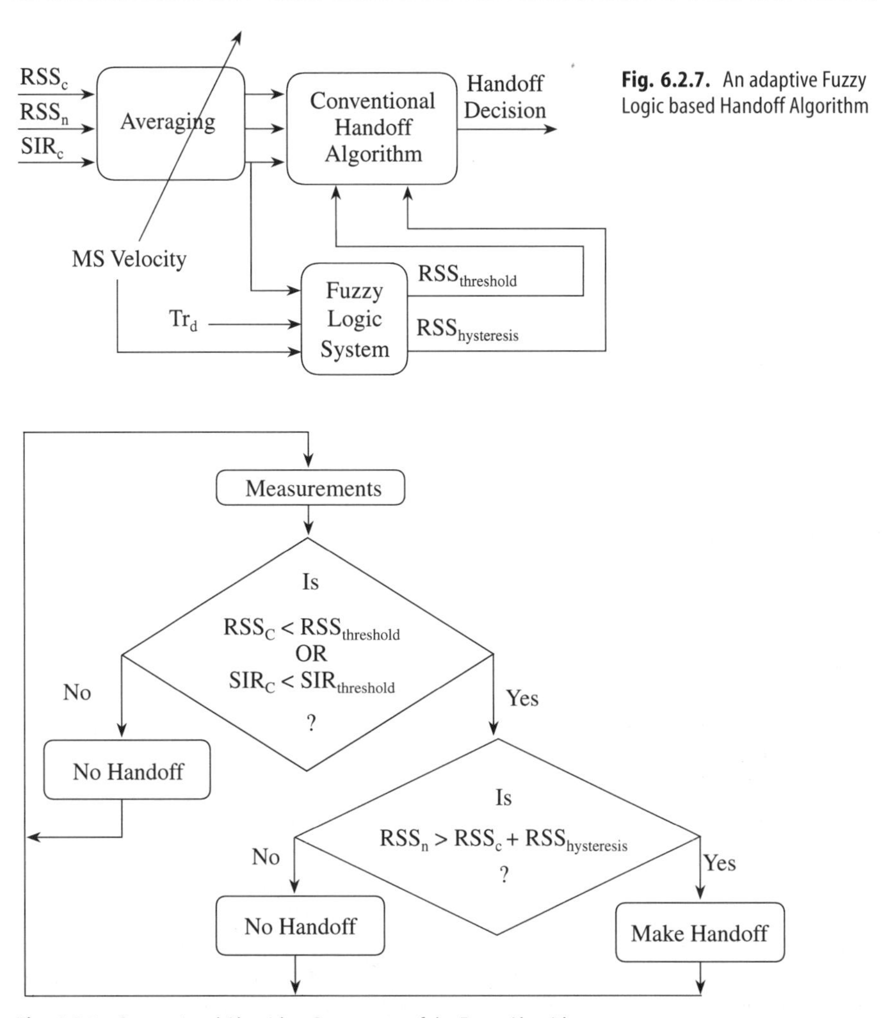

Fig. 6.2.7. An adaptive Fuzzy Logic based Handoff Algorithm

Fig. 6.2.8. Conventional Algorithm Component of the Fuzzy Algorithm

between the *BS* and the *MS* can be considered good, and no handoff takes place. If RSS_c or SIR_c is weak, and RSS_n exceeds RSS_c by the amount of $RSS_{hysteresis}$, handoff is made.

The handoff parameters, $RSS_{threshold}$ and $RSS_{hysteresis}$, are adapted according to several quantities that characterize the dynamics of cellular environment. Several measurements such as RSS_c, RSS_n, and SIR_c are averaged according to the velocity adaptive averaging mechanism. The MS velocity and the traffic difference, Tr_d (i.e., the difference in the number of calls in the current *BS* and the neighboring *BS*) are not averaged, because their most recent instantaneous values are of interest.

The inputs to the *FLS* are SIR_c, Tr_d ($= Tr_c - Tr_n$), and *MS* Velocity. The outputs of the *FLS* are $RSS_{threshold}$ and $RSS_{hysteresis}$. A fuzzy logic rule base is created based

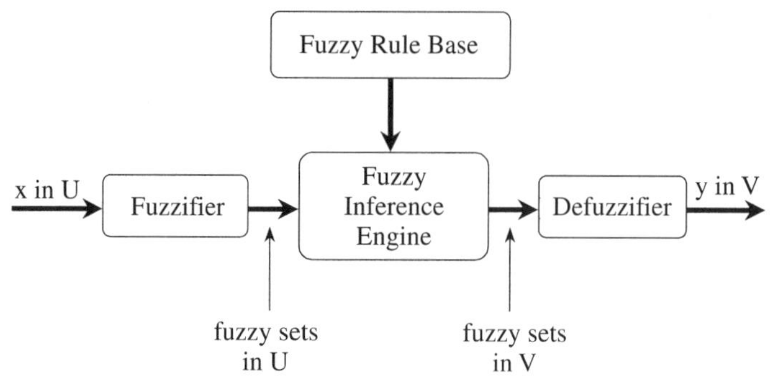

Fig. 6.2.9. Elements of the Mamdani Fuzzy Logic System

on the known sensitivity of the algorithm parameters to factors such as interference and traffic. The structure of the FLS used here is the Mamdani FLS [42], which is depicted in Fig. 6.2.9.

Components of the FLS are singleton fuzzifier, product operation fuzzy implication for fuzzy inference, and center average defuzzifier. The fuzzy rule base is summarized in Table 6.2.1.

Each input fuzzy variable is assigned one of the three fuzzy sets, "High," "Normal," or "Low." Each output variable is assigned one of the seven fuzzy sets, "Highest," "Higher," "High," "Normal," "Low," "Lower," or "Lowest." The membership functions of fuzzy variables are illustrated in Fig. 6.2.10. In Fig. 6.2.10, the universe of discourse for SIR_c is from 14 dB to 22 dB. The fuzzy set "High" for SIR_c is from 18 dB to 22 dB with the maximum degree of membership at 22 dB. The "Normal" fuzzy set of SIR_c is from 14 dB to 22 dB with the maximum degree of membership at 18 dB, and the "Low" fuzzy set of SIR_c is from 14 dB to 18 dB with the maximum degree of membership at 14 dB. The parameters of the fuzzy variables that can be tuned are centers of the Gaussian membership functions, spreads of the membership functions, and ranges of the universes of discourse. Equal weight is given to the input fuzzy variables to facilitate the creation of the fuzzy logic rule base. The sensitivity of the input fuzzy variable to the output of the *FLS* can be controlled by changing the universe of discourse. Furthermore, the addition of more fuzzy sets in a given universe of discourse provides improved resolution and better sensitivity control. The universe of discourse for each input fuzzy variable is classified into three fuzzy sets to keep the complexity of the fuzzy logic rule base low. The universe of discourse for the output fuzzy variable was divided into seven regions, so that appropriate weight can be given to the different combinations of the input fuzzy sets.

Two examples of the fuzzy rule base are given below, to highlight the philosophy that underlines the fuzzy logic rule base summarized in Table 6.2.1.

Example 1. Consider Rule Number Nineteen. SIRc is "Low," "Trd is "High," and MS Velocity is "High." These conditions indicate that handoff should be encouraged as much as possible. To make fastest handoff, the RSS threshold is increased to the highest value, and the RSS hysteresis is decreased to the lowest value.

Table 6.2.1. Fuzzy Logic Rule Base

Rule Number	SIR_c	Tr_d	*MS Velocity*	$\Delta RSS_{threshold}$	$\Delta RSS_{hysteresis}$
1	High	High	Low	High	Low
2	High	High	Normal	Normal	Normal
3	High	High	High	Low	High
4	High	Normal	Low	Normal	Normal
5	High	Normal	Normal	Low	High
6	High	Normal	High	Lower	Higher
7	High	Low	Low	Low	High
8	High	Low	Normal	Lower	Higher
9	High	Low	High	Lowest	Highest
10	Normal	High	Low	Higher	Lower
11	Normal	High	Normal	High	Low
12	Normal	High	High	Normal	Normal
13	Normal	Normal	Low	High	Low
14	Normal	Normal	Normal	Normal	Normal
15	Normal	Normal	High	Low	High
16	Normal	Low	Low	Normal	Normal
17	Normal	Low	Normal	Low	High
18	Normal	Low	High	Lower	Higher
19	Low	High	Low	Highest	Lowest
20	Low	High	Normal	Higher	Lower
21	Low	High	High	High	Low
22	Low	Normal	Low	Higher	Lower
23	Low	Normal	Normal	High	Low
24	Low	Normal	High	Normal	Normal
25	Low	Low	Low	High	Low
26	Low	Low	Normal	Normal	Normal
27	Low	Low	High	Low	High

Example 2. Now, consider Rule Number Nine. Since SIRc is "High", Trd is "Low," Since SIRc is "High," Trd is "Low," and MS velocity is "Low," handoff is discouraged to the maximum extent.

Influence of Multiple Criteria. If the majority of the input variables suggest encouraging a handoff, the threshold is increased, and hysteresis is decreased. On the other hand, if the majority of input variables suggest discouraging a handoff, the threshold is decreased, and hysteresis is increased. The extent to which the threshold and the hysteresis are changed depend on the number of agreements on a particular direction of the change in the threshold and hysteresis. Resolving conflicting criteria in accordance with the global system goals is an important benefit of fuzzy logic. As an example, consider Rule Number two. "High" SIRc discourages a handoff, while "High" Trd encourages a handoff, and MS velocity is neutral. Hence, the fuzzy logic rule makes the logical decision of keeping the threshold and hysteresis at nominal levels. The fuzzy logic based handoff algorithm of Fig. 6.2.6, is advantageous compared to conventional algorithms.

The proposed algorithm, exploits several attractive features of conventional algorithms. For example, the absolute and relative signal strength and SIR based

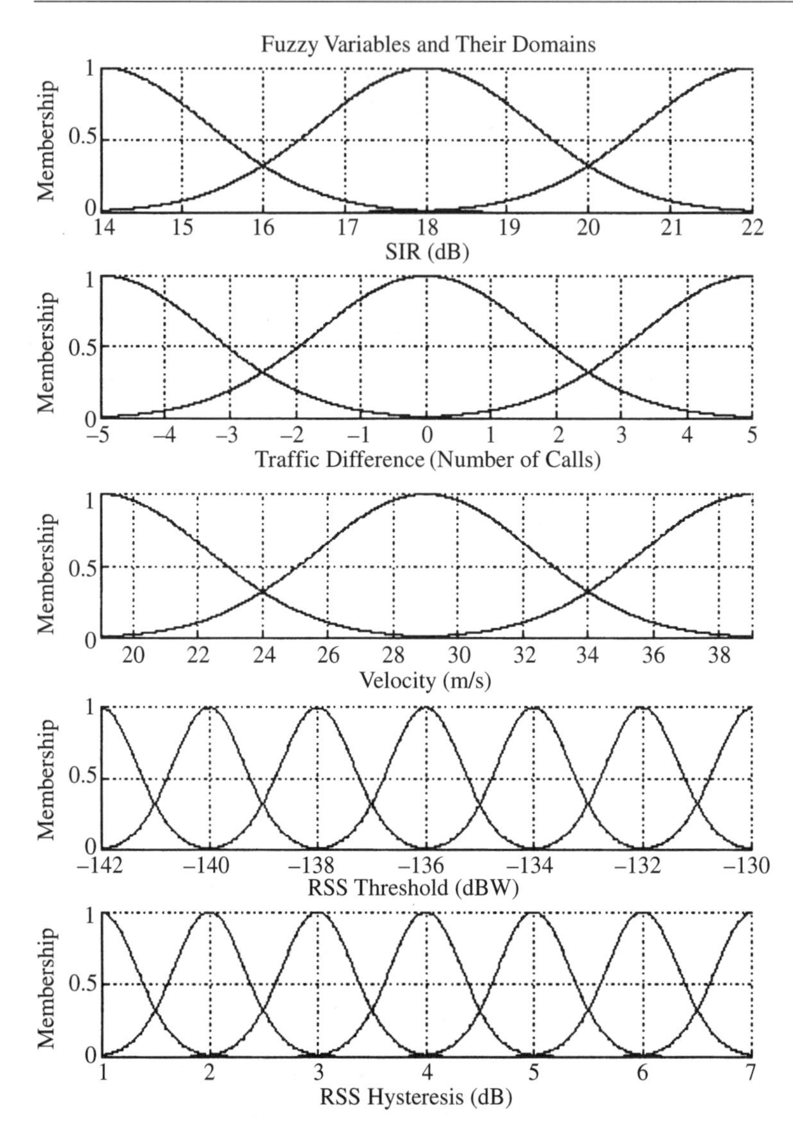

Fig. 6.2.10. Membership Functions for the Fuzzy Variables

algorithm that is part of the proposed algorithm leads to fewer handoffs, reduced ping-pong effect, and reduced unnecessary handoffs. The proposed algorithm complements the conventional algorithm.

The proposed algorithm is a multicriteria algorithm that yields better performance than single-criterion algorithms due to the additional flexibility and complementary nature of the criteria. A multi-criteria algorithm considers various measurements to make a decision consistent with the global system goals.

The proposed algorithm adapts to interference, traffic, and velocity. Interference adaptation improves RSS and SIR distribution, resulting in fewer dropped calls, better communication quality, and potentially lower power requirements. Traffic adaptation provides traffic balancing, reducing the call blocking probability, and handoff blocking probability. Velocity adaptation provides good performance at different speeds.

6.2.4.2 Training of a Neuro-Fuzzy System

The fuzzy logic based handoff algorithm represents a high performance algorithm with inherent parallelism. However, the fuzzy algorithm is much more complex than a conventional algorithm with a few binary IF-THEN rules. As the number of inputs to the FLS increases or as the universes of discourse for the fuzzy variables are divided into more fuzzy sets, the complexity of the FLS increases even further. The FLS complexity is two-fold, storage requirements and the number of computations that are performed every measurement sample period. To retain the high performance of an FLS but to reduce the overall complexity of the algorithm, a neurofuzzy system is proposed. The neurofuzzy system involves encoding of the FLS to simultaneously achieve high performance and low complexity. The mapping between the FLS inputs and outputs is encoded in the form of the parameters of the neural network.

1. Create a training data set that contains possible inputs and corresponding outputs. Different combinations of inputs are applied to the FLS, and the FLS outputs are observed. The FLS inputs and outputs become the data set for the training of the neural network.
2. Utilize suitable neural network architecture and a learning algorithm. Since the mapping between FLS inputs and outputs represents a function approximation problem, two suitable neural network architectures, a multi-layer perceptron (trained by the backpropagation algorithm) and radial basis function network, are utilized here.
3. Determine the input set and the target output set for the neural network. Since the mapping between the FLS inputs and outputs is static, the input and output data sets obtained in Step 1 can serve as the input set and target output set with appropriate scaling. The scaling ensures that the network parameters do not saturate.
4. Choose the configuration parameters (e.g., the number of neurons and the learning rate) and train the neural network. Various parameters that can be changed to obtain satisfactory performance include training time, pre-processing of inputs, neural network paradigm, and training parameters. After the neural network has been trained, the mapping between the FLS inputs and outputs is encoded in the form of the parameters of the neural network.

The complexity of a handoff algorithm can be analyzed in terms of storage requirements and computational requirements. At every sampling instant, the outputs of the adaptive mechanisms (FLS, MLP and RBFN) need to be calculated.

The output of an FLS, y, is given by:

$$y = \frac{\sum_{l=1}^{R} \bar{y}^l (\mu_{B^l}(\bar{y}^l))}{\sum_{l=1}^{R} (\mu_{B^l}(\bar{y}^l))} \tag{6.2.6}$$

where R is the number of rules, $\bar{y}^l$ is the center of the output fuzzy set, and $\mu_{B^l}(\bar{y}^l)$ is calculated as:

$$\mu_{B^l}(\bar{y}^l) = \prod_{i=1}^{m} \exp\left(-\left(\frac{x_i - \bar{x}_i^l}{\sigma_i^l} \right)^2 \right) \tag{6.2.7}$$

where m is the number of inputs to the FLS, $\bar{x}_i^l$ is the center of the fuzzy set for input i for rule l, and σ_i^l is the spread of the fuzzy set for input i for rule l. Since the FLS considered here is a single-output system, two FLSs are required to calculate two outputs. However, all the computations need not be carried out separately for these two FLSs since the membership values (i.e., $\mu_{B^l}(\bar{y}^l)$) are the same for a given input vector. Only $\bar{y}^l$ related calculations need to be carried out for individual outputs.

The output of an MLP is given by:

$$Y = W_2 \tanh(W_1 X + B_1) + B_2 \tag{6.2.8}$$

where W_1 is a hidden layer weight matrix of size $N \times m$, B_1 is a hidden layer bias matrix of size $N \times 1$, W_2 is an output layer weight matrix of size $p \times N$, B_2 is an output layer bias matrix of size $p \times 1$, X is an $m \times 1$ input vector, Y is a $p \times 1$ output vector, and "tanh" is the hyperbolic tangent function.

The output of an RBFN is given by:

$$Y = W_2 A_1 + B_2 \tag{6.2.9}$$

with

$$A_1 = radbas(dist(W_1, X) \times B_1). \tag{6.2.10}$$

Here, W_1 consists of centers of Gaussian functions and is of size $N \times m$, B_1 consists of spreads associated with Gaussian functions and is of size $N \times 1$, W_2 is an output layer weight matrix of size $p \times N$, B_2 is an output layer bias matrix of size $p \times 1$, X is an m $\times$ 1 input vector, Y is a $p \times 1$ output vector, and "dist" is the distance between X and each row of W_1. In other words,

$$dist(w, X) = \sqrt{\sum_{j=1}^{p} (w_j - X_j)^2}\ . \tag{6.2.11}$$

Here, w represents one row of W_1.

"Radbas" is the radial-basis function given by:

$$radbas(x) = \exp(-x^2). \tag{6.2.12}$$

The storage requirements of the FLS, MLP, and RBFN are derived next.

FLS Storage Requirements. For an FLS rule, m centers ($\bar{x}_i^l$) and m spreads (σ_i^l)for the antecedent part of the rule and p centers ($\bar{y}^l$) for the consequent part of the rule are required. Thus, for each rule, a total of $(2m + p)$ elements are required (Eqs. (6.2.6) and (6.2.7)). Since there are R rules in an FLS, a total of $(2m + p)R$ elements need to be stored for the FLS.

MLP Storage Requirements. For an MLP, $W_1, B_1, W_2,$ and B_2 are required. Since W_1 is of size $N \times m$ and B_1 is of size $N \times 1$, the number of elements for the first layer is $Nm + N$. Since W_2 is of size $p \times N$ and B_2 is of size $p \times 1$, the number of elements for the second layer is $pN + p$. The total number of elements are $Nm + N + pN + p = m\,(N + 1) + p\,(N + 1) = (m + p)\,(N + 1)$.

RBFN Storage Requirements. For an RBFN, the parameters $W_1, B_1, W_2,$ and B_2 have the same dimensions as in the case of an MLP. Hence, a total of $(m + p)$ $(N + 1)$ elements are needed for RBFN.

Table 6.2.2 summarizes the storage requirements for an FLS, MLP, and RBFN.

In this chapter, $R = 27$, $m = 3$, $p = 2$, $N = 5$ for BPNN, and $N = 8$ for RBFN. Table 6.2.3 gives the improvement in storage requirements for the neural techniques. The MLP and RBFN give an improvement of the factor 7.2 (i.e., 216/30) and 4.8 (i.e., 216/45) over the FLS for storage requirements.

The computational complexities of the FLS, MLP, and RBFN are derived next. The operations of addition and subtraction are grouped together and referred to as adds. Also, the operations of multiplication and division are grouped together and referred to as multiplies. Evaluations of functions (such as exponential, square-root) are referred to as functions.

FLS Computations. For each rule, the input vector is processed by Eq. (6.2.7). There are one subtraction, one division, one multiplication (squaring operation), and one function (exponential) evaluation for each element of the input vector X that has m elements. Thus, there are one add, two multiplies, and one function. For each rule, there are m adds, $2m$ multiplies, and m functions due to m inputs. The m exponential terms and $\bar{y}^l$ are multiplied, requiring m more multiplies. Hence, there are mR adds, $3mR$ multiplies, and mR functions for R

Table 6.2.2. Storage Complexity of Adaptation Mechanisms

System	Elements
FLS	(2m + p) R
MLP	(m + p) (N + 1)
RBFN	(m + p) (N + 1)

Table 6.2.3. Specific Examples of Storage Complexity

System	Elements
FLS	216
MLP (N = 5)	30
RBFN (N = 8)	45

rules. There are (R–1) additions (of $\bar{y}^l$ and $\mu_{B^l}(\bar{y}^l)$) in the numerator of Eq. (6.2.6) and one division of the numerator and denominator in Eq. (6.2.6). Thus, there are $Rm + (R - 1) = (m + 1)R - 1$ adds, $3mR + 1$ multiplies, and mR functions for one output calculation. For each additional output, there are R multiplies (of $\bar{y}^l$ and $\mu_{B^l}(\bar{y}^l)$) and (R-1) adds (of the terms $\bar{y}^l\ \mu_{B^l}(\bar{y}^l)$ and one numerator-denominator division). Hence, there are additional $(R + 1)$ multiplies and $(R - 1)$ adds for each additional output. If there are p outputs, $(R + 1)(p - 1)$ additional multiplies and $(R - 1)(p - 1)$ additional adds are required. Thus, the total number of adds is $(m + 1)R - 1 + (R - 1)(p - 1) = mR + R - 1 + Rp - R - p + 1 = mR + Rp - p$, the total number of multiplies is $3mR + 1 + (R + 1)(p - 1) = 3mR + 1 + Rp - R + p - 1 = (3m + p - 1)R + p$, and the total number of functions is mR.

MLP Computations. Each row of $W_1\ X$ multiplication requires m multiplies and $(m - 1)$ adds. Since there are N rows in W_1, a total of mN multiplies and $(m - 1)N$ adds are required. $W_1\ X$ and B_1 addition requires N more adds. Thus, $(m - 1)N + N = mN$ adds are required. Hence, mN multiplies, mN adds, and N functions (*tanh* calculations) are required to carry out a $tanh(W_1\ X + B_1)$ operation. The multiplication of W_2 and *tanh* terms is between the $p \times N$ and $N \times 1$ matrices, requiring pN multiplies and $(p - 1)N$ adds. The result of this multiplication is added to the $p \times 1$ matrix B_2, requiring additional p adds. Hence, the total number of adds are $mN + pN = (m + p)N$, and the total number of multiplies are $mN + pN = (m + p)N$.

RBFN Computations. There are m subtractions, m multiplies (squaring operation), $(m - 1)$ adds, and one function (square-root) for each row of W_1 in Eq. (6.2.11). Hence, there are $(2m - 1)$ adds, m multiplies, and one function for Eq. (6.2.11). For N rows of W_1, there are $(2m - 1)N$ adds, mN multiplies, and N functions. The distance function is $N \times 1$, and it is added to B_1 of size $N \times 1$. This requires N additions. Hence, there are $N(2m - 1) + N = 2mN$ adds, Nm multiplies, and N functions for the argument of "radbas" in Eq. (6.2.10). The $N \times 1$ matrix is processed by radial basis functions. In each radial function, there is one multiplication (squaring) and one function evaluation (exponential) operation. Thus, there are N more multiplies and N more function evaluations. Hence, there are $2mN$ adds, $Nm + N$ multiplies, and $2N$ functions for the calculation of A_1. The evaluation of Eq. (6.2.9) requires pN multiplies and pN adds. Hence, there are a total of $2mN + pN = N(2m + p)$ adds, $Nm + N + pN = (m + p + 1)N$ multiplies, and $2N$ functions.

Table 6.2.4 summarizes the computational requirements of the FLS, MLP, and RBFN.

Table 6.2.5 shows the improvement in computational requirements for the neural methods. There is an improvement of 8.8 (486/55 = 8.8) and 3.8 (486/128 = 3.8) for MLP and RBFN, respectively, over the FLS.

6.2.4.3 Performance Evaluation of the Neurofuzzy System Based Handoff Algorithm

Figure 6.2.11 shows the input data of the training data set. These different combinations of inputs are applied to the FLS, and the corresponding FLS outputs are calculated. The FLS outputs constitute the output data of the training data

Table 6.2.4. Computational Complexity of Adaptation Mechanisms

System	Multiplies	Adds	Function Evaluations
FLS	$3m + R + p$	$m + 2R + (p - 1)R$	mR
MLP	$N(m + p)$	$N(m - 1) + p(N - 1) + N + p$	N
RBFN	$N(p + m + 1)$	$N(p + 2m)$	$2N$

Table 6.2.5. Computational Complexity of Adaptation Mechanisms

System	Multiplies	Adds	Function Evaluations	Total Operations
FLS	272	133	81	486
MLP	25	25	5	55
RBFN	48	64	16	128

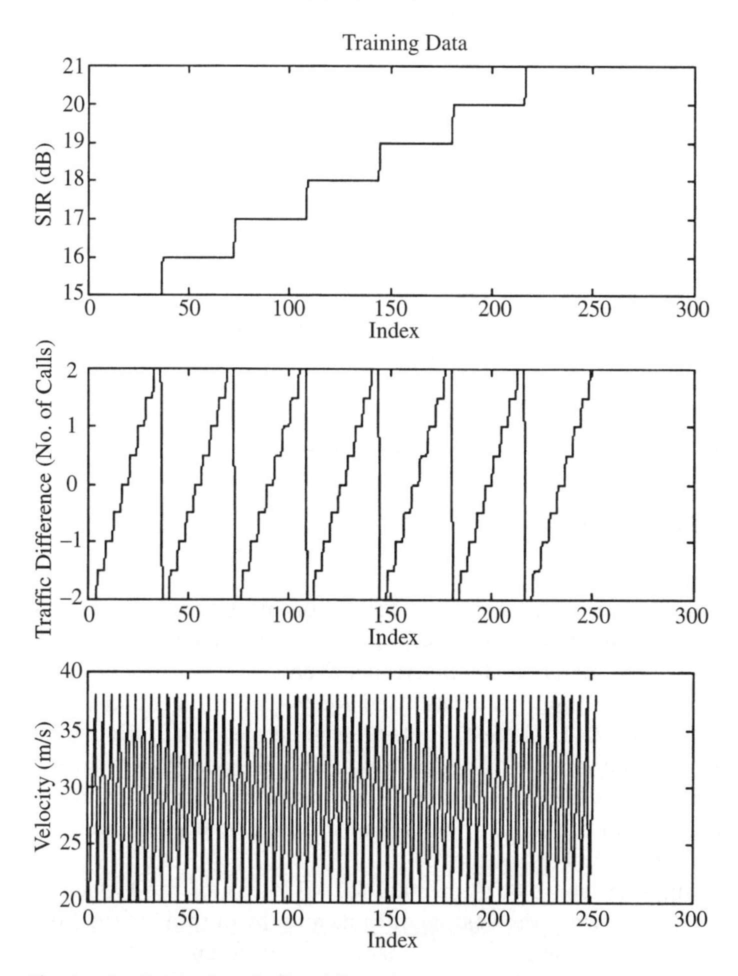

Fig. 6.2.11. Training Data for Neural Networks

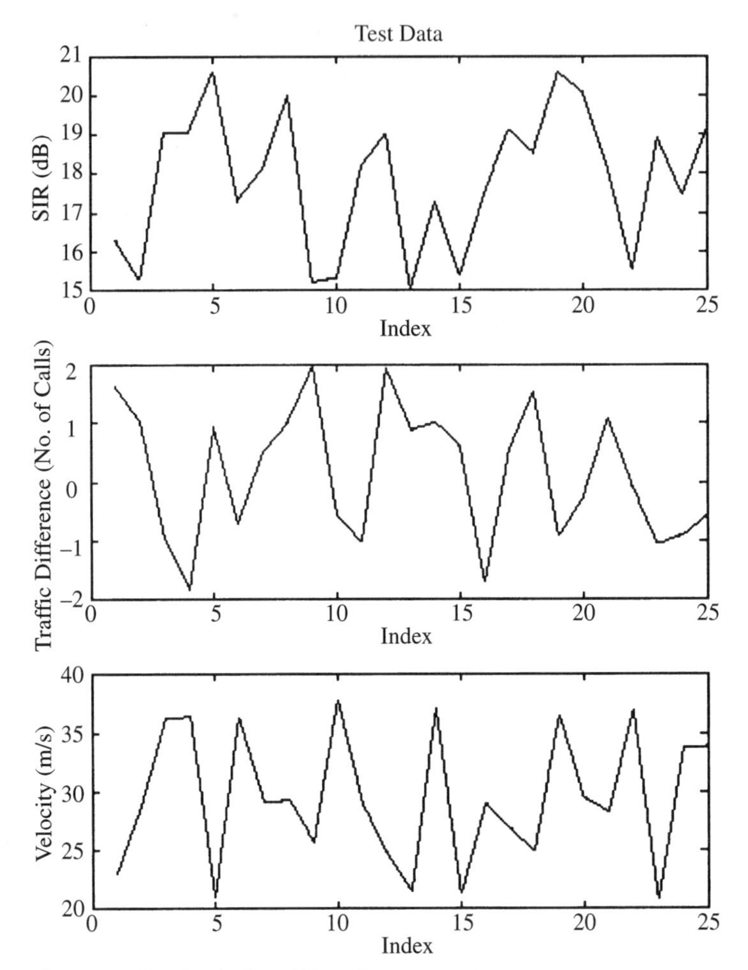

Fig. 6.2.12. Test Data for Neural Networks

set. This output data serves as the target or desired output when a neural network is trained. The ranges of SIR, traffic difference, and velocity are from 15 dB to 21 dB, −2 to 2 calls, and 20 m/s to 38 m/s, respectively.

Figure 6.2.12 shows the input test data used for testing the generalization property of the trained ANNs. The test inputs were generated randomly within the specified ranges of the variables. The corresponding FLS outputs are the desired outputs. Hence, if an ANN has learned the input-output mapping of the FLS well, the presentation of the test inputs shown here produces the outputs that are similar to the desired test outputs.

Table 6.2.6 summarizes some of the results for MLP. E_{train} is the Frobenius norm of the difference between the desired outputs and the outputs of the MLP for the training data. E_{test} is the Frobenius norm of the difference between the desired outputs and the outputs of the MLP for the test data. A different number

Table 6.2.6. Training and Test Results for MLP.

Number	Number of Hidden Layer Neurons	E_{train}	E_{test}
1	5	24.58	8.17
2	8	16.32	8.33
3	12	16.01	8.09
4	17	16.47	8.07
5	21	16.22	8.06

of hidden layer neurons were trained for different training times (5000 to 15000 epochs). One epoch is one pass through the training set. In general, more neurons can lead to an improved mapping. However, complexity increases as the number of neurons increases. The number of neurons is chosen to be eight as a compromise between the accuracy of generalization and complexity. Since the main interest is to represent the FLS with as few neurons as possible, a tradeoff between the number of neurons and the mapping accuracy must be achieved. For the application under consideration, the error performance of the MLP is quite acceptable.

Figure 6.2.13 shows the desired (or actual) test data and the MLP output data. The desired data and MLP predicted data are close to each other; however, they are not identical. This means that the ANN has learned most of the FLS mapping features, but it has not learned a perfect mapping. Hence, when the performances of the fuzzy logic based algorithm and neural algorithm are compared,

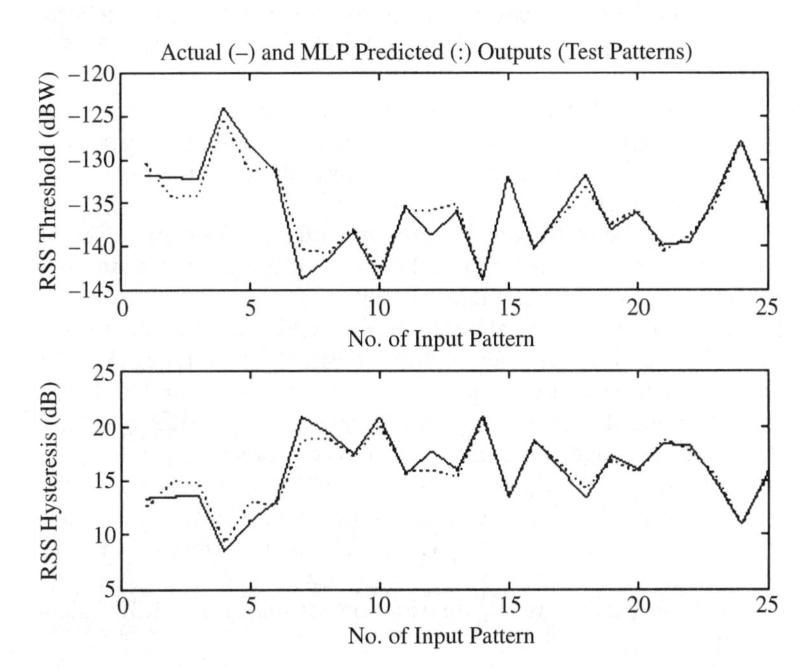

Fig. 6.2.13. MLP Test Data Performance

Table 6.2.7. Training and Test Results for RBFN

Number	Number of Radial Basis Functions	E_{train}	E_{test}
1	5	34.50	8.28
2	10	24.42	6.82
3	14	20.25	5.47
4	19	17.80	5.29
5	24	15.73	5.16

similar, but not identical, performances should be expected. The FLS mapping can be learned well by an ANN provided that appropriate processing is done and that a sufficient number of neurons are used. Since the goal here is to use fewer neurons, no attempt is made to obtain perfect FLS mapping.

The RBFN mapping performance is similar to the MLP training performance. The RBFN has learned the mapping well, though the mapping is not exact. Table 6.2.7 summarizes some of the results for RBFN.

Different spreads for radial basis functions and different numbers of radial basis functions were tried. In general, higher numbers of radial basis functions give an improved performance with the associated increase in complexity. The number of radial basis functions is chosen to be ten, as a compromise between accuracy of generalization and complexity. As in the case of MLP, the performances of the fuzzy logic based algorithm and neural algorithm can be expected to be similar but not identical.

Figure 6.2.14 shows the cumulative distribution function (CDF) of RSS for conventional, fuzzy logic (FL), and MLP algorithms. As expected, FL and MLP performances are similar.

Figure 6.2.15 shows the distribution of SIR for all the algorithms. Again, there is a close match between FL and MLP performance. The small discrepancy can be attributed to the stochastic nature of simulations and the approximate modeling of the FLS by the MLP.

Figure 6.2.16 shows the traffic distribution for different algorithms. The FL and MLP give similar traffic performances. Both the FL and MLP provide a two call improvement over the conventional algorithm.

Figure 6.2.17 illustrates the operating point for the FL and MLP algorithms. The symbols "*," "+," and "*x*" represent minimum (45 mph), average (65 mph), and maximum (85 mph) velocities, respectively. These symbols represent the FL operating points, while the encircled symbols represent the MLP operating points. Both FL and MLP algorithms give similar performance.

Similar simulation tests were performed for an RBFN based algorithm. The results of two representative simulations are shown here. Figure 6.2.18 shows the SIR distribution for FL and RBFN algorithms. The performances of these algorithms are similar.

Figure 6.2.19 shows that the traffic distribution for the FL and RBFN algorithms is similar.

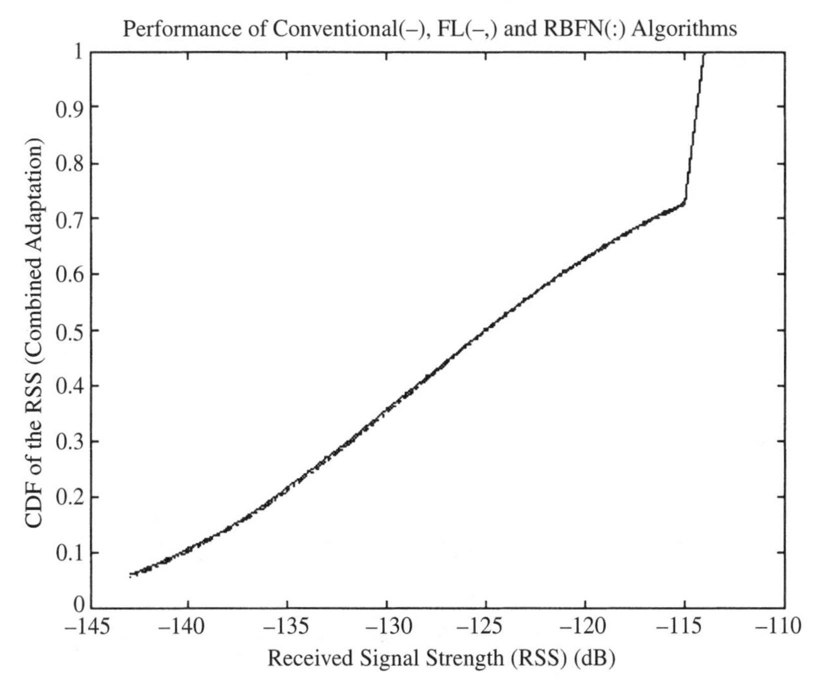

Fig. 6.2.14. Distribution of RSS for Conventional, Fuzzy, and MLP Algorithms

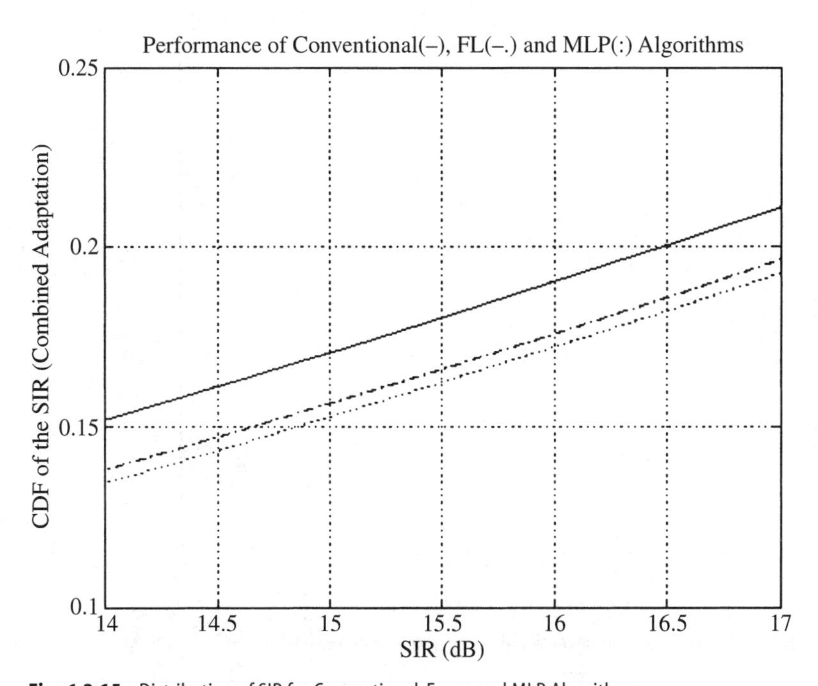

Fig. 6.2.15. Distribution of SIR for Conventional, Fuzzy, and MLP Algorithms

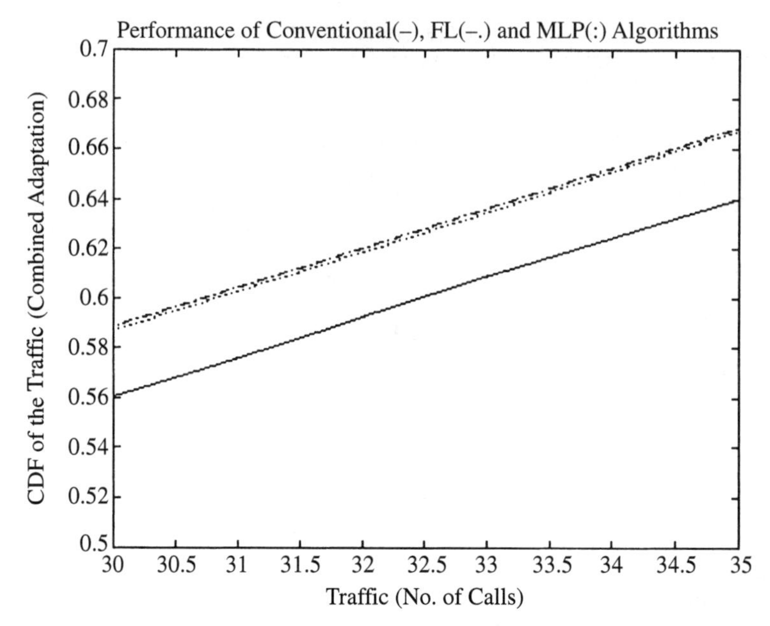

Fig. 6.2.16. Distribution of Traffic for Conventional, Fuzzy, and MLP Algorithms

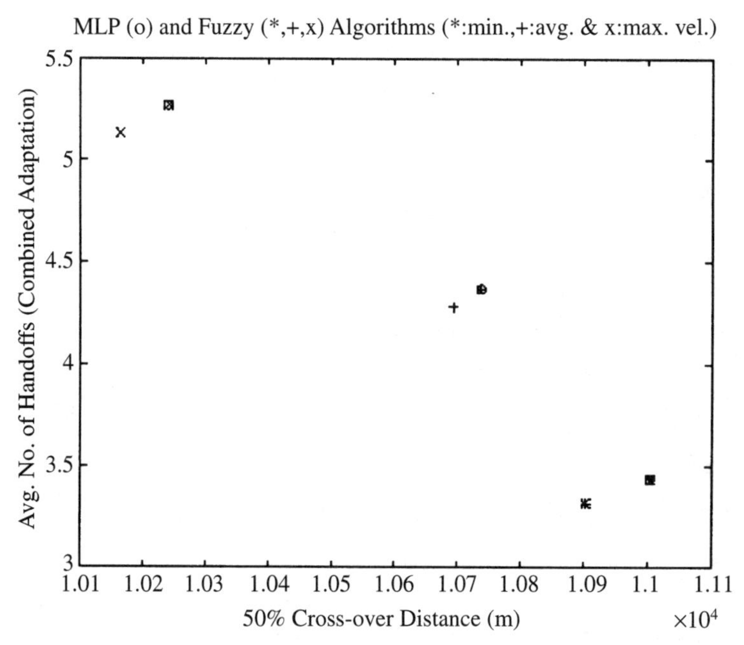

Fig. 6.2.17. Operating Points for Conventional, Fuzzy, and MLP Algorithms

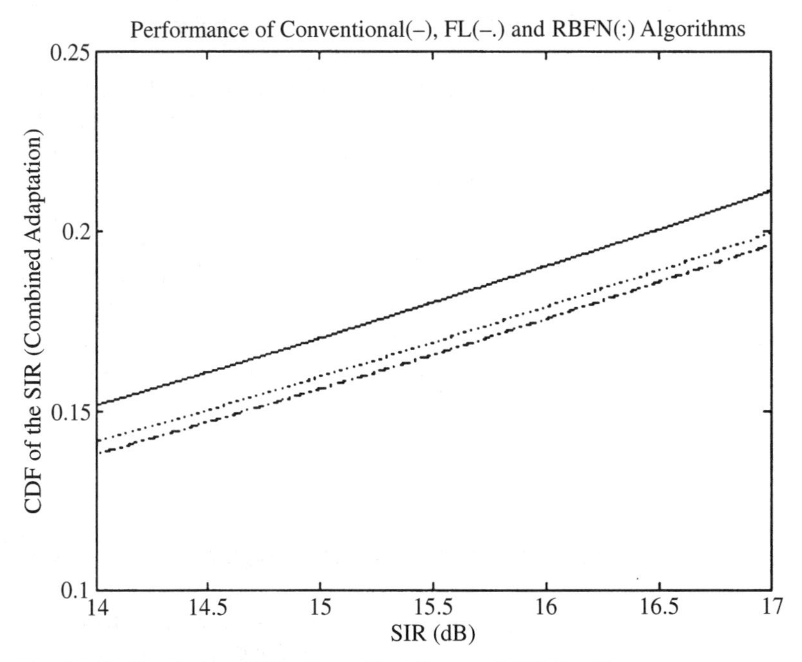

Fig. 6.2.18. Distribution of SIR for Conventional, Fuzzy, and RBFN Algorithms

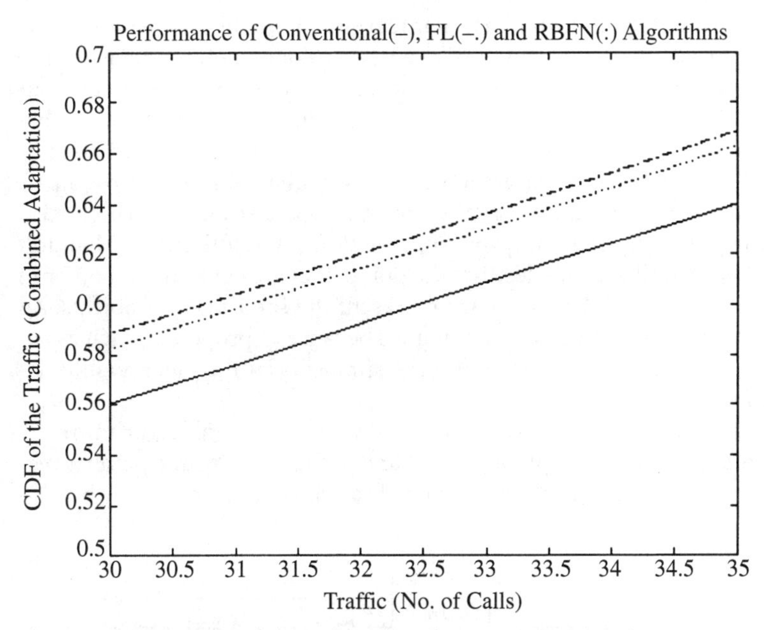

Fig. 6.2.19. Distribution of Traffic for Conventional, Fuzzy, and RBFN Algorithms

6.2.4.4 Summary of the Neuro-Fuzzy System Based Handoff Mechanism

An adaptive algorithm that encodes the working of an FLS is proposed. An FLS can be designed using known sensitivities of handoff parameters, but the FLS requires the storage of many parameters and needs a lot of computations. Several neural network paradigms such as an MLP and an RBFN, are universal approximators. The input-output mapping capability and compact data representation capability of these neural network paradigms are exploited to represent the FLS. The neural representation of the FLS provides an adaptive handoff algorithm that retains the high performance of the original fuzzy logic based algorithm and that has an efficient architecture for meeting storage and computational requirements. The analysis of the simulation results indicates that an adaptive multicriteria neural handoff algorithm performs better than a signal strength based conventional handoff algorithm and that the fuzzy logic based algorithm and neural network based algorithm perform similarly.

6.2.5 Pattern Recognition Based Neuro-Fuzzy System

Section 6.2.4 describes neural encoding of an FLS to achieve a high-performance handoff algorithm with lower complexity than the generic FLS based algorithm. Knowledge of the sensitivity of handoff parameters to the dynamics of cellular environment is used to create the rule base of the FLS. The mapping between the FLS inputs and outputs is then learned by the neural network. The neural network adapts the parameters of a conventional handoff algorithm to obtain high performance in a complex environment. The process of neural encoding of an FLS is generic and can generally be applied to any mechanism that utilizes an FLS. The problem of handoff can also be viewed as a pattern classification problem, in which the inputs to the handoff algorithm represents a pattern that is classified into a class representing a base station. Pattern classification (PC) is a convenient way of implementing a multi-criteria handoff algorithm. The PC utilizes the idea that the points that are close to each other in a mathematically defined feature space represent the same class of objects. An FLS can serve as a means of implementing PC because of properties such as capability of universal approximation and the similarity of the membership degree to the PC degree.

Figure 6.2.20 shows the block diagram of a PC based handoff algorithm.

Various measurements are preprocessed and are used to form a pattern vector. This pattern vector is classified into one of the classes. The class may be a BS

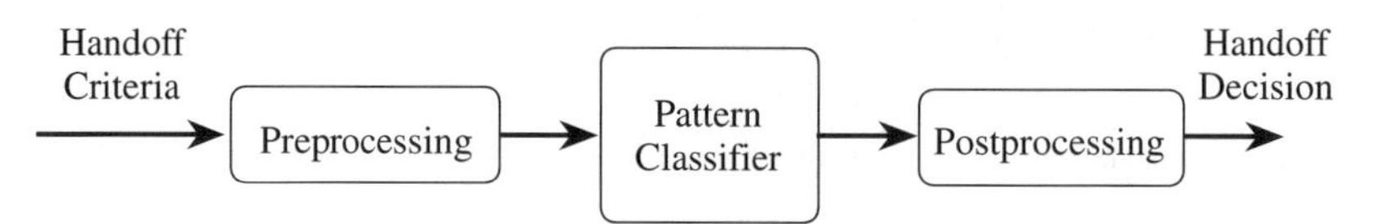

Fig. 6.2.20. Pattern Classification Based Handoff Algorithm

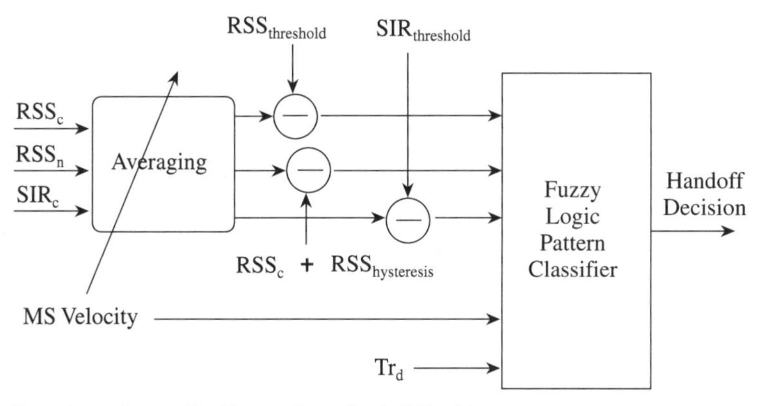

Fig. 6.2.21. Pattern Classification Based Handoff Algorithm

or the degree to which the MS belongs to a given BS. The output of the PC is post-processed to make a decision about handoff. The PC based handoff algorithm allows simultaneous consideration of multiple handoff criteria. The testing of a series of binary IF-THEN rules is replaced by a single operation of classification. Since the PC is a single output system, it can lead to reduction in storage and computational requirements. Adaptation can be obtained through suitable pre-processing of the handoff criteria.

A PC based neuro-fuzzy system for an adaptive handoff algorithm can be developed by the following procedure.

1. Design an FLS that implements PC.
2. Encode the FLS in the form of a neural network for reduced complexity.
3. Evaluate the performance of the neural encoded FLS.

Steps 2 and 3 can be executed similar to the procedure extensively described in Section 6.2.4. Step 1 is discussed in detail next.

The block diagram of an adaptive handoff algorithm that utilizes PC based FLS, is shown in Fig. 6.2.21.

The link measurements, RSS_c, RSS_n, and SIR_c, are averaged using a velocity adaptive averaging mechanism. *MS velocity* and traffic difference Tr_d are not averaged since their instantaneous values are of interest. The averaged RSS_c, RSS_n, and SIR_c are biased before forming a pattern to account for the thresholds. The FLPC assigns a class association degree to the input pattern. If this degree is greater than 5.5, handoff is made. The Mamdani FLS is used here as a basic architecture of the FLS. Two major stages of the PC based FLS design, are determination of the training data set and the operation of classification. These steps are described below.

6.2.5.1 Determination of the Training Data Set

The training data set should consist of representative patterns and the class association degrees, which is the degree to which a pattern belongs to a class.

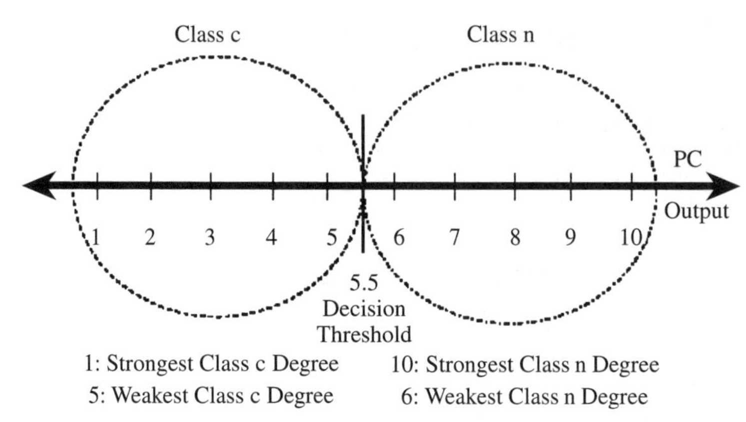

Fig. 6.2.22. The Concept of a Degree for the PC

The pattern vector consists of five elements, $RSS_c - RSS_{threshold}$, $RSS_n - (RSS_c + RSS_{hysteresis})$, $SIR_c - SIR_{threshold}$, $Tr_d = Tr_c - Tr_n$, and *MS velocity*. RSS_c is the received signal strength (*RSS*) from the current BS, $RSS_{threshold}$ is the *RSS* threshold, RSS_n is the received signal strength (*RSS*) from the neighboring (or candidate) BS, $RSS_{\text{hysteresis}}$ is the *RSS* hysteresis, SIR_c is the *SIR* of the current channel, $SIR_{threshold}$ is the *SIR* threshold, Tr_d is the traffic difference (i.e., the difference between the number of calls in the current cell (Tr_c) and the number of calls in the neighboring cell (Tr_n)).

Figure 6.2.22 illustrates the concept of the association degree for the PC. Class c denotes the class of the current BS, and class n denotes the class of the neighboring BS.

This figure is used with the following pieces of information to illustrate the creation of an appropriate training data set.

- If all the elements of the pattern vector suggest that handoff should not be made, the output of the PC is one. If all the elements of the pattern vector suggest that handoff should be made, the output of the PC is ten. If an element does not encourage or discourage handoff, its position is considered neutral.
- If the majority of the pattern elements favor a "No Handoff" decision, the output of the PC is in the range of one to five. Similarly, if the majority of the pattern elements favor a "Handoff" decision, the output of the PC is in the range of six to ten. The output value "1" indicates that the degree to which an MS belongs to class c is strongest, and the output value "5" indicates that the degree to which an MS belongs to class c is weakest. The output value "10" indicates that the degree to which an MS belongs to class n is strongest, and the output value "6" indicates that the degree to which an MS belongs to class n, is weakest.
- The output value of the PC depends on the net agreements between the elements of the pattern for a particular decision, "Handoff" or "No Handoff."

Table 6.2.8. PC Outputs for "No Handoff" Decision

Number	Net Agreements	PC Output
1	5	1
2	4	2
3	3	3
4	2	4
5	1	5

Table 6.2.9. PC Outputs for "Handoff" Decision

Number	Net Agreements	PC Output
1	5	10
2	4	9
3	3	8
4	2	7
5	1	6

- The training data set should cover the entire range of interest for all the variables. To minimize the number of patterns in the training data set, representative examples should be chosen carefully.

Examples for handoff decisions are given below.

Examples for a "No Handoff" Decision. If there are three agreements for "No Handoff" and two neutral positions, the net number of agreements is three and the PC output is three. If there are one agreement and four neutral positions, the net agreement is one and the output value is five. If there are four agreements and one disagreement, the net number of agreements is three and the output is three. Table 6.2.8 summarizes the PC outputs for the "No Handoff" decision scenario.

Examples for a "Handoff" Decision. If there are three agreements for "Handoff" and two neutral positions, the net number of agreements is three and the output is eight. If there are one agreement and four neutral positions, the net agreement is one and the output value is six. If there are four agreements and one disagreement, the net number of agreements is three and the output is eight. Table 6.2.9 summarizes the PC outputs for the "Handoff" decision scenario.

Several concepts of fuzzy logic were used to create a training data set. Each fuzzy variable (each element of the input pattern vector called $ip_j, j \in [1, 5]$) is divided into three fuzzy sets ("High" (H), "Medium" (M), and "Low" (L)). Two basic rules are used to derive a complete set of the rules, covering the entire region of interest:

Rule 1. If ip_1 is H, ip_2 is L, ip_3 is H, ip_4 is L, and ip_5 is L, the output is one;
Rule 2. If ip_1 is L, ip_2 is H, ip_3 is L, ip_4 is H, and ip_5 is H, the output is ten.

Since there are five fuzzy variables and three fuzzy sets, there are a total of $3^5 = 243$ rules.

6.2.5.2 Operation of Classification

If a pattern vector similar to one of the representative vectors is presented to the PC, the PC classifies it into the class associated with the closest stored pattern vector. The closeness can be quantified by the Euclidean distance between the stored patterns and the input pattern vector. The PC output indicates the degree to which a given pattern vector belongs to a class (or a BS).

6.2.6 Application of a Neuro-Fuzzy Handoff Approach to Various Cellular Systems

6.2.6.1 A Microcellular System

The propagation environment and handoff challenges are different in macrocells and microcells. A handoff algorithm with constant parameters, cannot perform uniformly well in various handoff scenarios encountered by an MS in an urban microcellular environment. A neurofuzzy system can be used as part of an adaptive handoff algorithm for enhanced performance in all the handoff scenarios of a microcellular system. Such application of a neurofuzzy system to a microcellular system is described here. The basic approach is identical to the neural encoding of the FLS discussed in Section 6.2.4 for a macrocellular system. The knowledge of a microcellular environment is used to create a rule base for the FLS.

Microcells increase system capacity at the cost of an increase in the complexity of resource management. In particular, the number of handoffs per call increases, and fast handoff algorithms are required to maintain an acceptable level of dropped call rate. Microcells impose distinct constraints on handoff algorithms due to the characteristics of their propagation environment. For example, a mobile station (MS) encounters a propagation phenomenon called *corner effect*, which demands a faster handoff. Figure 6.2.23 shows two generic handoff scenarios in microcells, a *line of sight (LOS) handoff* and a *non-line of sight (NLOS) handoff*. A LOS handoff occurs when the base stations (BSs) that serve an MS are LOS BSs before and after the handoff. When the MS travels from BS 0 to BS 2, it experiences a LOS handoff. A NLOS handoff occurs when one BS is a NLOS BS before the handoff, and the other BS becomes a NLOS BS after the handoff. When the MS travels from BS 0 to BS 1, it experiences a NLOS handoff. A good handoff algorithm performs uniformly well in both generic handoff scenarios. Important considerations for designing handoff algorithms for a microcellular system, are briefly described next.

Mobility and Traffic Characteristics. The MS speeds are lower, and the speed range is narrower compared to a macrocellular scenario. Traffic is normally allowed only along the streets.

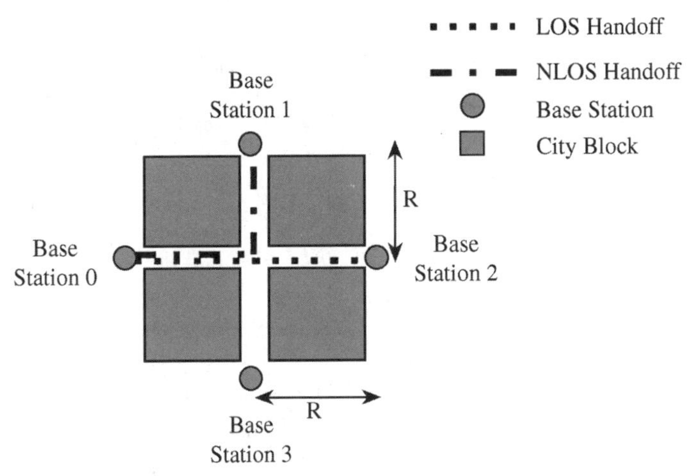

Fig. 6.2.23. Generic Handoff Scenarios in a Microcellular System

Propagation Features. The MS experiences the corner effect as discussed earlier. Field measurements have shown that the shadow fading intensity is lower in microcells than in macrocells.

Measurement Averaging. The averaging interval (or averaging distance) is shorter in microcells to respond to fast varying signal strength profiles. To provide sufficient averaging to counteract shadow fading effects, a sufficient number of samples are required, which may necessitate higher measurement sampling frequency.

Primary Handoff Requirements. A handoff algorithm should be fast and should minimize the number of handoffs. A fixed parameter handoff algorithm is suboptimal in a microcellular environment. For example, if hysteresis is large, it will cause a delay in NLOS handoff, increasing the probability of a dropped call. On the other hand, if hysteresis is small, it will increase the likelihood of the ping-pong effect. Since the situation of LOS or NLOS handoff cannot be known *a priori*, a proper tradeoff must be achieved between the LOS and NLOS handoff performance. In general, a large hysteresis gives good LOS handoff performance and poor NLOS handoff performance. A small hysteresis gives good NLOS handoff performance and poor LOS handoff performance.

Secondary Handoff Requirements. The algorithm should respond relatively faster to fast moving vehicles, should attempt to balance traffic, and should be adaptive to interference.

Figure 6.2.24 shows the block diagram of an adaptive handoff algorithm that uses two FLSs, primary FLS and secondary FLS. Both the FLSs can be transformed into a neural encoding based neurofuzzy system. This algorithm considers secondary handoff requirements whenever primary microcellular handoff objectives are not compromised. When an MS is near an intersection, there is a possibility of NLOS handoff, and any handoff parameter adaptation to obtain better performance in meeting secondary handoff objectives, can adversely

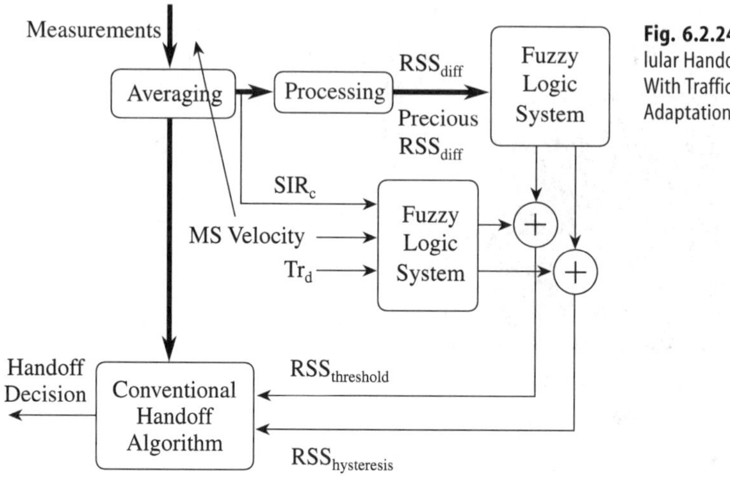

Fig. 6.2.24. A Microcellular Handoff Algorithm With Traffic and Mobility Adaptation

affect performance in meeting primary handoff objectives. Hence, the proposed algorithm switches on the secondary adaptation mechanism, only under LOS handoff type situations (i.e., when an MS is relatively far from an intersection). The vicinity of an MS to an intersection can be predicted based on the RSS difference between best BSs or the reliability of direction biasing.

The primary FLS provides base values of the handoff parameters, while the secondary FLS provides incremental variations in $RSS_{threshold}$ and $RSS_{hysteresis}$ to reflect the dynamics of traffic, interference, and mobility. The inputs to the secondary FLS are SIR_c, $Tr_d = Tr_c - Tr_n$, and MS velocity, and the outputs of the secondary FLS are incremental $RSS_{threshold}$ ($\Delta RSS_{threshold}$) and incremental $RSS_{hysteresis}$ ($\Delta RSS_{hysteresis}$). SIR_c is SIR of the current BS, and Tr_d is traffic difference (i.e., the difference in the number of calls in the current and the neighboring BS, $Tr_c - Tr_n$). *MS velocity* as an input to the FLS is the component of the MS velocity toward the serving BS. If the MS is moving toward the serving BS, the velocity is considered positive, and if the MS is moving away from the serving BS, the velocity is considered negative. Complete fuzzy logic rule bases for the primary and secondary FLSs are shown in Table 6.2.10 and Table 6.2.11. The geometry that underlines the philosophy behind the design of the FLS is illustrated in Fig. 6.2.24.

When an MS is relatively far from the intersection and close to a BS, the difference in RSS at the MS from the LOS BSs is high since the MS receives very high RSS from the closer BS and very low RSS from the far LOS BS. This situation is similar to a LOS handoff situation since the good handoff candidates are LOS BSs. Under these circumstances, it is advantageous to use high $RSS_{hysteresis}$ and low $RSS_{threshold}$ values to reduce the ping-pong effect. However, as an MS reaches the intersection, there is a likelihood of a NLOS handoff, and it is beneficial to use low $RSS_{hysteresis}$ and high $RSS_{threshold}$ to make a fast handoff in case a NLOS handoff is necessary. The intersection region is characterized by small (ideally, near zero) RSS differences. After an MS crosses an intersection, the RSS

Table 6.2.10. Fuzzy Logic Rule Base for a Microcellular Algorithm

Rule Number	*Current RSS Difference*	*Previous RSS Difference*	$RSS_{hysteresis}$	$RSS_{threshold}$
1	High	High	Lowest	Highest
2	High	Medium	Lower	Higher
3	High	Low	Low	High
4	Medium	High	Low	High
5	Medium	Medium	Medium	Medium
6	Medium	Low	High	Low
7	Low	High	Higher	Lower
8	Low	Medium	Higher	Lower
9	Low	Low	Highest	Lowest

Table 6.2.11. Secondary Rule Base for a Microcellular Algorithm

Rule Number	SIR_c	Tr_d	*MS Velocity*	$\Delta RSS_{threshold}$	$\Delta RSS_{hysteresis}$
1	High	High	Low	High	Low
2	High	High	Normal	Normal	Normal
3	High	High	High	Low	High
4	High	Normal	Low	Normal	Normal
5	High	Normal	Normal	Low	High
6	High	Normal	High	Lower	Higher
7	High	Low	Low	Low	High
8	High	Low	Normal	Lower	Higher
9	High	Low	High	Lowest	Highest
10	Normal	High	Low	Higher	Lower
11	Normal	High	Normal	High	Low
12	Normal	High	High	Normal	Normal
13	Normal	Normal	Low	High	Low
14	Normal	Normal	Normal	Normal	Normal
15	Normal	Normal	High	Low	High
16	Normal	Low	Low	Normal	Normal
17	Normal	Low	Normal	Low	High
18	Normal	Low	High	Lower	Higher
19	Low	High	Low	Highest	Lowest
20	Low	High	Normal	Higher	Lower
21	Low	High	High	High	Low
22	Low	Normal	Low	Higher	Lower
23	Low	Normal	Normal	High	Low
24	Low	Normal	High	Normal	Normal
25	Low	Low	Low	High	Low
26	Low	Low	Normal	Normal	Normal
27	Low	Low	High	Low	High

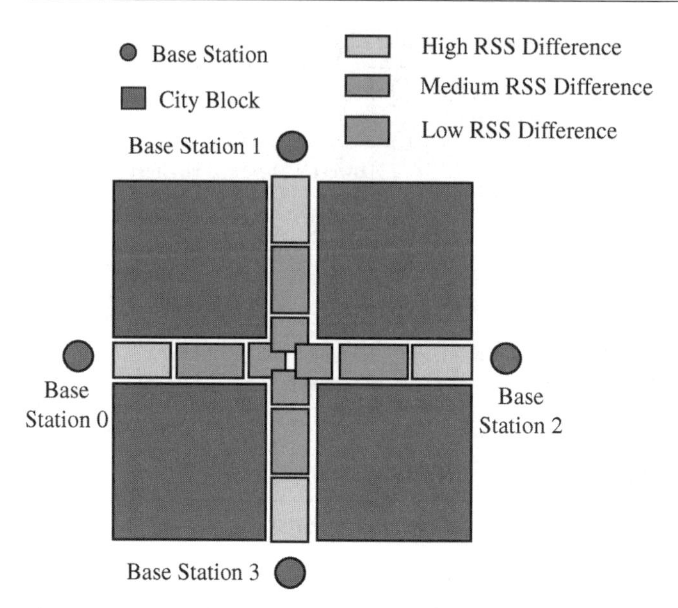

Fig. 6.2.25. Handoff Situations in a Microcellular System

difference keeps increasing, and this situation is similar to a LOS handoff scenario. Again, it is important to use high $RSS_{hysteresis}$ and low $RSS_{threshold}$ to reduce the ping-pong effect. Based on the knowledge of such propagation characteristics of a microcellular environment, a fuzzy logic rule base is created as shown in Table 6.2.10. *Current RSS Difference* and *Previous RSS Difference* are inputs to the rule base, and $RSS_{threshold}$ and $RSS_{hysteresis}$ are the outputs of the rule base. *Current RSS Difference* is the difference in RSS from two best BS at the current sample time, and *Previous RSS Difference* is the difference in RSS from two best BS at the previous sample time. Consider Rule 1. When *Current RSS Difference* and *Previous RSS Difference* are high, the MS is close to a BS, and, hence, $RSS_{threshold}$ is made lowest and $RSS_{hysteresis}$ is made highest to prevent the ping-pong effect. On the other hand, when *Current RSS Difference* and *Previous RSS Difference* are low, the MS is equally far from the BSs, and, hence, $RSS_{threshold}$ is made highest and $RSS_{hysteresis}$ is made lowest to make a fast handoff in potential NLOS handoff situations. The idea of using high hysteresis for LOS handoff situations and low hysteresis for NLOS handoff situations, conforms with the primary handoff objectives.

The philosophy behind the design of this rule base is explained next. When all the inputs suggest a change in the handoff parameters in the same direction (i.e., either increase or decrease), the parameters are changed to the maximum extent. For example, consider Rule 9. "High" SIR_c indicates that the quality of the current link is very good. "Low" Tr_d indicates that there are very few users in the current cell. "High" MS velocity indicates that the MS is moving toward the current BS at a high speed. All these secondary FLS inputs suggest that handoff

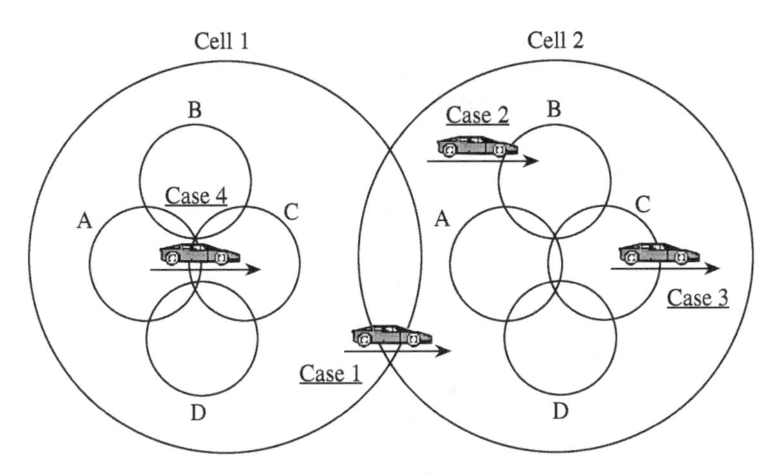

Fig. 6.2.26. Generic Handoff Scenarios in a Macrocell-Microcell Overlay System

from the current BS be discouraged. Hence, the FLS makes $\Delta RSS_{threshold}$ "Lowest" (making overall $RSS_{threshold}$ smaller) and $\Delta RSS_{hysteresis}$ "Highest" (making overall $RSS_{hysteresis}$ large).

6.2.6.2 An Overlay System

In an overlay cellular system, a large macrocell overlays clusters of small microcells. The overlay system is designed to balance system capacity and cost. The process of handoff is more complicated in an overlay system than in a pure macrocellular or microcellular system. First, handoff for an overlay system is described. The application of a neurofuzzy system to an overlay handoff is discussed later.

Figure 6.2.26 shows a macrocell-microcell overlay system and four generic handoff scenarios. Cell 1 and Cell 2 are macrocells that overlay clusters of microcells. A cluster of microcells consists of cells A, B, C, and D. Four generic types of handoffs are *macrocell to macrocell*, *macrocell to microcell*, *microcell to microcell*, and *microcell to macrocell*. When an MS travels from one macrocell to another (Case 1), a macrocell to macrocell handoff occurs. This type of handoff typically occurs near the macrocell borders. When an MS enters a microcell from a macrocell (Case 2), a macrocell to microcell handoff occurs. Even though the signal strength from the macrocell is usually greater than the signal strength from the microcell (due to relatively higher macrocell BS transmit power), this type of handoff is made to utilize the microcell connection that is economical, power efficient, and spectrally efficient and that generates less interference. When an MS leaves a microcell (Case 3) and enters a macrocell, a microcell to macrocell handoff is made to save the call since the microcell can no longer provide a good quality communication link to the MS. When an MS travels from one microcell to another (Case 4), a microcell to microcell handoff is made to reduce power requirements and get a better quality signal. An overlay system achieves a balance between maximizing the system capacity and minimizing the cost. Microcells cover areas with high traffic intensities while macrocells provide wide area coverage. Small size cells can provide very high capacity, but lead to an expensive system due to infrastructure costs. An overlay system is more complex than a pure macrocell or microcell system. Important considerations for designing efficient handoff algorithms for overlay systems, are outlined below:

Service. An attempt should be made to maximize the microcell usage, since the microcell connection has a low cost due to the better frequency reuse factor and low transmit power requirements. However, far regions should be served by macrocells for a better quality communication link. Microcell overflow traffic should be handled by macrocells.

Mobility. High speed vehicles should be connected to macrocells to reduce the handoff rate and the associated network load. This will also enable a handoff algorithm to perform uniformly well for line of sight (LOS) and non line of sight (NLOS) handoffs in microcells. The handoff parameters can now be optimized for LOS handoff situations, since the requirement of a very fast handoff for a typical NLOS situation can be easily avoided by connecting high speed users to macrocells.

Propagation Environment. In an overlay system, a user experiences both macrocell and microcell environments as the user travels across macrocells and microcells. Different fading intensities (e.g., low and high) exist in macrocells and microcells.

Resource Management. Resource management in an overlay system is a difficult task. One of the crucial issues is an optimum distribution of channels between macrocells and microcells.

Specific Handoff Requirements. A handoff algorithm should perform uniformly well in the four generic handoff situations described earlier. The algorithm should attempt to achieve the goal of an overlay system (i.e., the balance between the microcell usage and network load). The algorithm should balance traffic in the cells.

Figure 6.2.27 shows the block diagram of the proposed generic algorithm for an overlay system. The FLS shown as part of Fig. 6.2.27 can be replaced by a neurofuzzy system using the neural encoding approach of Section 6.2.4, or the pattern classification based approach of Section 6.2.5. Handoff criteria are averaged according to a velocity adaptive averaging mechanism. The *Handoff Initiation Mechanism* compares the *RSS* of the current BS (RSS_c) to a fixed threshold (RSS_{th}). If the current BS cannot provide sufficient RSS, the handoff process is initiated. A fuzzy logic system (FLS) serves as a *Cell Selection Mechanism* to determine the best macrocell "*x*" and the best microcell "*y*" as potential handoff candidates.

The inputs of the FLS are RSS_{micro} (*RSS* from the microcell BS), Tr_d (traffic difference or the difference in the number of calls in the microcell and in the macrocell), and *MS Velocity*. The output of the FLS is the *Cell Selection Index*, which indicates the degree to which a given user belongs to a microcell or a macrocell. A high value of the Call Selection Index indicates that the MS should

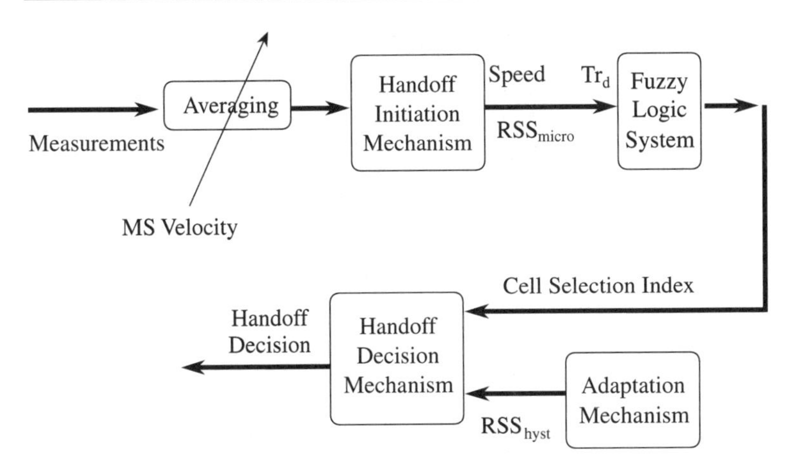

Fig. 6.2.27. Block diagram of a Generic Overlay Handoff Algorithm

be connected to a microcell, and a low value of the Call Selection Index indicates that the MS should be connected to a macrocell. Table 6.2.12 shows the entire rule base. Assume that RSS_{micro} is "Low" Tr_d is "High" and *MS velocity* is "High". These conditions indicate that the call should be encouraged to connect to a macrocell as much as possible; this is rule number nineteen. To indicate the highest degree of confidence for a macrocell connection, cell selection index is made lowest. Now consider rule nine. Since RSS_{micro} is "High" Tr_d is "Low" and *MS velocity* is "Low" the call should be encouraged to connect to a microcell as much as possible. This is done by making the cell selection index highest. If the output of the FLS is greater than zero, a microcell is selected for the communication with the MS; otherwise, a macrocell is selected.

The parameter adaptation for the proposed generic overlay algorithm is explained next.

Macrocell to Macrocell Handoff. If the currently serving cell is a macrocell and the candidate cell is also a macrocell, an incremental hysteresis $\Delta hyst_{macro}$ is found as shown in Fig. 6.2.27. If the MS is moving toward both the current and the candidate BSs or moving away from the BSs, a fixed hysteresis h_{macro} is used. If the MS is moving toward the current BS and moving away from the candidate BS, a handoff is discouraged by increasing the hysteresis value by an amount Δh_{macro}. However, if the MS is moving toward the candidate BS and away from the current BS, handoff is encouraged by decreasing the hysteresis value by an amount Δh_{macro}. The overall adaptive hysteresis is $RSS_{effective} = RSS_{hyst} + \Delta hyst_{macro}$ where RSS_{hyst} is the nominal value of the RSS hysteresis. This adaptation of hysteresis is based on direction biasing and helps reduce the ping-pong effect between two macrocells. A handoff is made to the macrocell "x" if RSS_x exceeds RSS_c by an amount $RSS_{effective}$.

Macrocell to Microcell Handoff. If the currently serving cell is a macrocell and the candidate cell is a microcell, an incremental hysteresis is taken as

Table 6.2.12. Rule Base for Cell Selection

Rule Number	RSS_{micro}	Tr_d	*MS Velocity*	Cell Selection Index
1	High	High	High	Low
2	High	High	Normal	Normal
3	High	High	Low	High
4	High	Normal	High	Normal
5	High	Normal	Normal	High
6	High	Normal	Low	Higher
7	High	Low	High	High
8	High	Low	Normal	Higher
9	High	Low	Low	Highest
10	Normal	High	High	Lower
11	Normal	High	Normal	Low
12	Normal	High	Low	Normal
13	Normal	Normal	High	Low
14	Normal	Normal	Normal	Normal
15	Normal	Normal	Low	High
16	Normal	Low	High	Normal
17	Normal	Low	Normal	High
18	Normal	Low	Low	Higher
19	Low	High	High	Lowest
20	Low	High	Normal	Lower
21	Low	High	Low	Low
22	Low	Normal	High	Lower
23	Low	Normal	Normal	Low
24	Low	Normal	Low	Normal
25	Low	Low	High	Low
26	Low	Low	Normal	Normal
27	Low	Low	Low	High

Δh_{micro}. The overall adaptive hysteresis is $RSS_{effective} = RSS_{hyst} - \Delta h_{micro}$. This adaptation of hysteresis is based on the idea of encouraging handoffs to the microcells to increase microcell usage. A handoff is made to the microcell "y" if RSS_y exceeds RSS_{th} by an amount $RSS_{effective}$.

Microcell to Microcell Handoff. If the currently serving cell is a microcell and the candidate cell is also a microcell, an incremental hysteresis $\Delta hyst_{micro}$ is found as shown in Fig. 6.2.29. If the MS is moving toward the current BS and away from the candidate BS, a handoff is discouraged by increasing the hysteresis value by an amount Δh_{micro}. However, if the MS is moving toward the candidate BS and away from the current BS, a handoff is encouraged by decreasing the hysteresis value by an amount Δh_{micro}. The overall adaptive hysteresis is $RSS_{effective} = RSS_{hyst} + \Delta hyst_{micro}$. This adaptation of hysteresis is based on direction biasing and helps reduce the ping-pong effect between two microcells. A handoff is made to the microcell "y" if RSS_y exceeds RSS_c by an amount $RSS_{effective}$.

Microcell to Macrocell Handoff. If the currently serving cell is a microcell and the candidate cell is a macrocell, an incremental hysteresis is taken as Δh_{macro}. The overall hysteresis is $RSS_{effective} = RSS_{hyst} - \Delta h_{macro}$. This adaptation of hystere-

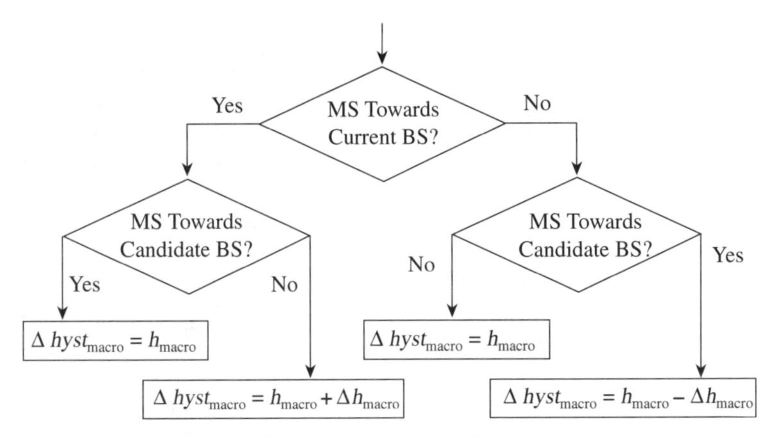

Fig. 6.2.28. Adaptive Handoff Parameters for a Current Macrocell Connection

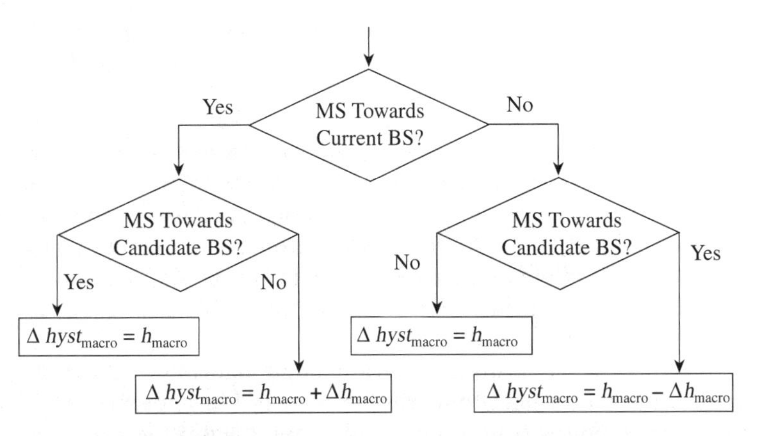

Fig. 6.2.29. Adaptive Handoff Parameters for a Current Microcell Connection

sis is based on the idea of encouraging handoffs to the macrocells from microcells to save the call since the microcell coverage area is limited and since the call may be dropped if handoff is not made early enough. A handoff is made to the macrocell "x" if RSS_x exceeds RSS_{th} by an amount $RSS_{effective}$.

6.2.6.3 A System with Soft Handoff Support

High performance soft handoff algorithms need to consider several aspects specific to soft handoff situations. A good soft handoff algorithm, attempts to achieve a balance between the quality of the signal and the associated cost. In general, the greater the number of BSs involved in soft handoff, the better the quality of the signal due to increased diversity gain and the higher the degree to which the network resources are consumed. Important considerations for designing soft handoff algorithms are described below.

Cellular System Layouts. Soft handoff may be implemented in a macrocellular, microcellular, or overlay system. Traffic, mobility, and propagation environment in these distinct system deployment scenarios, should be considered.

Primary Soft Handoff Requirements. The handoff algorithm should try to maximize signal quality and minimize the number of BSs involved in soft handoff. The number of Active Set (i.e., the set that contains the list of BSs in soft handoff) updates should be minimized, to reduce the network load.

Secondary Soft Handoff Requirements. The algorithm should correspond to vehicle speed. The algorithm should attempt to balance traffic.

Reference [43] analyzes performance of IS-95 soft handoff algorithm. A *simplified* soft handoff mechanism of IS-95 adapted from [43], is briefly summarized here. The IS-95 system implements MAHO. The MS measures pilot (E_c/I_o) (ratio of the chip energy and interference energy) of surrounding cells (or sectors). The MS communicates with members of the Active Set (AS) on both forward and reverse link. In other words, members of the AS transmit the same information signal to the MS, and receive the same information signal from the MS. On the forward link, the RAKE fingers of the MS demodulates signals from a maximum of three strongest AS members. On the reverse link, the signals received at the AS sectors of the same cell are combined using maximal ratio combining, while the best signal is selected form multiple BSs that have their sectors as members of the AS (an example of the implementation of selection diversity). If the MS measures a pilot that has (E_c/I_o) greater than a threshold, T_ADD, it is made a member of the AS. If the pilot strength of an AS member remains below a threshold, T_DROP, for T_TDROP seconds, it is removed from the AS. Thus, sectors with pilots that are received with strong (E_c/I_o) at the MS form the AS. Such SHO management ensures that reasonably strong signals, that can provide diversity gain, are utilized.

Figure 6.2.30 shows the block diagram of a handoff algorithm with traffic and mobility adaptation. This algorithm uses the primary FLS to provide adaptive handoff parameters to the conventional soft handoff algorithm and a secondary FLS to meet the secondary handoff goals. The input to the primary FLS, is the RSS at the MS, at the previous sample time. The outputs of the primary FLS are adaptive handoff parameters, $RSS_{threshold}$ and $RSS_{hysteresis}$. Inputs to the secondary FLS are Tr (traffic or number of users in a cell) and MS velocity (component of the velocity toward the BS). If the MS is moving toward a BS, the velocity is considered positive, and if the MS is moving away from the BS, the velocity is considered negative. The output of the secondary FLS is a preselection index, that indicates the degree to which the BS is a good candidate for the Active Set, if traffic and mobility were the only considerations.

The complete fuzzy logic rule bases for the primary and the secondary FLSs, are shown in Table 6.2.13 and Table 6.2.14.

The design philosophy for the primary FLS is explained next. When an MS is relatively close to a BS, the *RSS* at the MS is very high, and there is no need to initiate soft handoff since the current communication link already has sufficiently high RSS. To discourage any BS from becoming a member of the current Active Set (that consists of only the current BS), $RSS_{threshold}$ is set very high and

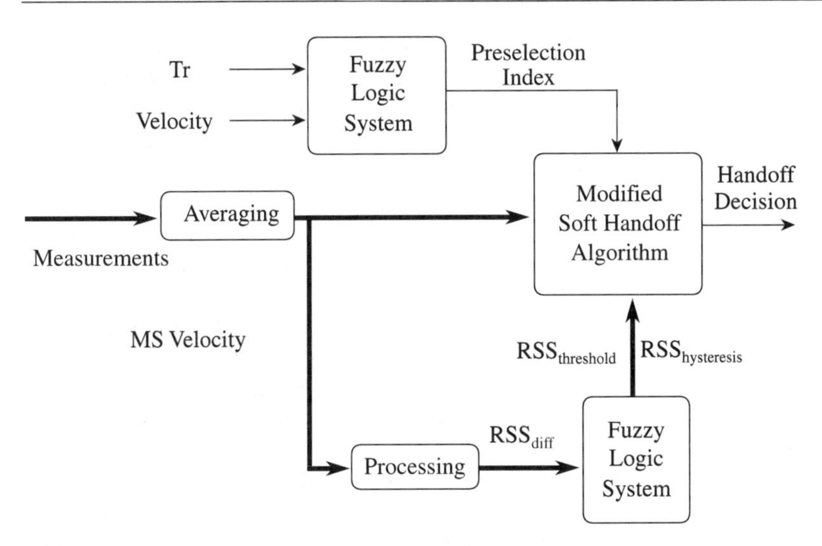

Fig. 6.2.30. A Soft Handoff Algorithm with Traffic and Mobility Adaptation

Table 6.2.13. Basic Soft Handoff Rule Base

Rule Number	RSS	$RSS_{threshold}$	$RSS_{hysteresis}$
1	Very High	Very High	Very Low
2	High	High	Low
3	Medium	Medium	Medium
4	Low	Low	High
5	Very Low	Very Low	Very High

Table 6.2.14. Rule Base for Traffic and Mobility Adaptation

Rule Number	Traffic (Tr)	Velocity	Preselection Index
1	High	High	Medium
2	High	Medium	Low
3	High	Low	Very Low
4	Medium	High	High
5	Medium	Medium	Medium
6	Medium	Low	Low
7	Low	High	Very High
8	Low	Medium	High
9	Low	Low	Medium

$RSS_{hysteresis}$ is set very low. When the MS is far from the neighboring BSs and the MS is connected to only one BS, RSS is very low, and, hence, soft handoff should be initiated to increase overall signal strength. Soft handoff can now be encouraged by setting $RSS_{threshold}$ very low and $RSS_{hysteresis}$ very high. When the MS is far from the neighboring BSs but the MS is connected to more than one BS, RSS may be high or low depending upon the number of BSs and the quality of existing MS-BS connections. If RSS is low, the number of BSs involved in soft handoff can be kept the same by using nominal values of $RSS_{threshold}$ and $RSS_{hysteresis}$, or it can be increased by setting $RSS_{threshold}$ low and $RSS_{hysteresis}$ high depending upon the overall quality of the existing links. If RSS is high, the number of BSs involved in soft handoff can be kept the same by using nominal values of $RSS_{threshold}$ and $RSS_{hysteresis}$ or, it can be decreased by setting $RSS_{threshold}$ low and $RSS_{hysteresis}$ high. Based on such knowledge, a fuzzy logic rule base shown in Table 13 is created.

Consider Rule 7 of the secondary FLS. The traffic in the cell is "Low" (meaning that there are very few users in the cell), and the velocity is "High" (meaning that the MS is moving at a high velocity toward the BS). This situation suggests that the BS under consideration should be encouraged to the maximum extent, to become a member of the Active Set. This would help achieve traffic balancing and reduce the number of Active Set updates.

Now consider Rule 3. The traffic in the cell is "High" (meaning that there are many users in the cell), and the velocity is "Low" (meaning that the MS is moving at a high velocity away from the BS). This situation suggests that the BS under consideration should be discouraged to the maximum extent from becoming a member of the Active Set. This would again help achieve traffic balancing and reduce the number of Active Set updates. The modified soft handoff algorithm is similar to the conventional soft handoff algorithm, but it selects the BSs according to the priority suggested by the preselection index. Note that the primary objectives have not been compromised, since the adaptive handoff parameters are supplied by the primary FLS.

6.2.7 Conclusion

A handoff algorithm is one of the critical design components of a cellular communication system. An efficient handoff algorithm leads to a high-performance system and helps increase the capacity and the quality of service of a cellular system. Conventional handoff algorithms cannot effectively utilize several pieces of knowledge about the behavior of a cellular environment. This chapter exploits two generic neurofuzzy approaches, neural encoding approach and pattern recognition approach, to design a high-performance and low-complexity adaptive handoff algorithm. Several specific cellular deployment scenarios such as macrocells, microcells, overlay cells, and systems with soft handoff support are considered.

References

[1] Liodakis G, Stavroulakis P (1994) A Novel Approach in Handover Initiation for Microcellular Systems, *Proc. 44th IEEE VTC*, pp. 1820–1823.

[2] Lee W C Y (1993) Mobile Communications Design Fundamentals, 2nd ed. John Wiley & Sons Inc.

[3] Pollini G P (1996) Trends in Handover Design, IEEE Communications Magazine, pp. 82–90.

[4] Anagnostou M E, Manos G C (1994) Handover related performance of mobile communication networks, Proc. 44th IEEE VTC, pp. 111–114.

[5] Frech E A, Mesquida C L (1989) Cellular Models and Handoff Criteria, Proc. 39th IEEE VTC, pp. 128–135.

[6] Mende W R (1990) Evaluation of a Proposed Handover Algorithm for the GSM Cellular System, Proc. 40th IEEE VTC, pp. 264–269.

[7] Munoz-Rodriguez D, Cattermole K W (1987) Multicriteria for Handoff in Cellular Mobile Radio, IEE Proc., Vol. 134, pp. 85–88.

[8] Senarath G H, Everitt D (1994) Comparison of Alternative Handoff Strategies for Microcellular Mobile Communication Systems, Proc. 44th IEEE VTC, pp. 1465–1469.

[9] Kanai T, Furuya Y (1988) A Handoff Control Process for Microcellular Systems, Proc. 38th IEEE VTC, pp. 170–175.

[10] Falciasecca G, Frullone M, Riva G, Serra A M (1989) Comparison of Different Handover Strategies for High Capacity Cellular Mobile Radio Systems, Proc. 39th IEEE VTC, pp. 122–127.

[11] Corazza G E, Giancristofaro D, Santucci F (1994) Characterization of Handover Initialization in Cellular Mobile Radio Networks, Proc. 44th IEEE VTC, pp. 1869–1872.

[12] Rappaport S S (1993) Blocking, Handoff and Traffic Performance for Cellular Communication Systems with Mixed Platforms," IEE Proceedings-I, Vol. 140.

[13] Ott G D (1977) Vehicle Location in Cellular Mobile Radio Systems, Vol. VT-26, No. 1, IEEE Trans. Veh. Tech., Vol. VT-26, No. 1, pp. 43–36.

[14] Lee W C Y (1995) *Mobile Cellular Telecommunications*, 2nd *ed.* McGraw Hill.

[15] Holtzman J M, Sampath A (1995) Adaptive Averaging Methodology for Handoffs in Cellular Systems, IEEE Trans. on Veh. Tech., Vol. 44, No. 1, pp. 59–66.

[16] Sampath A, Holtzman J M (1994) Adaptive Handoffs Through Estimation of Fading Parameters, Proc. ICC.

[17] Dassanayake P (1993) Effects of Measurement Sample on Performance of GSM Handover Algorithm, Electronic Letters, Vol. 29, pp. 1127–1128.

[18] Dassanayake P (1994) Dynamic Adjustment of Propagation Dependent Parameters in Handover Algorithms, Proc. 44th IEEE VTC, pp. 73–76.

[19] Rolle G (1986) The Mobile Telephone System C 450- a First Step Towards Digital, Proc. Second Nordic Seminar.

[20] Chuah C N, Yates R D, Goodman D J (1995) Integrated Dynamic Radio Resource Management, Proc. 45th IEEE VTC, pp. 584–88.

[21] Austin M D, Stuber G L (1994) Velocity Adaptive Handoff Algorithms for Microcellular Systems, IEEE Trans. Veh. Tech., Vol. 43, No. 3, pp. 549–561.

[22] Kawabata K, Nakamura T, Fukuda E (1994) Estimating Velocity Using Diversity Reception, Proc. 44th IEEE VTC, pp. 371–74.

[23] Austin M D, Stuber G L (1994) Directed Biased Handoff Algorithm for Urban Microcells, Proc. 44th IEEE VTC, pp. 101–5.

[24] Chuah C N, Yates R D (1995) Evaluation of a Minimum Power Handoff Algorithm, Proc. IEEE PIMRC, pp. 814–818.

[25] Chia S T S, Warburton R J (1990) Handover Criteria for a City Microcellular Radio Systems, Proc 40th IEEE VTC, pp. 276–281.

[26] Asawa M, Stark W E (1994) A Framework for Optimal Scheduling of Handoffs in Wireless Networks," Proc. IEEE Globecom," pp. 1669–1673.

[27] Kelly O E, Veeravalli V V (1995) A Locally Optimal Handoff Algorithm, Proc. IEEE PIMRC, pp. 809–813.
[28] Rezaiifar R, Makowski A M, Kumar S (1995) Optimal Control of Handoffs in Wireless Networks, Proc. 45th IEEE VTC, pp. 887–91.
[29] Kapoor V, Edwards G, Sankar R (1994) Handoff Criteria for Personal Communication Networks, pp. 1297–1301, Proc. ICC.
[30] Munoz-Rodriguez D, Moreno-Cadenas J A, Ruiz-Sanchez M C, Gomez-Casaneda F (1992) Neural Supported Handoff Methodology in Microcellular Systems, Proc. 42nd IEEE VTC, Vol. 1, pp. 431–434.
[31] Kinoshita Y, Itoh T (1993) Performance Analysis of a New Fuzzy Handoff Algorithm by an Indoor Propagation Simulator, Proc. 43rd IEEE VTC, pp. 241–245.
[32] Maturino-Lozoya H, Munoz-Rodriguez D, Tawfik H (1994) Pattern Recognition Techniques in Handoff and Service Area Determination, Proc. 44th IEEE VTC, Vol. 1, pp. 96–100.
[33] Munoz-Rodriguez D (1987) Handoff Procedure for Fuzzy Defined Radio Cells, Proc. 37th IEEE VTC, pp. 38–44.
[34] Vijayan R, Holtzman J M (1993) Sensitivity of Handoff Algorithms to Variations in the Propagation Environment, Proc. 2nd IEEE Intl. Conf. on Universal Personal Communications.
[35] Zhang N, Holtzman J (1994) Analysis of Handoff Algorithms using both Absolute and Relative Measurements, Proc. 44th IEEE VTC, Vol. 1, pp. 82–86.
[36] Viterbi A J, Viterbi A J, Gilhousen K S, Zehavi E (1994) Soft Handoff Extends CDMA Cell Coverage and Increases Reverse Link Capacity, IEEE JSAC, Vol. 12, No. 8, pp. 1281–1288.
[37] Rappaport T S (1996) Wireless Communications, Prentice-Hall Inc.
[38] Berg J E, Bownds R, Lotse F (1992) Path Loss and Fading Models for Microcells at 900 MHz, Proc. 42nd IEEE VTC, pp. 666–671.
[39] Gudmundson M (1991) Correlation Model for Shadow Fading in Mobile Radio Systems, Electronic Letters, Vol. 27, No. 23, pp. 2145–2146.
[40] Tripathi N D (1997) Generic Adaptive Handoff Algorithms Using Fuzzy Logic and Neural Networks, Ph. D. Dissertation, Virginia Tech.
[41] Tripathi N D, Reed J H, VanLandingham H F (2001) Radio Resource Management in Cellular Systems, Kluwer Academic Publishers.
[42] Mamdani E H (1974) Applications of Fuzzy Algorithms for Simple Dynamic Plant, Proc. IEE, Vol. 121, No. 12, pp. 1585–1588.
[43] Chheda A (1999) A Performance Comparison of the DS-CDMA IS-95B and IS-95A Soft Handoff Algorithms, IEEE VTC.

6.3 Handoff Based Quality of Service Control in CDMA Systems Using Neuro-Fuzzy Techniques

Bongkarn Homnan and Watit Benjapolakul

6.3.1 Introduction

There are many approaches for controlling Quality of Service (QoS). The QoS in this chapter is divided into two aspects, as follows:

Trunk-Resource Efficiency: the considered parameters include Trunk-Resource Efficiency (*TRE*), new call blocking probability (P_B), and handoff call blocking probability (P_{HO}).

Call quality: the forward-link bit energy to noise power spectral density ratio (E_b/N_o) and outage probability (P_{out}), are used to evaluate the system performance in this aspect. Note that P_{out} is defined as the probability that E_b/N_o is less than the assigned threshold [1, 2].

In this chapter, adapted add threshold (*T_ADD*) and adapted drop threshold (*T_DROP*) in SHO process are emphasized for controlling *TRE*, P_B, and P_{HO} or E_b/N_o and P_{out}.

Normally, handoff is an essential mechanism for maintaining the QoS during high fading environment or during changing serving Base Station (BS). Handoff, can be considered as an aspect of channel assignment depending on the assigned handoff thresholds in each BS. If the handoff thresholds are changed dynamically, the cellular system, of course, behaves as a dynamic channel assignment system. The concept of dynamic handoff thresholds can be applied to SHO, which is a kind of handoff by adaptation of *T_ADD* and *T_DROP*. SHO is used in Code Division Multiple Access (CDMA) mobile communication system. It is a diversity handoff, because each Mobile Station (MS) in SHO Area (SHA) uses two or more traffic channels from different BSs at the same time. Thus, when TRE is considered, SHO is not efficient due to higher P_B and $P_{HO,}$ when compared with FDMA (Frequency Division Multiple Access) or TDMA (Time Division Multiple Access) mobile communication systems which use only one channel during handoff (hard handoff, HHO).

Many SHO algorithms have been proposed for higher system performance as follows:

Chen (1994) [3], proposed the adaptive traffic load shedding scheme for reducing P_B and P_{HO} at high traffic load. This scheme allows heavily loaded BSs to dynamically shrink their coverage areas, while less loaded adjacent BSs increase their coverage areas to support extra traffic by transmitting higher power to those MSs that handoff from the adjacent heavily loaded BSs. The disadvantage is that the BS has to request the adjacent BSs to increase their coverage areas before it can reduce its coverage area. The BS cannot increase or reduce its pilot strength independently and it is not easy to rapidly react to the variations of traffic load [4, 5].

Hwang et al. (1997) [6], proposed to allow the handoff thresholds (*T_ADD* and *T_DROP*) to vary dynamically according to the traffic density in each cell. Only two fixed values of each threshold were assigned. The comparison of system performance with the conventional IS-95A SHO [3] observed only P_{out}, thus it was not sufficient in comparing all aspects of system performance which can be described by many indicators.

Jeon et al. (1997) [4], proposed a new channel assignment scheme for reducing P_B and P_{HO} in BS by increasing the value of *T_DROP* when the traffic channel is not available. The disadvantage of this method is that the BS must calculate mean and variance of the total received signal power. This is not practical in a real system. Moreover, they proposed to use only two fixed values of *T_DROP*. Therefore, it was a quite coarse adaptation of threshold, in following wide range of offered traffic load.

Worley and Takawira (1998) [5], proposed to have the upper and lower thresholds of the MS transmitted power define the conditions, in deciding when to handoff. Generally, the BS has no information about the MS transmitted power, so in this method, the MS must send this information to the BS to compare it with the defined thresholds. Because of the fact that the frequency ranges used in forward link and reverse link are not the same, when the pilot signal strengths of neighboring cells are stronger than *T_ADD* but the MS transmitted power is weaker than its lower threshold power, SHO will not occur. On the contrary, when the pilot signal strengths of neighboring cells are weaker than *T_DROP* but the MS transmitted power is still strong, MS still uses channel from that BS. This is inefficient in comparison to IS-95A SHO.

IS-95B/cdma2000 SHO [7, 8] utilizes three additional parameters: *SOFT_SLOPE*, *ADD_INTERCEPT*, and *DROP_INTERCEPT* for adjusting threshold dynamically. In this chapter, IS-95B/cdma2000 SHO will also be compared with the proposed method. Note that the IS-95B/cdma2000 SHO algorithm in this chapter is partly different from that considered in [9], but conforms to TIA/EIA/IS-95B [7] and TIA/EIA/IS-2000-5 [8]. In addition, in this chapter, SHO algorithms are based on statistical modeling with Poisson arrival process and exponential holding time process, while Chheda's proposed simulations [9] are not. Therefore, a better view of QoS-controlling SHO comparison can be expected.

There are some researchers applying fuzzy logic theory in handoff process, such as Kinoshita (1992, 1995) [10, 11]. They applied fuzzy inference in learning to know the cell boundary [10] and increasing the number of inference rules for softer decision [11], but emphasized on HHO in indoor area. Homnan and Benjapolakul (1998) [12], proposed the fuzzy inference system (FIS) by using the signal strength which MS receives and the distance between MS and BS as the inputs, while the output is the defined value for deciding handoff. This work was also applied for HHO.

Homnan et al. (2000), proposed the concept of dynamic handoff thresholds by using fuzzy inference system-based soft handoff (FIS SHO) [13–15], which consisted of inference rules of human-oriented information. Moreover, the number of the remaining channels (CH_{rm}) and the number of pilots in active set (no_{BS}) were, for the first time, introduced inputs. Thus, this method, which will be referred to as "the 2-input FIS SHO" throughout this chapter, is different from those of [10–12]. The advantages of using no_{BS} and CH_{rm}, were that they could adapt to SHO thresholds properly because no_{BS} represents the number of pilots in active set (AS) directly and infers the pilot signal strength, the distance between MS and BS, and E_b/N_o MS receives, while CH_{rm} informs the remaining status of traffic channels in each BS and infers traffic load and interference in each BS. This FIS SHO method can support more users than the conventional SHO methods (IS-95A SHO and IS-95B/cdma2000 SHO). In other words, it gives higher *TRE*. Unfortunately, at high traffic load, it gives lower E_b/N_o. However, the E_b/N_o values are still acceptable.

In this chapter, two new practical methods applied to the conventional SHO based on Simple Step Control (SSC) method and neuro-fuzzy inference system with gradient descent (GD) method, referred to as N-FIS&GD SHO are considered. These methods utilize a trade off between *TRE* (P_B, P_{HO}) and E_b/N_o (P_{out}) in case *TRE* or E_b/N_o is not guaranteed in the previous proposed method of FIS SHO [13–15]. Therefore, a proper input, i.e., E_b/N_o is introduced into the control plant to solve this problem.

The motivation to propose our algorithm is based on the following considerations. SSC is a simple method that can be used to control a parameter, by letting that parameter swing around the threshold. This method is similar to the step power control algorithm [1, 7, 8, 16–18]. The advantage of SSC is that its complexity is low, because it only compares the controlled parameter with the required value in order to give the new appropriate controlled value. For the combination of FIS and GD (FIS&GD), FIS&GD method was previously proposed to control call blocking probability in hierarchical cellular systems [19]. However, the N-FIS&GD SHO method in this chapter is proposed to control *TRE* and E_b/N_o.

When QoS can be controlled, it means that the system performance or SHO performance can be controlled in the direction of requirements. In other words, *TRE* (P_B, P_{HO}) or E_b/N_o (P_{out}), can be improved depending on the requirements. In practice, all of the proposed methods: SSC SHOs (*TRE*-controlling SHO based on SSC, E_b/N_o-controlling SHO based on SSC), the 3-input N-FIS SHO proposed in this chapter by adding another input "E_b/N_o" to the 2-input FIS SHO (E_b/N_o-controlling SHO based on N-FIS) and N-FIS&GD SHOs (*TRE*-controlling SHO based on N-FIS&GD, E_b/N_o-controlling SHO based on N-FIS&GD), can be implemented at base station controller solely by changing software and increasing the size of the memories.

This chapter is divided into 7 sections. Section 6.3.2 describes the classification of the problems and performance indicators. Section 6.3.3 describes an overview of IS-95A and IS-95B/cdma2000 SHOs. Section 6.3.4 describes SSC, while section 6.3.5 describes FIS SHO and N-FIS&GD SHO. Section 6.3.6 describes the system model, computer simulation and results. Conclusions are presented in the last section.

6.3.2 Classification of the Problems and Performance Indicators

The problems are classified into two main groups as follows.

I) *TRE*-controlling SHO
- **a) *TRE*-controlling SHO based on SSC**
- **b) *TRE*-controlling SHO based on N-FIS&GD**

II) E_b/N_o-controlling SHO
- **a) E_b/N_o-controlling SHO based on SSC**
- **b) E_b/N_o-controlling SHO based on FIS**
- **c) E_b/N_o-controlling SHO based on N-FIS&GD**

Thus, there are 5 cases for consideration, as outlined above. Note that *TRE*-controlling SHO based on FIS is not studied, because the 2-input (CH_{rm}, no_{BS}) FIS SHO – which is previously proposed in [13–15] – aimed to adjust T_ADD and

T_DROP in order to support more users especially at high traffic load. For this reason, *TRE* is higher according to higher traffic load. In addition, CH_{rm} and no_{BS} are not appropriate inputs to the FIS for adaptation of *T_ADD* and *T_DROP* in case of *TRE* control because the value of *TRE* is inversely proportional to the expected value of no_{BS} because $TRE = 1/no_{BS}$. So, it seems contradictory that no_{BS} is used for controlling itself by adaptation of *T_ADD* and *T_DROP* derived from CH_{rm} and no_{BS}. In this case, it is difficult to control *TRE*. In order to control *TRE*, the Gradient method (GD) is introduced to be used with N-FIS, to produce a controlled parameter which may be different from the instantaneous value in the system. This difference term is used in the *TRE*-controlling process. However, E_b/N_o-controlling SHO can use only FIS (case II.b) for controlling E_b/N_o, by adding E_b/N_o itself as one of the inputs, because when *T_ADD* and *T_DROP* are lower, the system gives higher E_b/N_o. In addition, E_b/N_o is normally controlled by the power control process. The value of E_b/N_o is within a small range, which differs from that of *TRE* derived from only FIS, which is within a larger range (0–100%).

The performance indicators are categorized into two types [16].

Call quality indicator. Average forward-link E_b/N_o and P_{out} for a given system load are selected for this issue. The values of P_{out} correspond to those of E_b/N_o, because P_{out} is calculated from E_b/N_o, as shown in Eq. (6.3.1) [2].

$$P_{out} = \Pr\left(\frac{E_b}{N_o} < 5dB\right) \tag{6.3.1}$$

Good resource allocation indicators. There are 5 parameters as follows:

T_c — *carried traffic:* the traffic that system can support.

P_B — *new call blocking probability:* the probability that a new call is blocked.

P_{HO} — *handoff call blocking probability:* the probability that all channels are occupied in the handoff target cell during a handoff procedure.

NO_{BS} — *the expected number of pilots in the AS:* NO_{BS} can be found by averaging no_{BS}. This is a measure of system resource utilization.

TRE — *Trunk-Resource Efficiency:* the expected system "efficiency", where efficiency is 1/(size of the AS). TRE is equal to 100% for HHO and *TRE* is less than 100% for SHO.

NO_{update} — *the expected percentage of pilot update in the AS:* a measure of network loading.

6.3.3 An Overview of IS-95A and IS-95B/cdma2000 SHOs

In an MS, there are four channel sets: the active set (AS), the candidate set (CS), the neighbor set (NS), and the remaining set (RS). An example of IS-95A SHO process, as shown in Fig. 6.3.1, can be explained as follows [17]:

1) When the pilot signal strength in the NS which MS receives, exceeds *T_ADD*, MS sends Pilot Strength Measurement Message (PSMM) to BS and transfers the pilot to the CS.

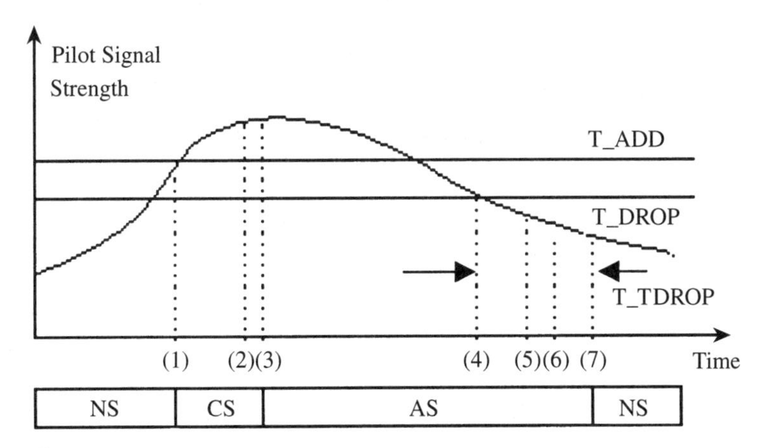

Fig. 6.3.1. IS-95A SHO process

2) BS sends Extended Handoff Direction Message (EHDM), General Handoff Direction Message (GHDM) or Universal Handoff Direction Message (UHDM) to MS.
3) MS transfers the pilot to the AS and sends Handoff Completion Message (HCM) to BS.
4) When the pilot signal strength in the AS drops below *T_DROP*, MS sends PSMM to BS.
5) BS receives the PSMM.
6) BS sends EHDM, GHDM, or UHDM, to MS.
7) MS moves the pilot from the AS into the NS and sends HCM to BS.

The SHO process is finished after the signal strength of a pilot in the AS has dropped below *T_DROP*, for longer than *T_DROP* seconds.

There are three additional parameters: *SOFT_SLOPE*, *ADD_INTERCEPT*, and *DROP_INTERCEPT* used in IS-95B/cdma2000 SHO algorithm [7,8].

In Fig. 6.3.2, P_1 is the signal strength of any pilot to be dropped from being an active pilot of an MS, and P_2 is the signal strength of any pilot to be added as an active pilot of an MS. An example of IS-95B/cdma2000 SHO process can be described as follows:

1) When the signal strength of a pilot in NS (P_2) – which MS receives – exceeds *T_ADD*, MS transfers this pilot to the CS.
2) When the signal strength of this pilot (P_2) – which is now in the CS – exceeds $[(SOFT\_SLOPE/8)*10*10\log_{10}(P_1)+ADD\_INTERCEPT/2]$,
MS sends PSMM to the BS. Note that P_1, at this point, is the signal strength of an active pilot in the AS, as shown in the Fig. 6.3.2.
3) Upon receiving EHDM, GHDM, or UHDM, MS transfers the pilot having signal strength P_2 to the AS, and sends HCM to the BS.
4) When P_1 drops below
$[(SOFT\_SLOPE/8)*10*10\log_{10}(P_2)+DROP\_INTERCEPT/2]$,
MS starts the handoff drop timer.

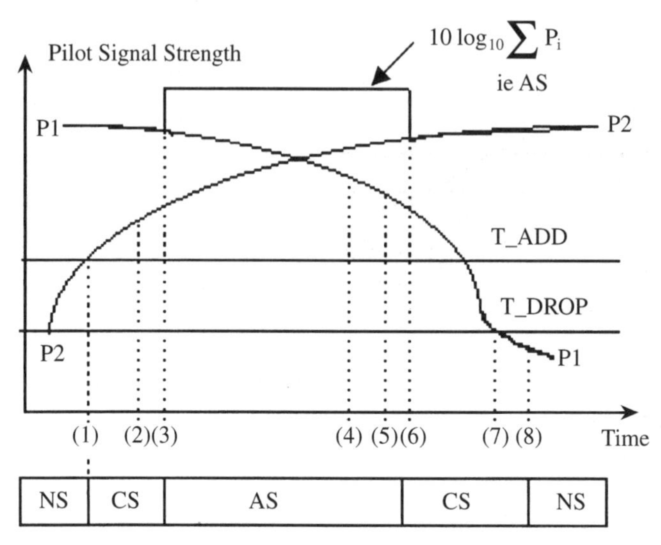

Fig. 6.3.2. IS-95B/cdma2000 SHO process

5) When the handoff drop timer expires, MS sends PSMM to the BS.
6) Upon receiving EHDM, GHDM, or UHDM, MS transfers the pilot having signal strength P_1 to the CS and sends HCM to the BS.
7) When P_1 drops below *T_DROP*, MS starts the handoff drop timer.
8) Finally, when the handoff drop timer expires, MS moves the pilot having signal strength P_1, from the CS to the NS.

6.3.4 Simple Step Control Algorithm (SSC)

There is only one input to the proposed SSC algorithm. The outputs of the algorithm are new *T_DROP* and new *T_ADD,* which are the thresholds for SHO decision.

6.3.4.1 TRE-Controlling SHO Based on SSC (case I.a)

no_{BS} is used for the input to the SSC algorithm. The algorithm can be explained step by step, as follows.

1) The requirement of TRE is set, and the approximated required no_{BS} ($no_{BS}^{required}$) can be calculated by $1/TRE(\%)*100$.
2) If no_{BS} of each MS in SHA is higher than $no_{BS}^{required}$, *T_DROP* of each MS will be set 0.5 dB higher.
3) If no_{BS} of each MS in SHA is lower than $no_{BS}^{required}$, *T_DROP* of each MS will be set 0.5 dB lower.
4) *T_ADD* of each MS is defined as *T_DROP* + 2 dB.

6.3.4.2 E_b/N_o-controlling SHO based on SSC (case II.a)

The concept is similar to that in section 6.3.4.1, but the parameters to be controlled are changed from TRE to E_b/N_o instead. The algorithm can also be explained step by step, as follows:

1) The requirement of E_b/N_o ($E_b/N_o(req)$) is set.

2) If E_b/N_o of each MS in SHA is higher than $E_b/N_o(req)$, *T_DROP* of each MS will be set 0.5 dB higher.

3) If E_b/N_o of each MS in SHA is lower than E_b/N_o (req), *T_DROP* of each MS will be set 0.5 dB lower.

4) *T_ADD* of each MS is defined as *T_DROP* + 2 dB.

6.3.5 FIS SHO and FIS&GD SHO

6.3.5.1 FIS SHO

There are three important sub-procedures in the previous proposed FIS SHO algorithm [13–15]: Fuzzification, Fuzzy Inference, and Defuzzification [21–23], as shown in Fig. 6.3.3. The inputs of the previous proposed SHO algorithm are no_{BS} and CH_{rm} of the serving BS. The outputs of the algorithm are new *T_DROP* and new *T_ADD* [13,14,15]. The inputs and output of FIS SHO can be considered as linguistic variables with the fuzzy sets as follows:

Linguistic variables	Fuzzy sets
1. no_{BS}	∈ {Low, Medium, High}
2. CH_{rm}	∈ {Low, Medium, High}
3. *T_DROP*	∈ {Low, Medium, High}

Obviously, Homnan *et al.* [13] adapted the SHO thresholds depending on the instantaneous traffic load of each BS by using CH_{rm} as one of the two inputs to the FIS SHO. In addition, the problem of pilot overuse encountered in IS-95A SHO were reduced, by using no_{BS} as the other one of the two inputs. However, P_{out} was not guaranteed.

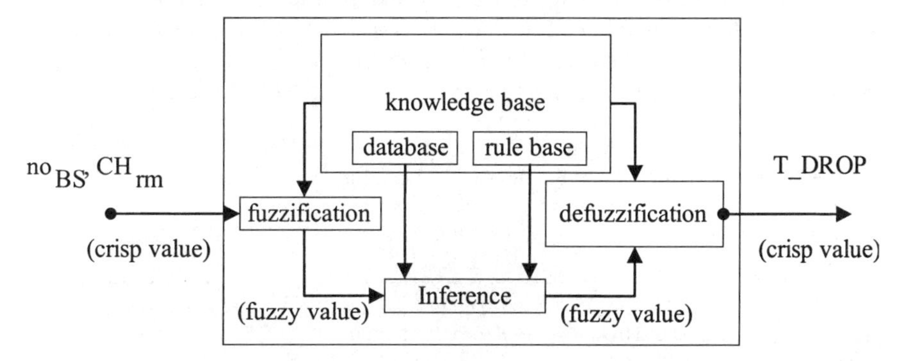

Fig. 6.3.3. FIS SHO (the 2-input FIS SHO) control plant

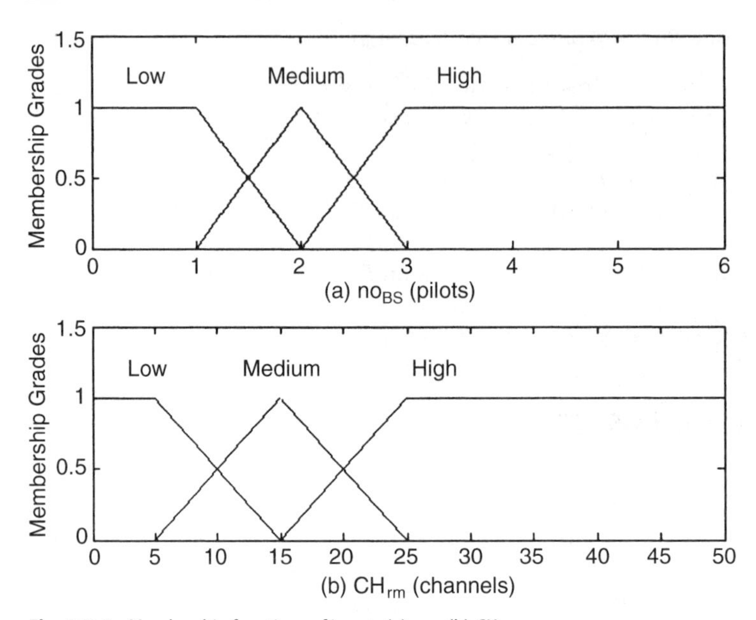

Fig. 6.3.4. Membership functions of inputs (a) no_{BS} (b) CH_{rm}

A. Fuzzification. Both of the crisp inputs are changed to be fuzzy inputs in this sub-procedure, before used in a rule base by the membership functions in Fig. 6.3.4, which are the databases in the knowledge base of FIS SHO control plant, as shown in Fig. 6.3.3.

B. Fuzzy Inference. This sub-procedure uses "if-then" rules relevant to human-oriented information to properly adapt *T_DROP*, for example:

" If no_{BS} is *A* and CH_{rm} is *B* then *T_DROP* is *C*."

where each of *A*, *B*, or *C* is a term (Low, Medium, or High) in the fuzzy sets defined above.

Rule base

There are 9 rules for determining 1 output from 2 inputs with 3 terms in each fuzzy set, as shown in Table 6.3.1. *T_DROP* is based on CH_{rm} which informs the status of the capacity of each BS including traffic loading, and no_{BS} which infers the pilot signal strength, the distance from BS, and E_b/N_o MS receives.

Table 6.3.1. Rules for the 2-input FIS threshold adaptation

CH_{rm} / no_{BS}	Low	Medium	High
Low	*T_DROP*: High	*T_DROP*: Medium	*T_DROP*: Low
Medium	*T_DROP*: High	*T_DROP*: Medium	*T_DROP*: Low
High	*T_DROP*: High	*T_DROP*: High	*T_DROP*: Medium

The MS which only has one pilot in the AS, will not use the rules in Table 6.3.1. That is, its *T_ADD* and *T_DROP* are the same as the initial assigned values.

Aggregation and Composition in max-min composition, considered in [21–23], are the procedures for reasoning [21], as described in the following:

Aggregation: Consider "if-then" conditions for each case of all inference rules such as

let: (no_{BS} = High)	: P
(CH_{rm} = Low)	: Q
(T_DROP = High)	: R
grade of membership function	: μ

then $\mu_R = \min(\mu_P, \mu_Q)$ is the fuzzy value of (T_DROP = High).

Composition: Use the results obtained from the aggregation procedure to compute the value of each case of *T_DROP*, for example, consider some cases $(1, 2, \ldots, m)$ in Table 6.3.1 that give the results of (T_DROP = High) (in this case, m = 4), then the final fuzzy value of (T_DROP = High) equals to $\max(\mu_{R1}, \mu_{R2}, \mu_{R3}, \ldots, \mu_{Rm})$. The final fuzzy values of (T_DROP = Medium) and (T_DROP = Low), can be calculated by the same procedure as that of (T_DROP = High).

C. Defuzzification. To extract a crisp value that represents a fuzzy set, is the defuzzification. In this chapter, the selected defuzzification scheme is weighted average formula (WAF) [22, 24, 25], as shown in Eq. (6.3.2).

$$z_{WAF} = \frac{\sum_{i=1}^{n} \mu_i Z_i^*}{\sum_{i=1}^{n} \mu_i} \qquad (6.3.2)$$

Where n is the number of membership functions for defuzzification. Define Z as a universe of discourse, for example, in Fig. 6.3.5, $-16 \le Z \le -12$, $z = \{z \mid z \in Z\}$, $\mu(z)$ as the aggregated output membership function, μ^* as the maximum value of membership function, then $Z^* = \{z$ at the position of each membership function $\mid \mu(z) = \mu^*\}$.

T_ADD for each MS is assigned as *T_DROP* + 2 dB for fixed SHO window size, in order to neglect the effect of window size.

6.3.5.2 E_b/N_o-Controlling SHO Based on FIS (case II.b)

A. Fuzzification. In case II.b, E_b/N_o is the additional input to the 2-input FIS SHO control plant, whose inputs are no_{BS} and CH_{rm} [12–14] to improve call quality. The membership functions of E_b/N_o as depicted in Fig. 6.3.6, are added into the database of the 2-input FIS SHO. The values of membership function parameters are designed to obtain a higher value of E_b/N_o. Because the quality of traffic channels (E_b/N_o) is not guaranteed in the previous study (the 2-input FIS SHO), E_b/N_o is added as another input to the 2-input FIS SHO control plant to improve the quality, by modifying and adding more inference rules, as shown in Table 6.3.2.

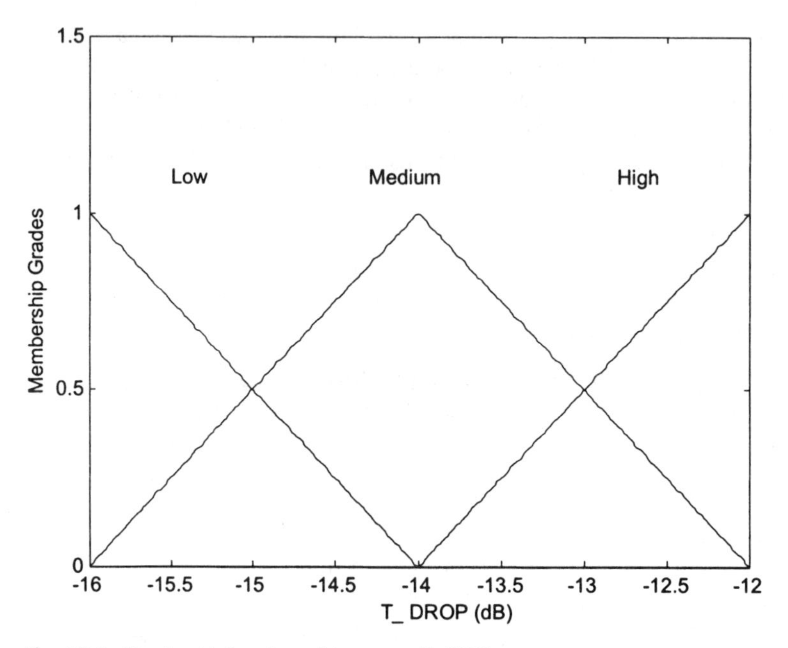

Fig. 6.3.5. Membership functions of the output (*T_DROP*)

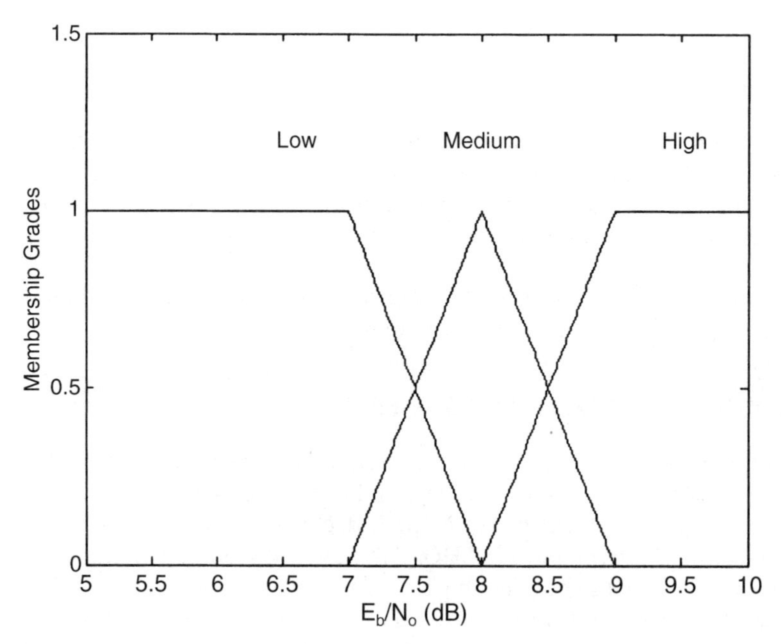

Fig. 6.3.6. Membership functions of E_b/N_o

Table 6.3.2. Rules for the 3-input FIS SHO threshold adaptation {Low: L, Medium: M, High: H}.

No.	E_b/N_o	no_{BS}	CH_{rm}	*T_DROP*
1	L	L	L	***M (H)***
2	L	L	M	***L (M)***
3	L	L	H	L (L)
4	L	M	L	***M (H)***
5	L	M	M	***L (M)***
6	L	M	H	L (L)
7	L	H	L	***M (H)***
8	L	H	M	***L (H)***
9	L	H	H	***L (M)***
10	M	L	L	H (H)
11	M	L	M	M (M)
12	M	L	H	L (L)
13	M	M	L	H (H)
14	M	M	M	M (M)
15	M	M	H	L (L)
16	M	H	L	H (H)
17	M	H	M	***M (H)***
18	M	H	H	***L (M)***
19	H	L	L	H (H)
20	H	L	M	***H (M)***
21	H	L	H	***M (L)***
22	H	M	L	H (H)
23	H	M	M	***H (M)***
24	H	M	H	***M (L)***
25	H	H	L	H (H)
26	H	H	M	H (H)
27	H	H	H	M (M)

B. Fuzzy Inference. In this sub-procedure of case II.b, if-then rules are modified to be:

"If $\boldsymbol{E_b/N_o}$ **is** ***A*** and no_{BS} is *B* and CH_{rm} is *C* then *T_DROP* is *D*."

where each of *A*, *B*, *C*, and *D* is a term (Low, Medium, or High) in the fuzzy sets. So, there are 3*3*3 = 27 rules in rule base calculated from 3 fuzzy subsets of 3 inputs as shown in Table 6.3.2.

The outcomes of *T_DROP* in the brackets in Table 6.3.2, were proposed in [13–15]. This will be compared with the proposed method in this chapter. The **bold** *italic letters* in Table 6.3.2 represent the rules that give different outcomes of *T_DROP* in the proposed method, compared with those in [13–15]. There are 13 rules that are different from those of the 2-input FIS SHO. The new linguistic values of *T_DROP* which are different from the case of 2-input FIS SHO, are obviously assigned in the case of low E_b/N_o.

Note that all of these rules are used when MS is only in the SHA. For those MSs which are not in SHA, they use the conventional SHO [17].

C. Defuzzification. The defuzzification is the WAF scheme.

6.3.5.3 *TRE*-Controlling SHO Based on N-FIS&GD (case I.b)

The FIS&GD method was used by Lo [19], in order to control call blocking probability for hierarchical cellular systems. However, the proposed N-FIS&GD method, based on the FIS&GD method, is used in this chapter to control *TRE* and E_b/N_o. One of the differences between FIS&GD SHO (Fig. 6.3.7 (a)) and FIS SHO (Fig. 6.3.3) is that the GD method is introduced for usage instead of the defuzzification sub-procedure. Consequently, the Sugeno's position-gradient type reasoning method is applied to derive $t\_drop_r(t+1)$, the adaptive drop threshold of each rule of each MS at time $t+1$, and then to obtain $T\_DROP(t+1)$ of each MS. However, the inputs and knowledge base (9 inference rules), are identical to those of the 2-input FIS SHO [13–15].

The $t\_drop_r(t+1)$ can be expressed as [26]

$$t\_drop_r(t+1) = \kappa T\_DROP(t) + \rho_r(t+1) \tag{6.3.3}$$

where
r: rule
κ: a constant to maintain the stability of the GD method
ρ: adjustment parameter for $t\_drop_r(t+1)$.

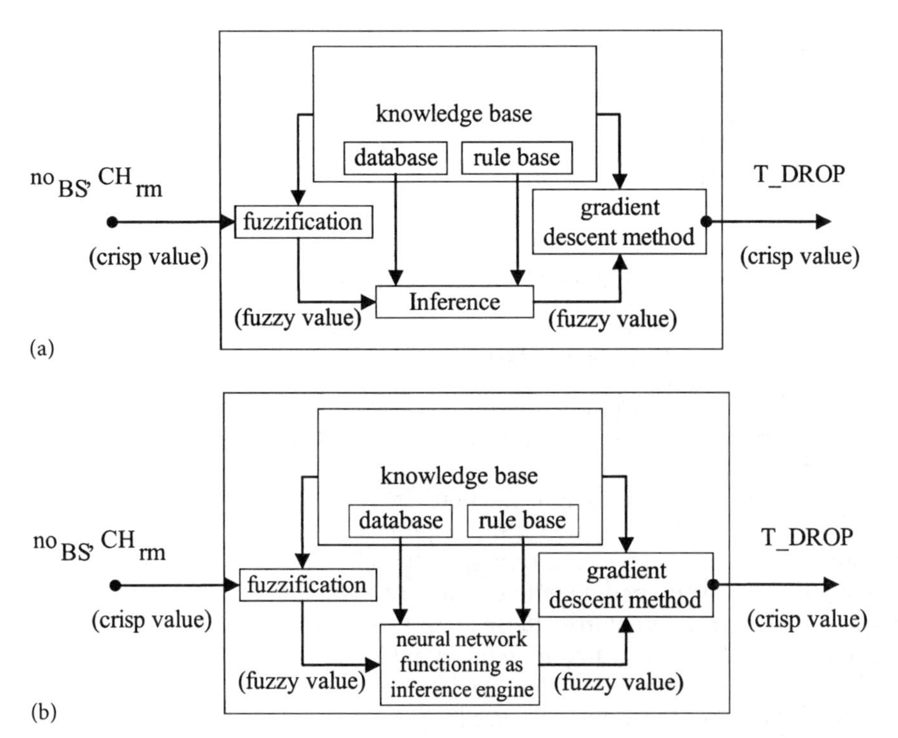

Fig. 6.3.7. (a): FIS&GD SHO control plant. (b) N-FIS&GD SHO control plant. (c) N-FIS&GD SHO control plant with neural network functioning as inference engine and GD method functioning as defuzzification

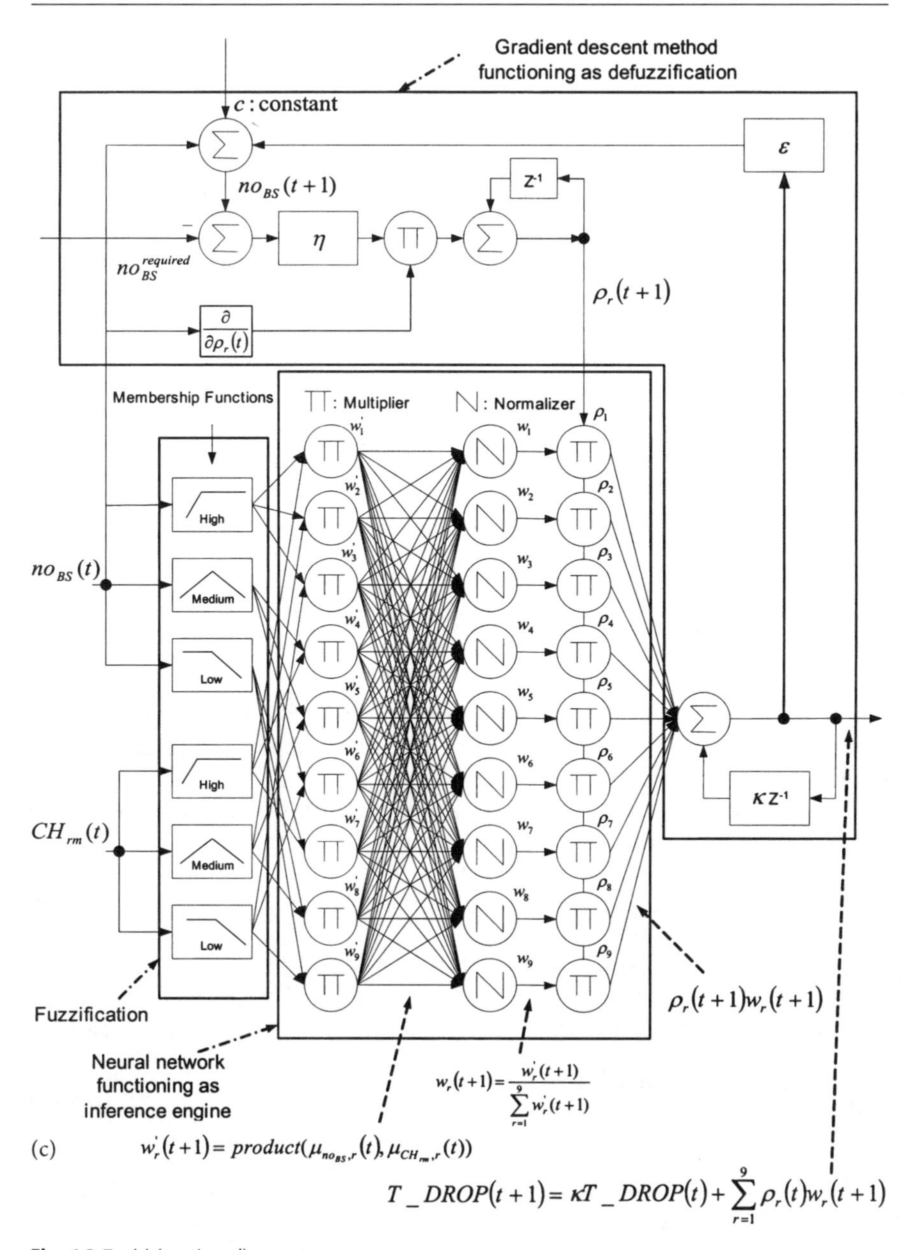

Fig. 6.3.7. (c) (continued)

Then $T\_DROP(t+1)$ is expressed by:

$$T\_DROP(t+1)=\sum_{r=1}^{9}t\_drop_r(t+1)w_r(t+1) \tag{6.3.4a}$$

$$\begin{aligned}T\_DROP(t+1)&=\sum_{r=1}^{9}\kappa T\_DROP(t)w_r(t+1)+\sum_{r=1}^{9}\rho_r(t+1)w_r(t+1)\\&=\kappa T\_DROP(t)\sum_{r=1}^{9}w_r(t+1)+\sum_{r=1}^{9}\rho_r(t+1)w_r(t+1)\\&=\kappa T\_DROP(t)+\sum_{r=1}^{9}\rho_r(t+1)w_r(t+1)\end{aligned} \tag{6.3.4b}$$

where $w_r(t+1)$ is the weighting factor for the output variable of rule r, defined in Eq. (6.3.5b) which normalize $w_r'(t+1)$ derived from the product of fuzzy values (μ), as shown in Eq. (6.3.5a).

$$w_r'(t+1)=product(\mu_{no_{BS},r}(t),\mu_{CH_{rm},r}(t)) \tag{6.3.5a}$$

$$w_r(t+1)=\frac{w_r'(t+1)}{\sum_{r=1}^{9}w_r'(t+1)} \tag{6.3.5b}$$

Note that the weighting factor in the neural network obtained from Eq. (6.3.5b) is different from that obtained from the max-min composition of FIS SHO, as explained in section 6.3.5.1B.

The parameter $\rho_r(t+1)$ in Eq. (6.3.3) or (6.3.4b), is derived from the GD method, where the error function at time t is defined as:

$$erf(t)=\frac{1}{2}(no_{BS}(t+1)-no_{BS}^{required})^2 \tag{6.3.6}$$

Then, $\rho_r(t+1)$ is given by

$$\rho_r(t+1)=\rho_r(t)+\Delta\rho_r(t+1) \tag{6.3.7a}$$

$$\rho_r(t+1)=\rho_r(t)+\eta\frac{\partial erf(t)}{\partial\rho_r(t)} \tag{6.3.7b}$$

$$\rho_r(t+1)=\rho_r(t)+\eta(no_{BS}(t+1)-no_{BS}^{required})\frac{\partial no_{BS}(t)}{\partial\rho_r(t)} \tag{6.3.7c}$$

where η is an adaptation gain which must be properly chosen.

The change of $no_{BS}(t+1)$ during $(t, t+1]$, denoted by $\Delta no_{BS}(t+1)$, would be varied in accordance with $T\_TDROP(t+1)$, so that $no_{BS}(t+1)$ can be kept around $no_{BS}^{required}$ to fulfil *TRE* requirement. $\Delta no_{BS}(t+1)$ can be approximated to be linearly proportional to $T\_DROP(t+1)$. Thus, the $no_{BS}(t+1)$ can be expressed as:

$$no_{BS}(t+1)=no_{BS}(t)+\Delta no_{BS}(t+1) \tag{6.3.8a}$$

$$no_{BS}(t+1) \approx no_{BS}(t) + \varepsilon T\_DROP(t+1) + c \tag{6.3.8b}$$

where ε is an empirical value and c is a constant value. Using z-transform, Eq. (6.3.8b) can be rewritten as:

$$(1-z^{-1})\frac{\partial no_{BS}(t)}{\partial \rho_r(t)} \approx \varepsilon \frac{\partial T\_DROP(t)}{\partial \rho_r(t)} \tag{6.3.9}$$

where z^{-1} denotes one delay unit of an SHO interval time. Then $\frac{\partial no_{BS}(t)}{\partial \rho_r(t)}$ can be expressed as:

$$\frac{\partial no_{BS}(t)}{\partial \rho_r(t)} \approx \gamma \frac{\partial no_{BS}(t-1)}{\partial \rho_r(t-1)} + \varepsilon \frac{\partial T\_DROP(t)}{\partial \rho_r(t)} \tag{6.3.10}$$

where γ is set to a value close to but less than 1 to avoid marginal stability of this gradient evolution.

From Eq. (6.3.4b), Eq. (6.3.10) can be rewritten as:

$$(1-\kappa z^{-1})\frac{\partial T\_DROP(t)}{\partial \rho_r(t)} = w_r(t) \tag{6.3.11}$$

and $\frac{\partial T\_DROP(t)}{\partial \rho_r(t)}$ can be expressed as:

$$\frac{\partial T\_DROP(t)}{\partial \rho_r(t)} = \kappa \frac{\partial T\_DROP(t-1)}{\partial \rho_r(t-1)} + w_r(t) \tag{6.3.12}$$

For each control cycle of FIS&GD SHO, the result of $\frac{\partial T\_DROP(t)}{\partial \rho_r(t)}$ in Eq. (6.3.12) is substituted into Eq. (6.3.10), in order to get $\frac{\partial no_{BS}(t)}{\partial \rho_r(t)}$ term and then $\frac{\partial no_{BS}(t)}{\partial \rho_r(t)}$ and ($no_{BS}(t+1) - no_{BS}^{required}$ are substituted into Eq. (6.3.7 c) to obtain $\rho_r(t+1)$ term. Finally, $\rho_r(t+1)$, and $w_r(t+1)$ from Eq. (6.3.5b) are used in Eq. (6.3.4b) to get the $T\_DROP(t+1)$.

The structure of N-FIS&GD SHO control plant, is shown in Fig. 6.3.7 (b). In Fig. 6.3.7 (c), the neural network with 9 input nodes and 9 output nodes is designed to perform the fuzzy inference function, according to 9 fuzzy values obtained from the fuzzification sub-procedure needed to be the inputs to the fuzzy inference function and 9 fuzzy values obtained from the fuzzy inference function needed to be the inputs to the gradient descent method, functioning as defuzzification sub-procedure. The structure of neural network shown in Fig. 6.3.7 (c) is of the "single layer perceptron" type. However, this neural network can be designed to have any number of hidden layers depending on the complexity of the fuzzy inference function required.

6.3.5.3 E_b/N_o-Controlling SHO Based on N-FIS&GD (case II.c)

Case II.c extends from case II.b in all parts, except the defuzzification part which is substituted by the GD method, and the fuzzy inference part which is substituted by the neural network. There are 3 inputs which differ from the 2 inputs of case I.b, hence, there are 27 inference rules instead of 9 rules as described in the section for case I.b. Thus, the neural network in this case has 27 input nodes and 27 output nodes designed according to the same criteria as in case I.b. Again, this neural network can also be designed to have any number of hidden layers, depending on the complexity of the fuzzy inference function required.

It is observed from the 2-input and 3-input FIS SHO algorithms, that when the number of the inputs is increased, the number of the fuzzy inference rules will also be increased (from 9 rules to 27 rules as shown in Table 6.3.1 and 6.3.2). In addition, if the number of the fuzzy values in each fuzzy set is also increased, the number of the fuzzy inference rules will acceleratively be increased. For example, if the number of the inputs to the FIS is increased from 2 inputs to 4 inputs and the number of the fuzzy values in each fuzzy set is increased from 3 values to 5 values, the number of the fuzzy inference rules will be acceleratively increased from 9 rules to 625 rules! That is, the complexity of the FIS is dramatically increased. In this case, the neuro-fuzzy inference system using the neural network functioning as the fuzzy inference (mapping the inputs of the fuzzy inference rules to the outputs of the fuzzy inference rules), can decrease the complexity. The strategy is to train the neural network with all inference rules until it can give the outputs, generally, as close as needed to the required fuzzy value. The time required to train the neural network may be very long but when the training process is finished, the neural network will be able to function as the fuzzy inference engine, very quickly. This is the most advantageous point of the neuro-fuzzy inference system over the fuzzy inference system, especially when the fuzzy inference system is expected to subtly control any process with fast response.

Eqs. (6.3.4 a) and (6.3.4 b) are rewritten for this case as:

$$T\_DROP(t+1) = \sum_{r=1}^{27} t\_drop_r(t+1)w_r(t+1) \tag{6.3.13a}$$

$$T\_DROP(t+1) = \kappa T\_DROP(t) + \sum_{r=1}^{27} \rho_r(t+1)w_r(t+1) \tag{6.3.13b}$$

$w_r(t+1)$ can be found by including fuzzy value (μ) of E_b/N_o into Eq. (6.3.5 a) and then normalize $w_r'(t+1)$, as shown in Eqs (6.3.14a) and (6.3.14b).

$$w_r'(t+1) = product\,(\mu_{no_{BS},r}(t), \mu_{CH_{rm},r}(t), \mu_{E_b/N_o,r}(t)) \tag{6.3.14a}$$

$$w_r(t+1) = \frac{w_r'(t+1)}{\sum_{r=1}^{27} w_r'(t+1)} \tag{6.3.14b}$$

The error function in this case can be shown as:

$$erf(t) = \frac{1}{2}(E_b / N_o(t+1) - E_b / N_o(required))^2 \tag{6.3.15}$$

The remaining sub-procedures of the control process are the same as those of case I.b. $\Delta E_b/N_o\,(t+1)$ is also assumed to be linearly proportional to *T_DROP* $(t + 1)$.

The N-FIS&GD method of case II.c is similar to that of case I.b but the E_b/N_o is added to the system as the input and the controlled parameter. Therefore, E_b/N_o is considered in some processes as shown in Eqs (6.3.14a) and (6.3.15).

6.3.6 System Model, Computer Simulation and Results

6.3.6.1 System Model [13–15]

The MSs in the system, are assumed to be perfectly reverse power controlled. There are 19 hexagonal cells with one center cell and two tiers of cells. The radius of each cell is 3000 meters, with an omni-directional antenna. The attenuation due to shadowing is a random variable with log-normal distribution, zero mean, 8 dB standard deviation, and 50% correlation of shadowing between cells. Path loss exponent is selected to be 4 [27, 28].

The call arrival process is assumed to be a Poisson process and arrive uniformly in coverage area. Holding time is exponentially distributed with a mean of 120 seconds over the coverage area. In addition, an MS moves in uniformly distributed direction (0–2π). The initial velocity, is a Gaussian random variable with a mean of 40 km/h and a standard deviation of 10 km/h. Only the velocity within the range of [0, 60] km/h is chosen. The MS is assumed to change its velocity at random intervals, which are exponentially distributed [4, 5], with a mean of 30 seconds. The updated and previous speeds are approximately 30% uniformly correlated. The new direction is 30% correlated with the previous one and the new angle is assumed to be a uniformly distributed random variable.

6.3.6.2 Computer Simulation and Results

The values of parameters for simulation are as follows: [9, 13–15, 29–31]

1) Initial *T_ADD* (dB)	–13
2) Initial *T_DROP* (dB)	–15
3) *T_TDROP* (seconds)	5
4) *T_COMP**0.5 (dB)	1
5) The number of traffic channels/cell	50
6) Voice activity factor	0.4
7) Orthogonal factor	0.8
8) Processing gain	128
9) Maximum BS power (watt)	5
10) Required E_b/N_o (dB)	7

The pilot, paging and synchronization channel power percentages with respect to the total BS transmitted power, are set up as follows:

11) Pilot channel power (%) 15
12) Paging channel power (%) 12
13) Synchronization channel power (%) 1.5

The values of κ, η, γ, and ε are set as 0.99, 0.01, 0.99, and 0.01, respectively. Note that *T_COMP* is the comparison threshold for transferring a pilot from the CS to the AS. The step size of *T_COMP* is 0.5 dB [32].

The results are illustrated in the same sequence, as those in the section of problem classification section (section 6.3.2). The comparative performances among various SHO algorithms at high traffic load (50 erlang), are shown in Table 6.3.3.

The values of all parameters in algorithms in Table 6.3.3 are compared with those of the reference values, which are defined as those of IS-95A SHO at 50 erlang traffic load, shown in **bold** in the last column of Table 6.3.3, except the values of E_b/N_o and P_{out}. The reference value of E_b/N_o is defined as 7 dB [32, 33], while the reference value of P_{out} is 0.1 [27]. The +/– symbol of each value, means that the value of the performance indicator of each algorithm, is higher/lower than that of the reference value. The **bold** numbers in columns 2 and 5 of Table 6.3.3, show the best specific performance indicators when compared with the other algorithms.

I) *TRE*-controlling SHO

All figures of the results of group I, consist of both algorithms (N-FIS&GD SHO and SSC SHO) with 3 assigned values of *TRE* (33.33%, 40% and 50%). All algorithms are compared with IS-95A SHO. N-FIS&GD SHO by "TRE (N-FIS&GD2)", while SSC SHO is represented by "TRE (SSC)". Note that, for example, TRE (N-FIS&GD2) = 50%, implies that the expected number of no_{BS} of all MSs in SHA is equal to 1/*TRE*(%)*100 = 2 pilots, at any given traffic load.

Fig. 6.3.8 shows the *TRE* in SHA of both algorithms. SSC SHO can control *TRE* better than N-FIS&GD SHO at low value of assigned *TRE*, while N-FIS&GD SHO can control *TRE* better than SSC SHO at high value of assigned *TRE*. The errors (the highest difference between the controlled value of *TRE* and the assigned value of *TRE*) of N-FIS&GD SHO are 5.5%, 5% and 1.6% for 33.33%, 40%, and 50% assigned values of *TRE*, respectively. The errors of SSC SHO are 0.94%, 2.7%, and 10.4% for 33.33%, 40%, and 50% assigned values of *TRE*, respectively. *Thus, at 50% assigned values of TRE, the N-FIS&GD SHO is preferred. In contrast, at 40% and 33.33% assigned values of TRE, the SSC SHO is preferred.*

In Fig. 6.3.9, the E_b/N_o of 33.33% and 40% assigned values of *TRE* (SSC SHO), are close to each other in the range of 10–40 erlang traffic load, but are different at 50 erlang traffic load. In contrast, the E_b/N_o of 50% assigned value of *TRE* (N-FIS&GD SHO), is lower than those of both 33.33% and 40% assigned values of *TRE* (SSC SHO). From Table 6.3.3, at 50 erlang traffic load, 33.33% assigned value of *TRE* (SSC SHO) gives higher E_b/N_o, than 40% assigned value of *TRE* (SSC SHO) and 50% assigned value of *TRE* (N-FIS&GD SHO), because the NO_{BS}

Table 6.3.3. Comparative performance among various SHO algorithms at 50 erlang traffic load. {Ref.: reference, erl: erlang, BS: base station, HO: handoff, +/– symbol means that the value of performance indicator is higher/lower than that of the Ref. value}

Group		**I**					**II**			
SHO Algorithms	IS-95A	**(I.a) SSC**		**(I.b) 2-input N-FIS&GD**	**IS-95B**	**2-input FIS**	**(II.a) SSC**	**(II.b) 3 input**	**(II.c) 3-input N-FIS&GD**	**Ref. value at 50 erl.**
Assigned *TRE*	–	**33.33%**	**40%**	**50%**	–	–	–	–	–	
***TRE* Error in SHA**	–	+0.94%	+2.70%	+1.60%	–	–	–	–	–	
E_b/N_o	**–1.05%**	–4.85%	–6.99%	–11.48%	–1.93%	–5.88%	–1.07%	–2.88%	–1.11%	**7 dB**
P_{out}	**–91.60%**	–70.10%	–58.20%	–24.40%	–87.30%	–65.20%	–90.90%	–81.80%	–89.70%	**0.1**
T_c	Ref.	+18.84%	+35.66%	**+53.65%**	+3.42%	+30.28%	–4.19%	+13.51%	–18.75%	**28.45 erl.**
P_B	Ref.	–17.86%	–37.79%	**–75.76%**	–38.58%	–25.59%	+15.94%	–9.52%	+32.05%	**0.35**
P_{HO}	Ref.	–25%	–44.73%	**–72.88%**	–44.66%	–36.49%	+9.72%	–15.29%	+46.21%	**0.28**
NO_{BS}	Ref.	–20.40%	–27.74%	–39.40%	–20.13%	–24.96%	–30.56%	–11.42%	+21.36%	**1.69 pilots**
TRE	Ref.	+15.15%	+22.71%	**+38.43%**	+14.90%	+19.66%	–0.80%	+7.62%	–10.40%	**59.11%**
NO_{update}	Ref.	–0.48%	–1.60%	**–3.77%**	+0.46%	–1.24%	–0.38%	–0.99%	+5.78%	**5.57%**
1-*way HO*	Ref.	+31.77%	+34.66%	+47.45%	+22.13%	+23.16%	–1.87%	+5.66%	+2.46%	**50.16%**
2-*way HO*	Ref.	–26.12%	–27.58%	–29.12%	–10.79%	–4.98%	+1.48%	–7.21%	–10.14%	**31.38%**
3-*way HO*	Ref.	–8.19%	–12.35%	–17.47%	–10.74%	–17.32%	+0.35%	–12.07%	–8.65%	**17.59%**
4-*way HO*	Ref.	+2.20%	+0.14%	–0.86%	–0.60%	–0.86%	–0.77%	–0.78%	+5.95%	**0.86%**

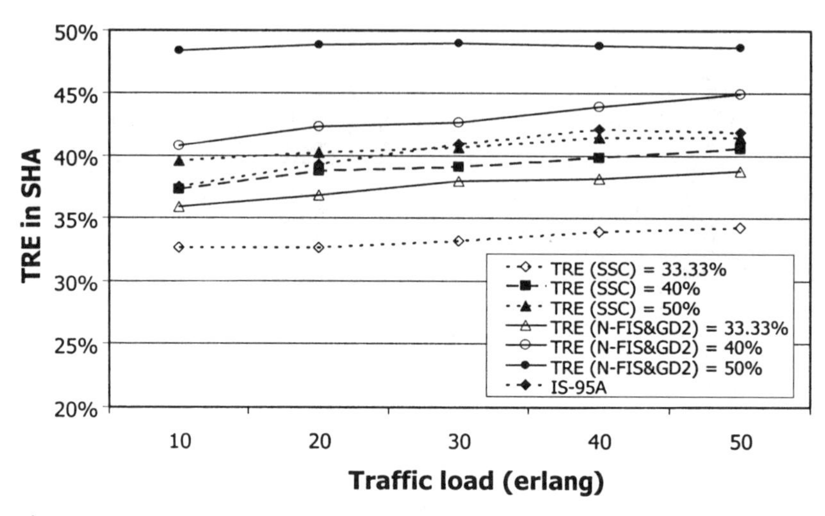

Fig. 6.3.8. *TRE* in SHA as a function of traffic load (Group I)

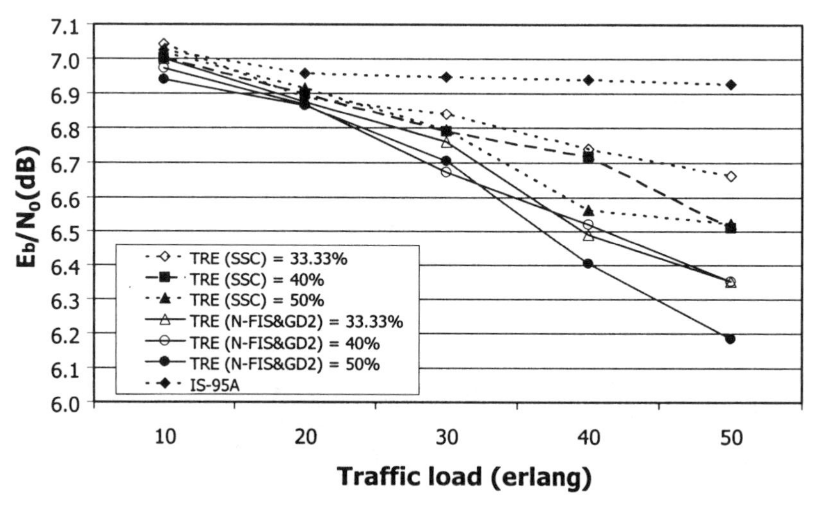

Fig. 6.3.9. E_b/N_o as a function of traffic load (Group I)

of 33.33% assigned value of *TRE* (SSC SHO) is higher than those of the other 2 methods (40% assigned value of *TRE* (SSC SHO) and 50% assigned value of *TRE* (N-FIS&GD SHO)), as also shown in Fig. 6.3.14.

The values of P_{out} correspond to those of E_b/N_o as shown in Eq. (6.3.1). The results of P_{out} are as shown in Fig. 6.3.10 and Table 6.3.3. The reference value of P_{out} in Table 6.3.3 is defined as 0.1 [27].

When the *TRE* is controlled to be high, the system can support more T_c, as shown in Figure 6.3.11 and Table 6.3.3, because the system can allocate resources with high efficiency, especially at 50% assigned value of *TRE* (N-

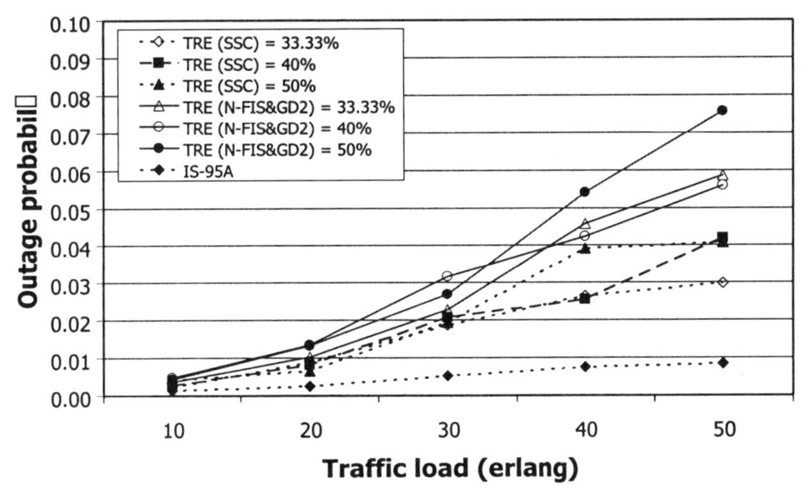

Fig. 6.3.10. P_{out} as a function of traffic load (Group I)

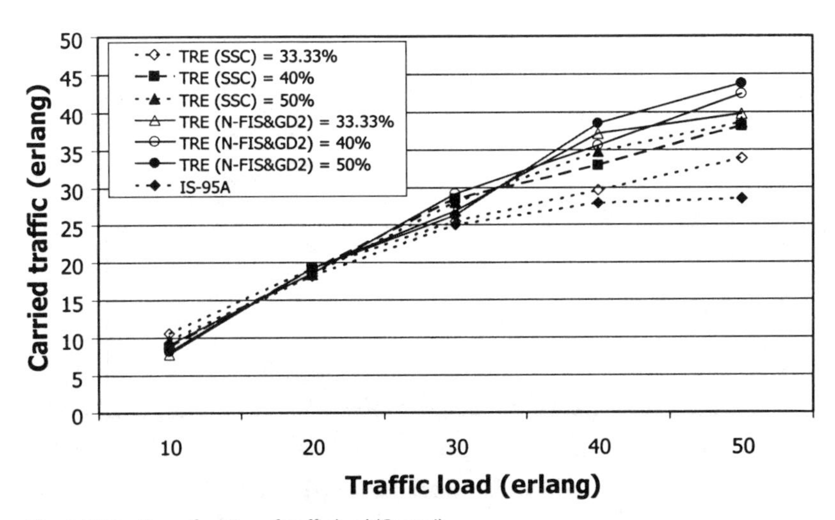

Fig. 6.3.11. T_c as a function of traffic load (Group I)

FIS&GD SHO). It can support carried traffic by 53.65% more than the reference value (28.45 erlang of IS-95A SHO) at 50 erlang traffic load.

From Fig. 6.3.12, all of the values of P_B of 33.33% assigned value of *TRE* (SSC SHO), are higher than those of 40% assigned value of *TRE* (SSC SHO), as well as, all of the values of P_B of 40% assigned value of *TRE* (SSC SHO), are higher than those of 50% assigned value of *TRE* (N-FIS&GD SHO). Obviously, 50% assigned value of *TRE* (N-FIS&GD SHO) gives P_B 75.76% less than the reference value (0.35 of IS-95A SHO) at 50 erlang traffic load. The values of P_{HO} are in the same way as those of P_B, as shown in Fig. 6.3.13 and Table 6.3.3.

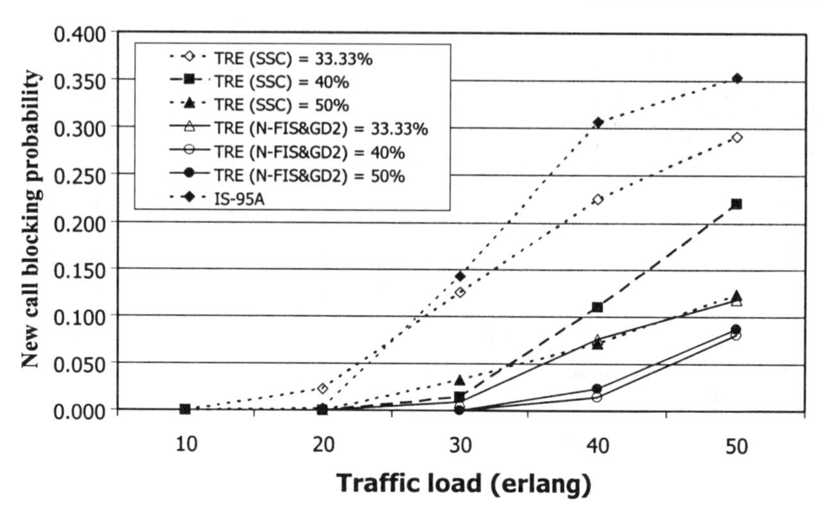

Fig. 6.3.12. P_B as a function of traffic load (Group I)

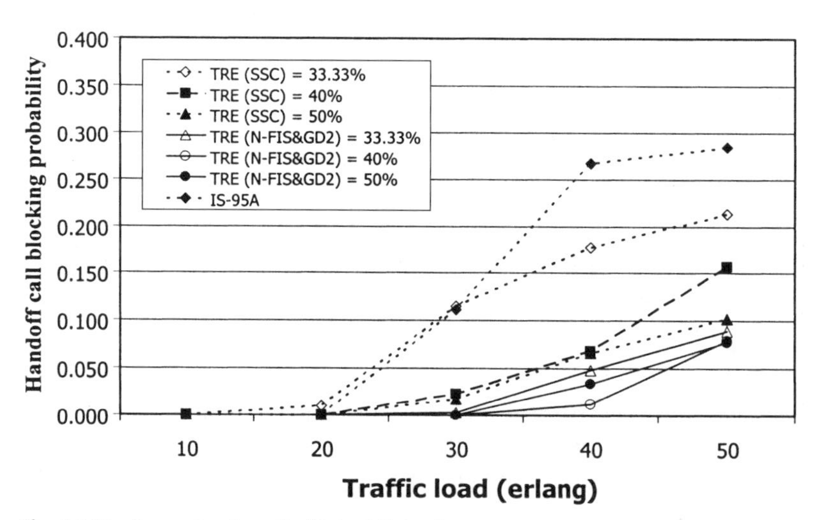

Fig. 6.3.13. P_{HO} as a function of traffic load (Group I)

*TRE*s in the whole cell of 33.33% and 40% assigned values of *TRE* using SSC SHO, and *TRE* in the whole cell of 50% assigned value of *TRE* using N-FIS&GD SHO, tend to be parallel to one another, as shown in Fig. 6.3.15. 50% assigned value of *TRE* (N-FIS&GD SHO) also gives the highest value (38.43% more than the reference value) of *TRE*s in the whole cell among all algorithms at 50 erlang traffic load, as shown in Table 6.3.3.

NO_{update} is the expected percentage of pilot update in the AS. From Fig. 6.3.16, NO_{update} of 50% assigned value of *TRE* (N-FIS&GD SHO), is the lowest among all cases. NO_{update} of 40% assigned value of *TRE* (SSC SHO), is lower than that of

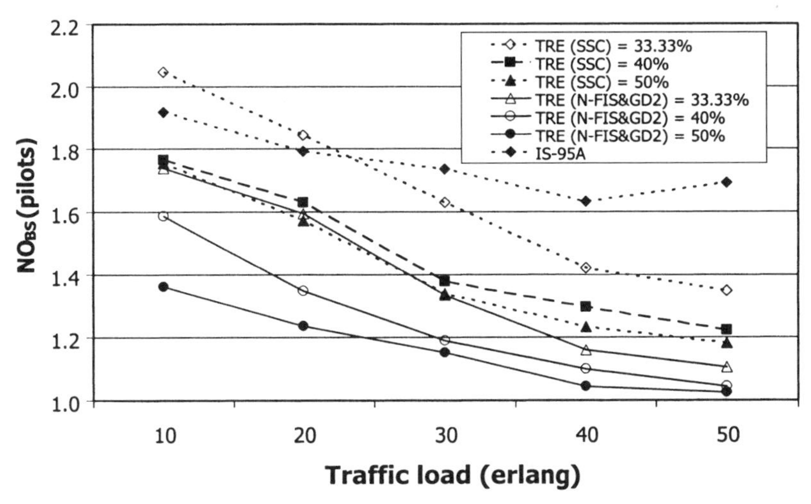

Fig. 6.3.14. NO_{BS} as a function of traffic load (Group I)

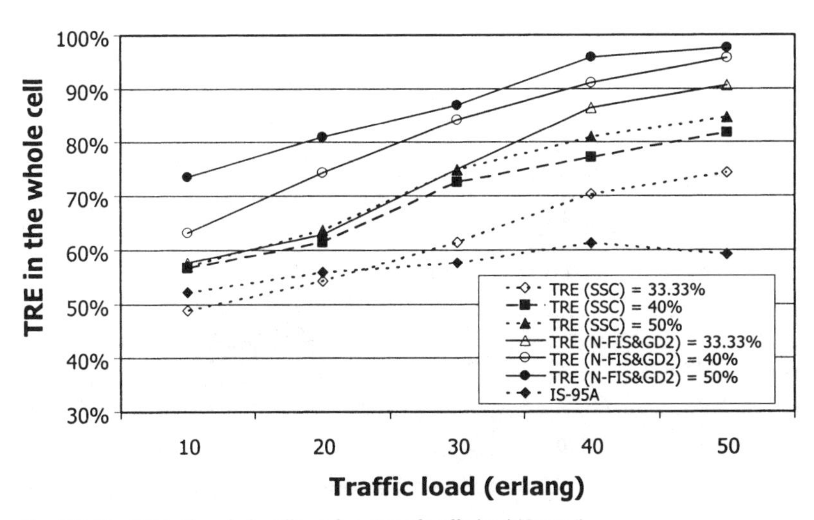

Fig. 6.3.15. *TRE* in the whole cell as a function of traffic load (Group I)

33.33% assigned value of *TRE* (SSC SHO). At 50% erlang traffic load, 50% assigned value of *TRE* (N-FIS&GD SHO), gives the lowest value of NO_{update} (3.77% less than the reference value) among all algorithms, as shown in Table 6.3.3.

The percentage of 1-*way* handoff of 33.33% assigned value of *TRE* (SSC SHO), is lower than those of the other assigned values of *TRE* SHOs, as shown in Fig. 6.3.17, while the percentages of 3-*way* and 4-*way* handoffs of 33.33% assigned value of *TRE* (SSC SHO), are higher than those of the other assigned values of *TRE* SHOs, as shown in Figs. 6.3.19 and 6.3.20. On the other hand, the percentages of 3-*way* and 4-*way* handoffs of 50% assigned value of *TRE* (N-FIS&GD

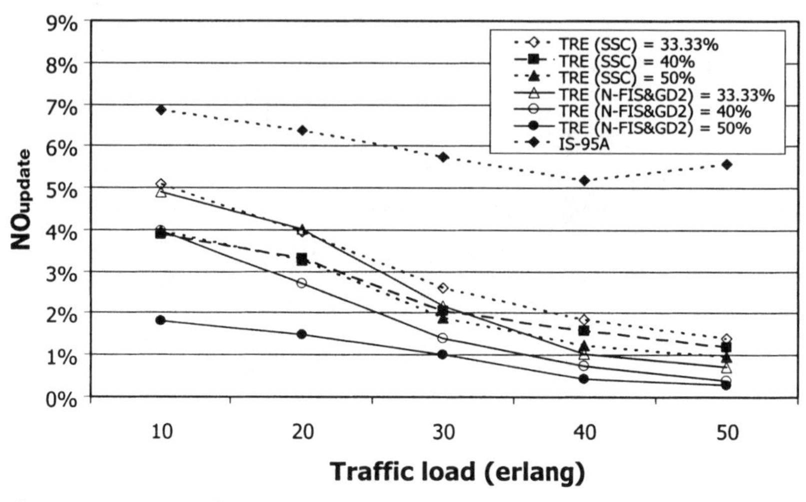

Fig. 6.3.16. NO_{update} as a function of traffic load (Group I)

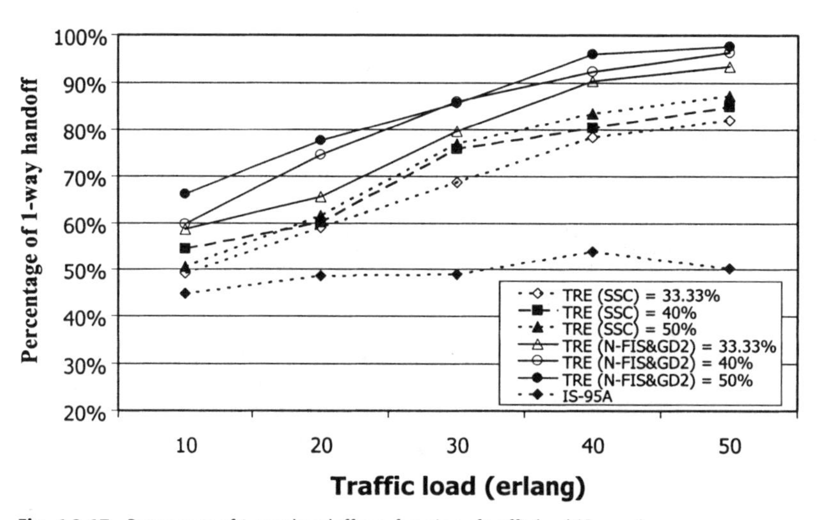

Fig. 6.3.17. Percentage of 1-*way* handoff as a function of traffic load (Group I)

SHO), are the lowest among all cases. In Fig. 6.3.18, the percentage of 2-*way* handoff of 50% assigned value of TRE (N-FIS&GD SHO), is higher than those of 33.33% and 40% assigned values of *TRE* (SSC SHO) in the range of 10–35 erlang traffic load, but lower in the range of 35–50 erlang traffic load. Table 6.3.3 also shows the percentages of 1-*way* to 4-*way* handoffs of all algorithms at 50 erlang traffic load. The percentages of 5-*way* to 6-*way* handoffs are not shown, since their values are very little compared with those of 1 to 4-*way* handoffs.

Average *T_DROP* of all cases, are shown in Fig. 6.3.21. Only three schemes are considered, due to their abilities of proper control: 50% assigned value of *TRE*

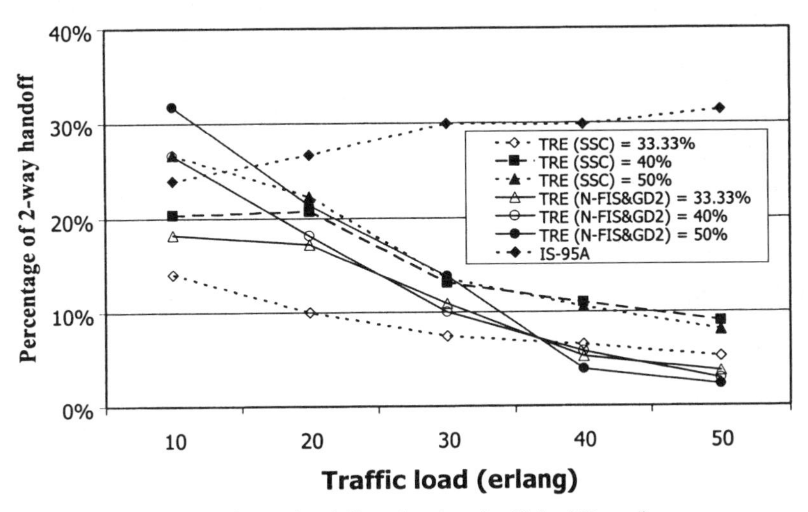

Fig. 6.3.18. Percentage of 2-*way* handoff as a function of traffic load (Group I)

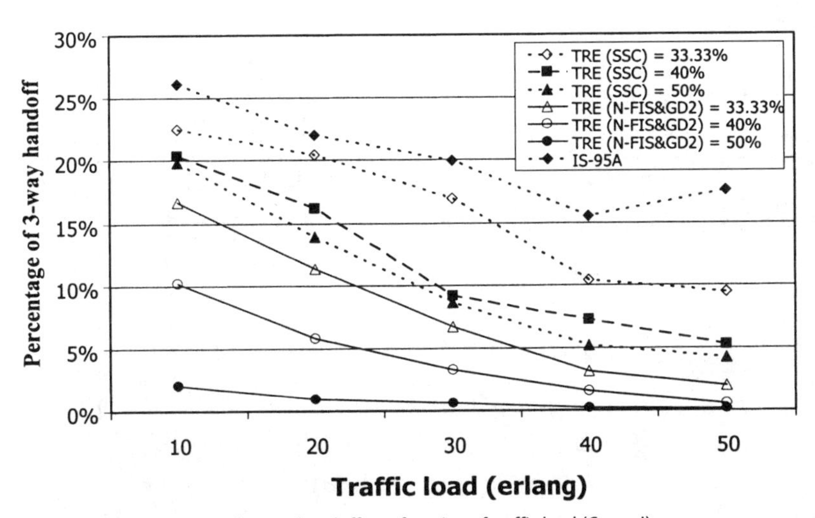

Fig. 6.3.19. Percentage of 3-*way* handoff as a function of traffic load (Group I)

(N-FIS&GD SHO), 40% assigned value of *TRE* (SSC SHO), and 33.33% assigned value of *TRE* (SSC SHO). For 50% assigned value of *TRE* (N-FIS&GD SHO), average *T_DROP* is high at low traffic load and then it decreases while the traffic load is higher, in order to control *TRE* with respect to the inference table of [13–15, 26] and GD method. In case of 33.33% assigned value of *TRE* (SSC SHO), the values of average *T_DROP* are lower than those of 50% assigned value of *TRE* (N-FIS&GD SHO), because 33.33% assigned value of *TRE* (SSC SHO) is intended to support more BSs or links than 50% assigned value of *TRE* (N-FIS&GD SHO). The values of average T_DROP of 40% assigned value of *TRE*

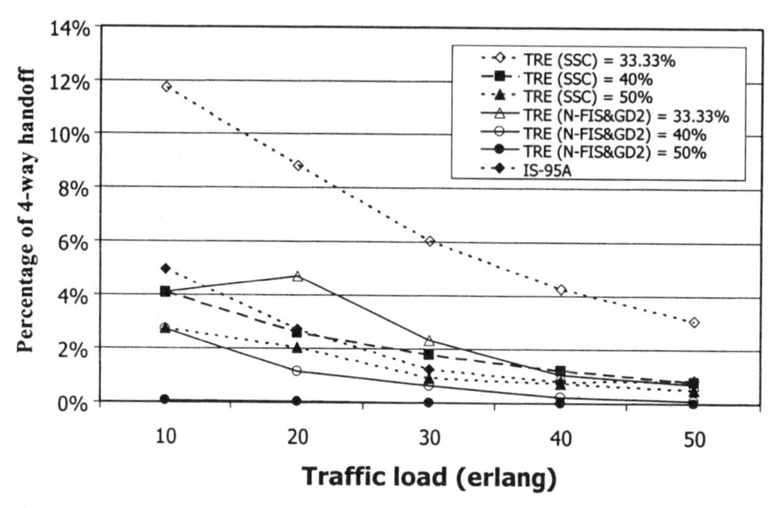

Fig. 6.3.20. Percentage of 4-*way* handoff as a function of traffic load (Group I)

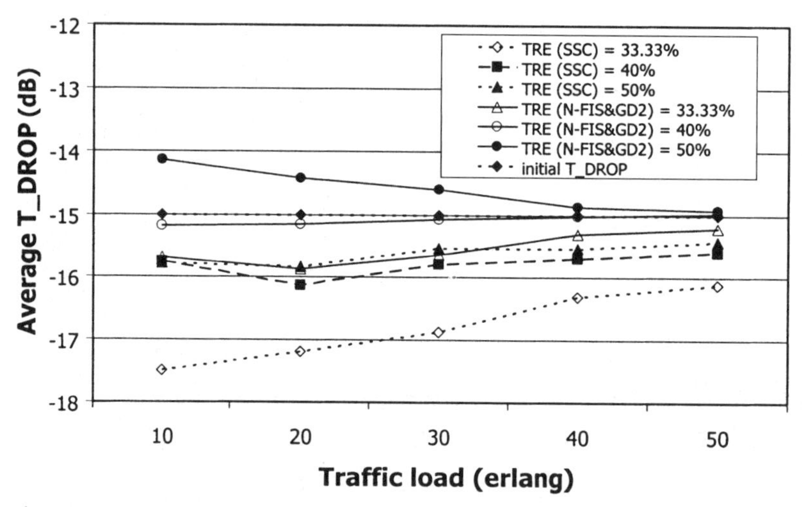

Fig. 6.3.21. Average *T_DROP* as a function of traffic load (Group I)

(SSC SHO), are between those of 50% assigned value of *TRE* (N-FIS&GD SHO) and 33.33% assigned value of *TRE* (SSC SHO).

The results show that *TRE* can efficiently be controlled by using N-FIS&GD SHO or SSC SHO, depending on the assigned value of *TRE*. At 50% assigned value of *TRE*, it is better to use the N-FIS&GD SHO and at 33.33% and 40% assigned values of *TRE*, it is better to use SSC SHO. The error between the controlled value of *TRE* and the assigned value of *TRE*, is within 3%. When *TRE* is controlled to a higher value, the system (with 50% assigned value of *TRE* (N-FIS&GD SHO)) can support higher number of users and give lower blocking probability, lower NO_{update} but quite higher outage probability (24.4% less than the reference value), as shown in Table 6.3.3. On the other hand, when *TRE* is controlled to a lower value, the system can support lower number of users, give higher blocking probability and higher NO_{update} as it should be, but can still keep low outage probability, for instance 70.1% and 58.2% less than the reference value for 33.33% and 40% assigned values of *TRE* (SSC SHO) respectively, as shown in Table 6.3.3. Thus, the appropriate operation in real system should be considered, according to the requirement of the system users.

II) E_b/N_o-controlling SHO

In all figures of the results of group II, there are 6 schemes. The notations in the figures are described as follows:

IS-95A: IS-95A SHO
IS-95B: IS-95B/cdma2000 SHO
Eb/No (FIS2): the 2-input FIS SHO with triangular membership functions and WAF defuzzification
Eb/No (FIS3): the 3-input FIS SHO with WAF defuzzification (case II.b).
Eb/No (SSC): SSC SHO (case II.a).
Eb/No (N-FIS&GD3): the 3-input N-FIS&GD SHO (case II.c).

The IS-95B and FIS SHOs are illustrated in this group, not in group I, because in group I, there have already been so many graphs shown in the Figures.

As shown in Fig. 6.3.22, E_b/N_o of Eb/No (FIS3), Eb/No (SSC), and Eb/No (N-FIS&GD3), are improved from that of Eb/No (FIS2) in the range of 30–50 erlang traffic load. Moreover, the values of E_b/N_o of Eb/No (SSC), are close to those of IS-95A SHO. At 50 erlang traffic load, the values of E_b/N_o of Eb/No (FIS3), Eb/No (SSC), Eb/No (N-FIS&GD3), and IS-95A SHO, are 2.88%, 1.07%, 1.11% and 1.05% less than the reference value respectively, as shown in Table 6.3.3. However, for thorough performance evaluation, the other indicators should also be considered.

P_{out} depends on E_b/N_o as mentioned earlier in Eq. (6.3.1). The results of P_{out} are shown in Fig. 6.3.23 and Table 6.3.3.

Figure 6.3.24 shows T_c of all schemes in group II. At 50 erlang traffic load, it obviously shows that T_c of Eb/No (FIS2) is the highest (30.28% more than the reference value, T_c of IS-95A SHO, at 50 erlang traffic load, as shown in Table 6.3.3). T_c of Eb/No (FIS3) is still higher than those of IS-95A and IS-95B SHOs (13.51% more than the reference value at 50 erlang traffic load), while T_c of Eb/No (SSC) and Eb/No (N-FIS&GD3) are lower than those of IS-95A and IS-95B SHOs.

The P_B and P_{HO}, are also important performance indicators. They are shown in Figs. 6.3.25 and 6.3.26, respectively. Both P_B and P_{HO}, of both Eb/No (FIS2) and IS-95B SHO, tend to give lower values than the other schemes, while those of Eb/No (N-FIS&GD3) are worse than the other schemes. The values of P_B and P_{HO} of Eb/No (FIS3), are between the highest and lowest values (9.52% and 15.29% less than the reference value at 50 erlang traffic load, respectively, as shown in Table 6.3.3).

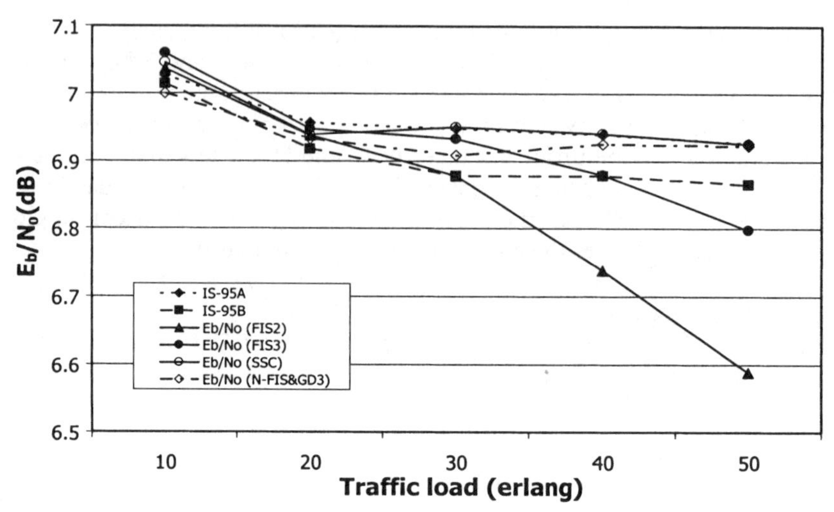

Fig. 6.3.22. E_b/N_o as a function of traffic load (Group II)

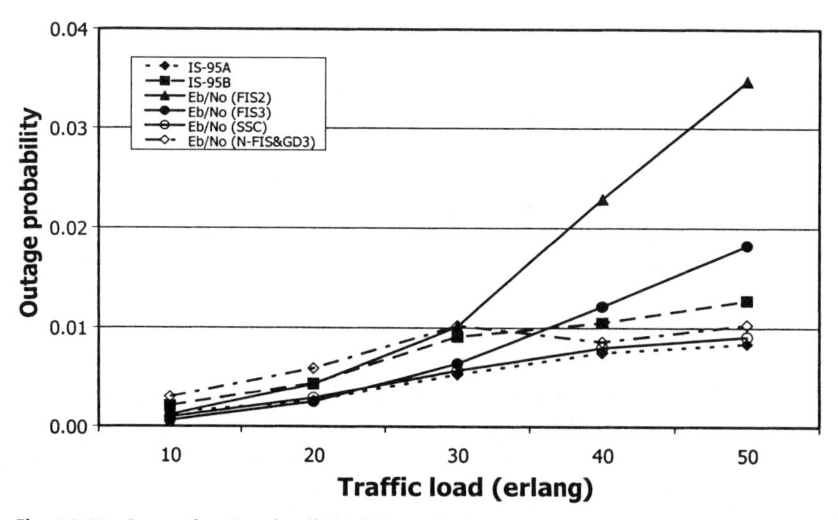

Fig. 6.3.23. P_{out} as a function of traffic load (Group II)

The *TRE* in the whole cell of Eb/No (FIS2) is still the highest among all schemes at high traffic load, while Eb/No (FIS3) gives the medium value of *TRE* at the highest values of traffic load, as shown in Fig. 6.3.28 and Table 6.3.3 (7.62% more than the reference value at 50 erlang traffic load). Obviously, Eb/No (N-FIS&GD3) gives different values of *TRE* to the other schemes. NO_{BS}, which is inversely proportional to *TRE*, is shown in Fig. 6.3.27.

It can be seen that NO_{update} of Eb/No (FIS3) is the same as that of Eb/No (FIS2), as shown in Fig. 6.3.29 and Table 6.3.3. In addition, NO_{update} of Eb/No (SSC) is also the same as IS-95A SHO, while the values of NO_{update} of IS-95B SHO

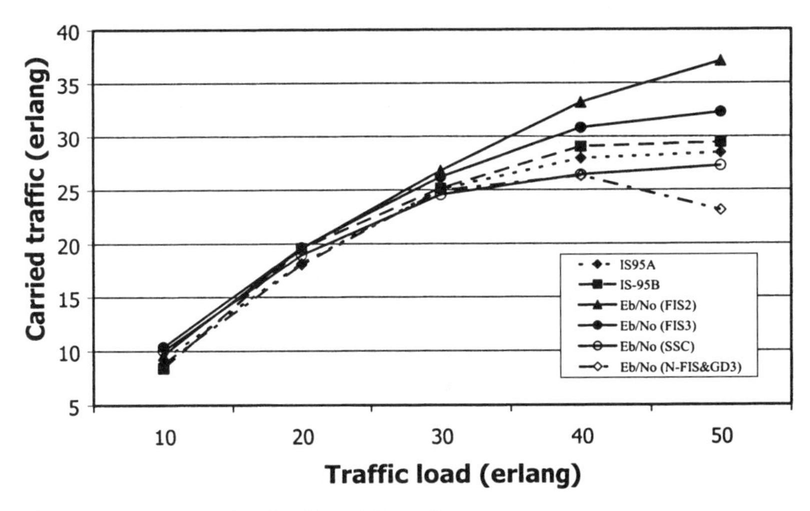

Fig. 6.3.24. T_c as a function of traffic load (Group II)

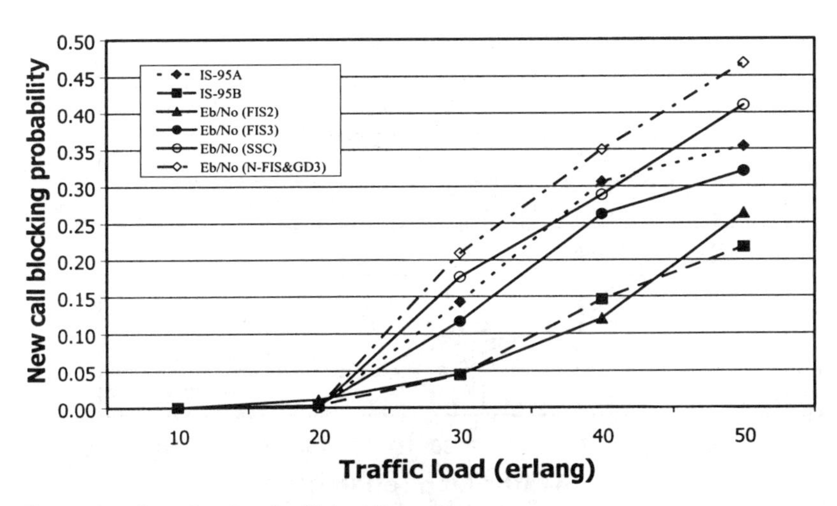

Fig. 6.3.25. P_B as a function of traffic load (Group II)

are quite higher than those of the other schemes, except those of Eb/No (N-FIS&GD3).

The percentages of 1-*way* to 4-*way* handoffs, are shown in Figs 6.3.30–6.3.33. The percentages of 5-*way* and 6-*way* handoffs are not shown, because of their very low values. The prominent result is the percentage of 4-*way* handoff of Eb/No (N-FIS&GD3), which is obviously different from the other schemes. Consequently, this scheme gives low *TRE* and high blocking. The Eb/No (FIS3), tends to give lower percentages of 1-*way* and 2-*way* handoffs, as shown in Figs. 6.3.30 and 6.3.31, but gives higher percentages of 3-*way* and 4-*way* handoffs as

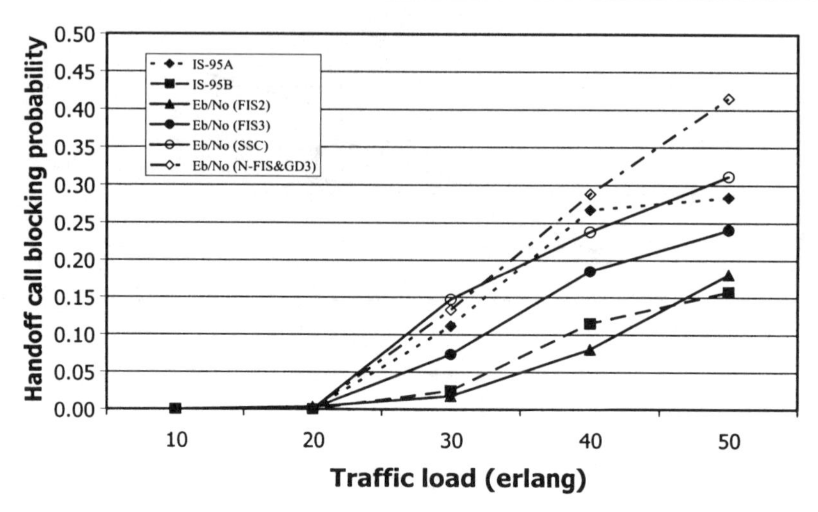

Fig. 6.3.26. P_{HO} as a function of traffic load (Group II)

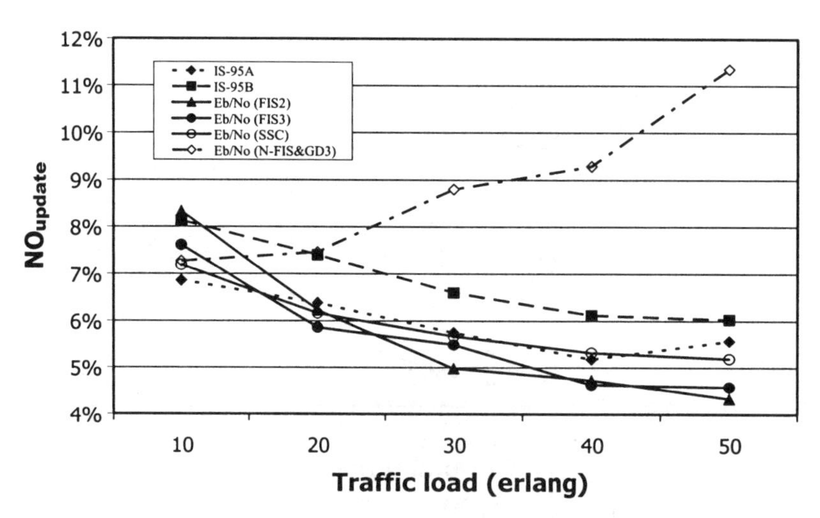

Fig. 6.3.27. NO_{BS} as a function of traffic load (Group II)

shown in Figs. 6.3.32 and 6.3.33, while keeping the call quality acceptable, as shown in Figs 6.3.22 and 6.3.23.

All performances shown above are the operation results of the average *T_DROP*, as depicted in Fig. 6.3.34. Obviously, the values of average *T_DROP* of Eb/No (SSC), are close to those of initial T_DROP, therefore all performances of Eb/No (SSC) are close to those of IS-95A SHO, as shown in Figs. 6.3.22–6.3.29. Furthermore, the values of average T_DROP of Eb/No (FIS3), which is extended from Eb/No (FIS2), are lower than those of Eb/No (FIS2), in order to improve the call quality. By the way, average T_DROP of Eb/No (N-FIS&GD3) is different

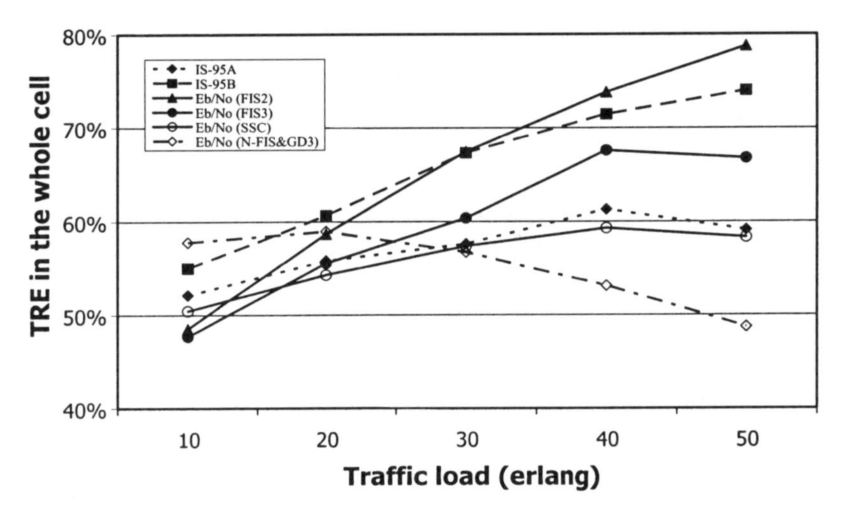

Fig. 6.3.28. *TRE* in the whole cell as a function of traffic load (Group II)

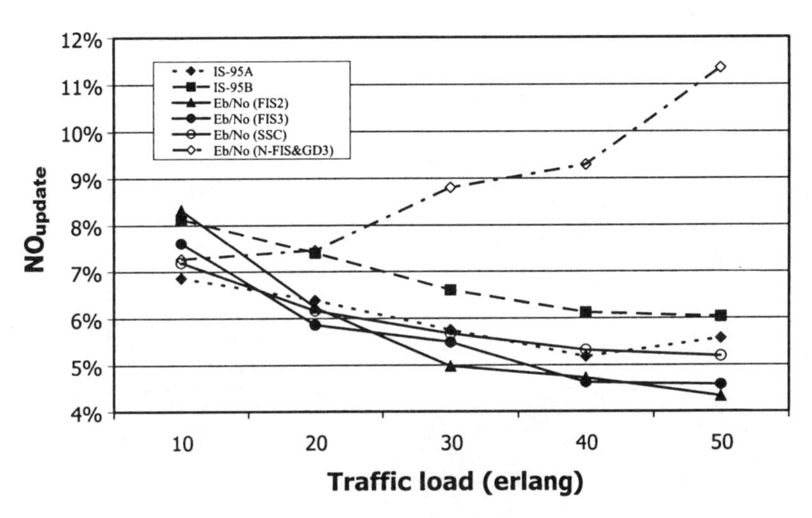

Fig. 6.3.29. NO_{update} as a function of traffic load (Group II)

from those of the others, because of the GD operation performance in controlling the call quality.

It can be seen from all of the results of group II, that E_b/N_o-controlling SHOs based on the 2-input FIS and the 3-input FIS are suitable to compromise the call quality with the call blocking probability, because these schemes can reduce call blocking probability when compared with IS-95A SHO, but still keep high call quality. However, the requirement depends on the system operators and/or users, because when high capacity and low blocking are required, Eb/No (FIS2) is preferable, but when high call quality is required, IS-95A SHO or

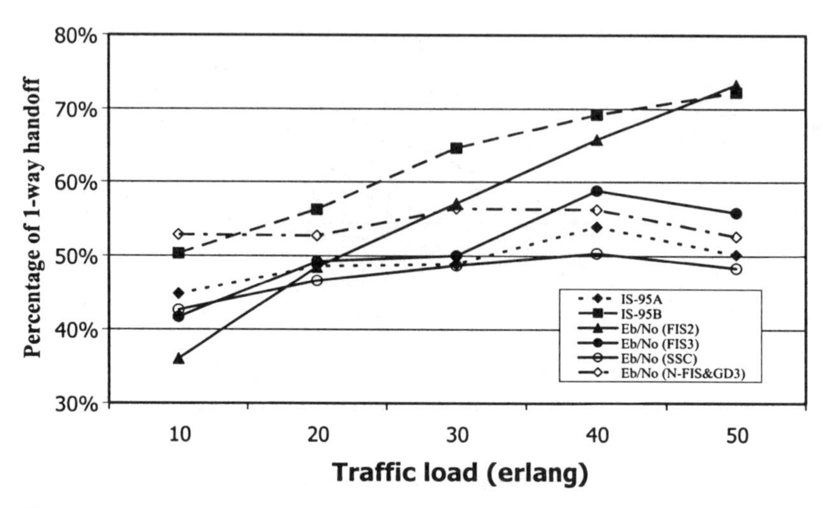

Fig. 6.3.30. Percentage of 1-*way* handoff as a function of traffic load (Group II)

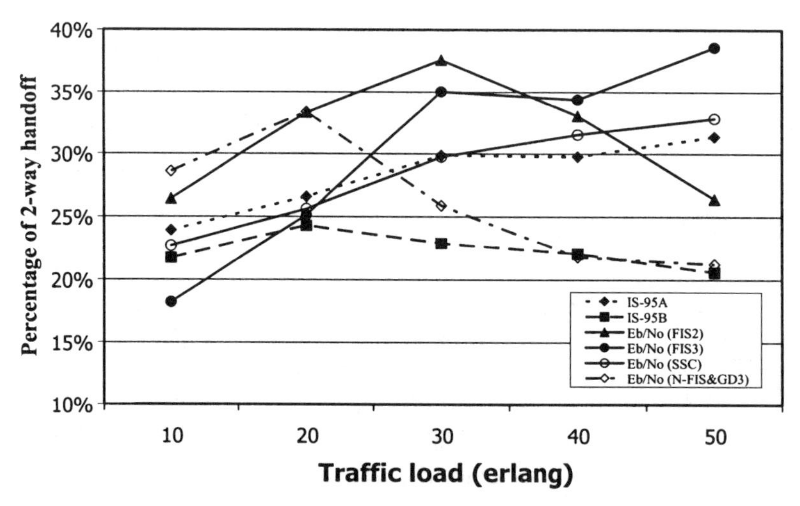

Fig. 6.3.31. Percentage of 2-*way* handoff as a function of traffic load (Group II)

Eb/No (SSC) is preferable. As mentioned earlier, Eb/No (FIS3) compromises Eb/No (FIS2) with IS-95A SHO or Eb/No (SSC). Moreover, Eb/No (FIS3) still gives low NO_{update}. On the other hand, Eb/No (N-FIS&GD3) gives low performance in all aspects except the call quality, as shown in Figs 6.3.22–6.3.29 and Table 6.3.3.

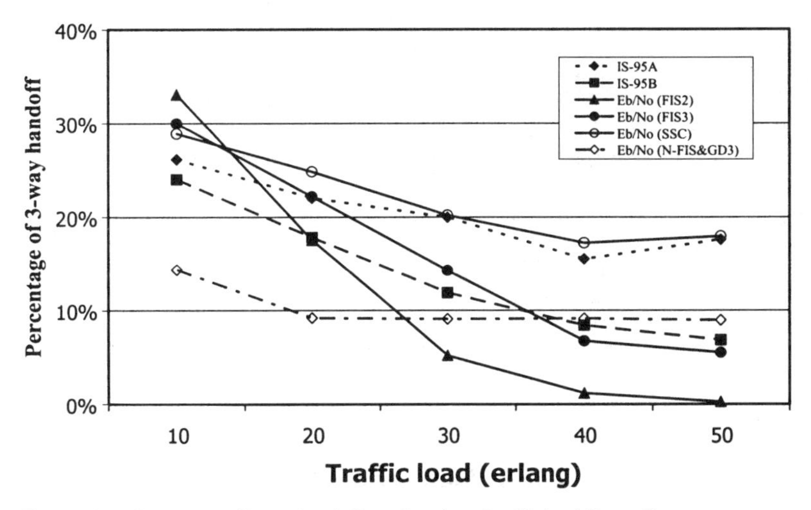

Fig. 6.3.32. Percentage of 3-*way* handoff as a function of traffic load (Group II)

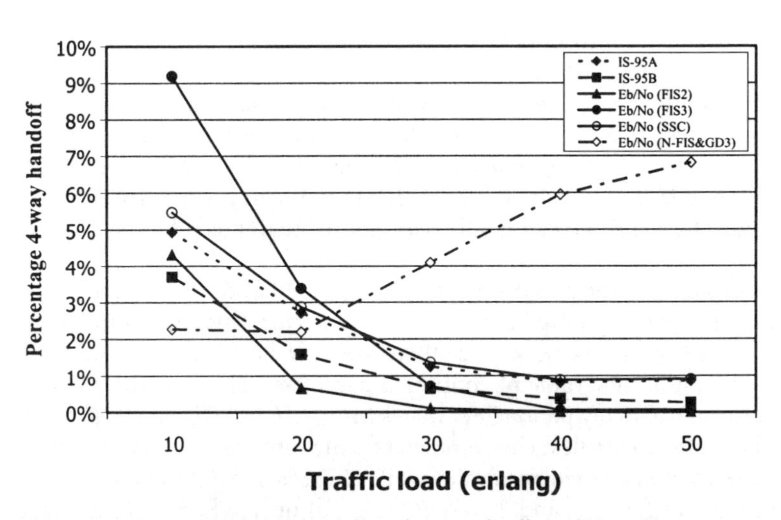

Fig. 6.3.33. Percentage of 4-*way* handoff as a function of traffic load (Group II)

6.3.7 Evaluation of Handoff as a Quality of Service Controller

In summary, *TRE*-controlling SHO and E_b/N_o-controlling SHO (which are considered as two kinds of *QoS*-controlling SHOs for this chapter), can improve the system performance. However, the selection of the method to be used depends on the environment and the requirement of the system operator and/or users. Each approach, can trade off Trunk-Resource Efficiency (TRE, P_B, P_{HO}) against the call quality (E_b/N_o, P_{out}). Because SSC SHOs including *TRE*-controlling SHO based on SSC method and E_b/N_o-controlling SHO based on SSC method, the 3-

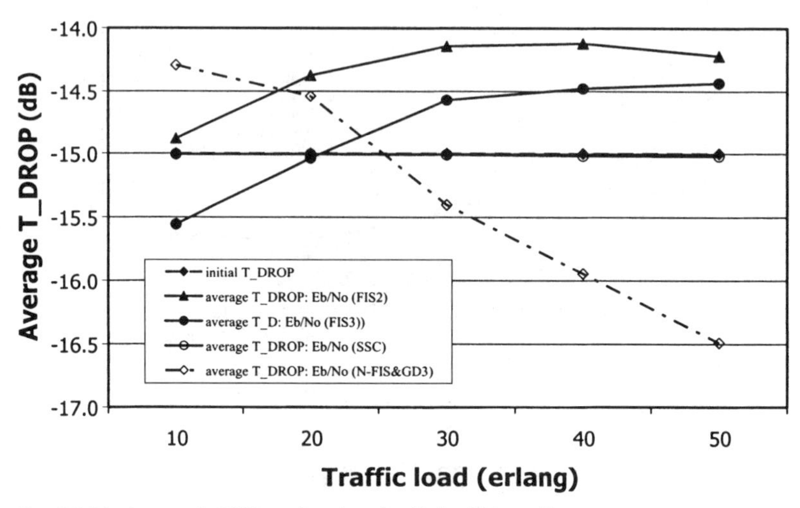

Fig. 6.3.34. Average *T_DROP* as a function of traffic load (Group II)

input FIS SHO (E_b/N_o-controlling SHO based on FIS), and N-FIS&GD SHOs including *TRE*-controlling SHO based on N-FIS&GD method and E_b/N_o-controlling SHO based on N-FIS&GD method, which are proposed in this chapter, have capability to control *QoS* (*TRE* and the call quality) as described in the introduction section, system performance or SHO performance can be controlled as required, while IS-95A and IS-95B/cdma2000 SHOs can not control the performance indicators and the call quality can not be guaranteed by the 2-input FIS SHO.

SSC SHO can control the parameter value (no_{BS} (case I.a) or E_b/N_o (case II.a)), by comparing it with the required value to obtain low error. This algorithm is very simple and has low complexity, because there are only step of comparison and step of assignment of new value of controlled parameter. The 3-input FIS SHO (case II.b) can control many parameters values (no_{BS}, CH_{rm}, E_b/N_o, and *T_DROP*), by using proper inference rules. The parameters values obtained in this algorithm are not compared with any required value while N-FIS&GD SHO can control parameters values (no_{BS}, CH_{rm}, and *T_DROP* (case I.b) or no_{BS}, CH_{rm}, E_b/N_o, and *T_DROP* (case II.c)), by using both inference rules and comparison between parameter value and its required value (no_{BS} (case I.b) or E_b/N_o (case II.c)). Thus, N-FIS&GD SHO has the most complexity among all the proposed SHOs.

From the results in sections 6.3.2 (I), (II) and Table 6.3.3 (at high traffic load), the advantages and disadvantages of all SHO algorithms are summarized below:

IS-95A SHO and E_b/N_o-controlling SHO based on SSC (case II.a) give high call quality but poor performance for the other indicators. Moreover, the performance of case II.a scheme, tends to be lower than that of IS-95A SHO.

IS-95B SHO gives a little higher T_c than IS-95A SHO and gives a higher performance of the other aspects than IS-95A SHO, except E_b/N_o and P_{out}.

The performances of ***the 2-input FIS SHO*** are close to those of ***TRE controlling SHO based on SSC (40% assigned value TRE),*** which emphasize the improvement of T_c, therefore all values of the performance indicators are better than IS-95A SHO, except E_b/N_o and P_{out}.

***TRE* controlling SHOs (case I.a and I.b),** which is an improvement on the 2-input FIS SHO, in case of *TRE*, can give better performance than the other algorithms, except the call quality when compared with IS-95A SHO. However, the call quality is still acceptable and can be varied, depending on the required assigned value of *TRE*.

E_b/N_o-controlling SHO based on N-FIS (case II.b) or *the 3-input FIS SHO,* is also an improvement on the 2-input FIS SHO or IS-95A SHO, in the aspect of call quality. Thus, this scheme gives higher E_b/N_o and lower P_{out} than the 2-input FIS SHO and gives E_b/N_o and P_{out} close to those of IS-95A SHO. On the other hand, it gives lower performance in the other aspects, when compared with the 2-input FIS SHO.

E_b/N_o-controlling SHO based on N-FIS&GD (case II.c) gives poor performance for all aspects, except the call quality, since this scheme is devised to emphasize E_b/N_o and P_{out}.

Note that when QoS is controlled, *TRE* (P_B, P_{HO}) and E_b/N_o (P_{out}) can be improved depending on the requirements. The fuzzy logic based handoffs that use no_{BS} as input, are robust to the environment or cell structure, because there is little need to measure the values of new environment parameters for the designed membership functions, such as distance or signal strength, as shown in the works of [10–12]. Moreover, SSC SHOs, the 3-input FIS SHO and N-FIS&GD SHOs can be implemented at base station controller, by only changing software and increasing the size of the memories.

References

[1] Viterbi A J (1995) Principles of Spread Spectrum Communication, Addison Wesley.

[2] Togo T, Yoshii I, Kohno R (1998) Dynamic Cell-Size Control according to Geographical Mobile Distribution in a DS/CDMA Cellular System, The 9th IEEE International Symposium on Personal, Indoor, and Mobile Radio Communications, pp. 677–681.

[3] Chen X H (1994) A Novel Adaptive Load Shedding Scheme for CDMA Cellular Mobile System, *Proc.* SINGAPORE ICC '94, Singapore, pp. 566–570.

[4] Jeon H G, Hwang S H, Kwon S K (1997) A Channel Assignment Scheme for Reducing Call Blocking Rate in a DS-CDMA Cellular System, IEEE 6th International Conference on Personal Comm Record, Vol. 2, pp. 637–641.

[5] Worley B, Takawira F (1998) Handoff Scheme in CDMA Cellular Systems, IEEE COMSIG' 98, South Africa, pp. 255–260.

[6] Hwang S H, Kim S L, Oh H S, Kang C E, Son J Y (1997) Soft Handoff Algorithm with Variable Thresholds in the CDMA Cellular Systems, IEEE Electronics Letters, Vol. 33, No. 19, pp. 1602–1603.

[7] TR45 TIA/EIA/IS-95B (1998) Mobile Station-Base Station Compatibility Standard for Dual-Mode Spread Spectrum Systems.

[8] TIA/EIA/IS-2000-5 (1999) Upper Layer (Layer 3) Signaling Standard for cdma2000 Spread Spectrum Systems.

[9] Chheda A (1999) A Performance Comparison of the CDMA IS-95B and IS-95A Soft Handoff Algorithms, IEEE VTC '99, 49th, Vol. 2, pp. 1407–1412.

[10] Kinoshita Y, Omata Y (1992) Advanced Handoff Control using Fuzzy Inference for Indoor Radio Sytems, IEEE 42nd VTC, Vol. 2, pp. 649–653.

[11] Kinoshita Y, Oku K (1995) Robustness Analysis of New Fuzzy Handover Control for Indoor Cellular, IEEE UPC, pp. 667–671.
[12] Homnan B, Benjapolakul W (1998) A Handover Decision Procedure for Mobile Telephone Systems Using Fuzzy Logic, *IEEE* APCCAS '98, pp. 503–506.
[13] Homnan B, Kunsriruksakul V, Benjapolakul W (2000) Adaptation of CDMA Soft Handoff Thresholds Using Fuzzy Inference System, IEEE ICPWC 2000, pp. 259–263.
[14] Homnan B, Kunsriruksakul V, Benjapolakul W (2000) Fuzzy Inference System based Adaptation of CDMA Soft Handoff Thresholds with Different Defuzzification Schemes, Proc. 5th CDMA International Conference, pp. 347–351.
[15] Kunsriruksakul V, Homnan B, Benjapolakul W (2001) Comparative Evaluation of Fixed and Adaptive Soft Handoff Parameters using Fuzzy Inference Systems in CDMA Mobile Communication Systems, IEEE VTC 2001, Greece, pp. 1077–1081.
[16] Wong D, Lim T J (1993) Soft Handoff in CDMA Mobile Systems, IEEE Personal Communications, pp. 6–17.
[17] TIA/EIA/IS-95A (1993) Mobile Station-Base Station Compatibility Standard for Dual-Mode Spread Spectrum Systems.
[18] Furukawa H (1998) Site Selection Transmission Power Control in DS-CDMA Cellular Downlink, IEEE ICUPC '98, pp. 987–991.
[19] Lo K R, Chang C J, Shung C B (1999) A QoS-Guaranteed Fuzzy Channel Allocation Controller for Hierachical Cellular Systems, IEEE VTC '99, pp. 2428–2432.
[20] Farinwata S S, Filev D, Langari R (2000) Fuzzy Control: Synthesis and Analysis, John Wiley & Sons.
[21] Jang J S R, Sun C T, Mizutani E (1997) Neuro-Fuzzy and Soft Computing, Prentice Hall.
[22] Klir G J, Yuan B (1995) Fuzzy Sets and Fuzzy Logic; Theory and Application, Prentice Hall.
[23] Altrock C V (1995) Fuzzy Logic and Neuro-Fuzzy Applications Explained, Prentice Hall.
[24] Saade J J, Diab H B (2000) Defuzzification Techniques for Fuzzy Controllers, IEEE Trans Syst, Man, Cybern, Vol. 30, pp. 223–229.
[25] Kandel A, Friedman M (1998) Defuzzification Using Most Typical Values, IEEE Trans Syst, Man, Cybern, Vol. 28, pp. 901–906.
[26] Homnan B, Benjapolakul W (2001) Trunk-Resource-Efficiency-Controlling Soft Handoff based on Fuzzy Logic and Gradient Descent Method, IEEE VTC 2001, Greece, pp. 1017–1021.
[27] Viterbi A J, Viterbi A M, Gilhousen K S, Zehavi E (1994) Soft Handoff Extends CDMA Cell Coverage and Increases Reverse Link Capacity, IEEE Journal on Selected Areas in Communications, Vol. 12, No. 8, pp. 1281–1288.
[28] Lee W C Y (1989) Mobile Cellular Telecommunication Systems, McGraw-Hill, NY.
[29] Yang J, Lee W C Y (1997) Design Aspects and System Evaluations of IS-95 based CDMA Systems, IEEE Universal Personal Comm Record, 6th, Vol. 2, pp. 381–385.
[30] Homnan B, Kunsriruksakul V, Benjapolakul W (2000) The Evaluation of Soft Handoff Performance between IS-95A and IS-95B/cdma2000, IASTED SPC 2000, Spain, pp. 38–42.
[31] Homnan B, Kunsriruksakul V, Benjapolakul W (2000) A Comparative Performance Evaluation of Soft Handoff between IS-95A and IS-95B/cdma2000, IEEE APCCAS '2000, China, pp. 34–37.
[32] Qualcomm (1992) The CDMA Network Engineering Handbook, Vol. 1.
[33] Viterbi A M, Viterbi A J (1993) Erlang Capacity of a Power Controlled CDMA System, IEEE Journal on Selected Areas in Communications, Vol. 11, No. 6, pp. 892–900.

7 An Application of Neuro-Fuzzy Systems for Access Control in Asynchronous Transfer Mode Networks

Watit Benjapolakul and Aimaschana Niruntasukrat

7.1 Introduction

Usage Parameter Control (UPC), or traffic policing in Asynchronous Transfer Mode (ATM) networks, aims at protecting the existing connection from violating the contract with Call Admission Control (CAC) in connection setup phase, in order to protect congestion and violation of the Quality of Service (QoS) of the other connections. UPC performs this, by having penalty measure to malicious users. Thus, UPC functions as the "police" in the ATM network. Many methods or mechanisms have been proposed to function as the traffic policing in ATM networks. These conventional mechanisms can be categorized into two trends: Leaky Bucket (LB) and window mechanisms.

The main objective of traffic policing is to police cell stream from traffic sources, that is whether the traffic characteristics violate the traffic descriptors namely Peak Cell Rate (PCR), Mean Cell Rate (MCR) and Maximum Burst Duration (MBD). Thus, the perfect traffic policing must be able to control all of the parameters. However, in practice, the policing of PCR and MCR of burst traffic sources simultaneously by the above conventional mechanisms may be difficult, because there is usually contradiction in parameter adjustment in the structures of mechanisms. For example, if the LB is required to police PCR, its counter size should be assigned to be small and its leak rate to be approximately the same as that of PCR, while MCR policing requires large counter size and the leak rate being approximately the same as that of MCR. This contradiction makes it impossible to apply only one mechanism to police traffic perfectly. Then, in subsequent researches, traffic policing was categorized into two trends of PCR policing and MCR policing. Although PCR is simple, PCR policing alone is not appropriate to ATM networks with high load. This is because fewer number of users will be supported by the network. On the other hand, MCR policing is more difficult but it can utilize the network resources more efficiently, thus the development of MCR policing to function solely, or function together with PCR policing for UPC, is the more appropriate trend.

The research work of [1] made performance comparison in MCR policing of LB with respect to jumping window, moving window, triggered jumping window and Exponentially Weighted Moving Average (EWMA) mechanisms. This work has found that LB and EWMA are mechanisms with long-term behavior most appropriate to UPC, i.e., they are robust to short-term statistical fluctuation of traffic sources, in the way that Cell Loss Ratio (CLR) are not made too high when compared with the other mechanisms. However, in the aspect of dynamic behavior, EWMA responds to the violation much slower than LB, in fact it is the slowest in responding to the violation. This makes LB the most popular traffic policing mechanism at present. Although conventional LB has the advantage of being simple in its structure, the research work of [2] analyzed its performance mathematically and the analysis has resulted in an approximate formula of CLR, originated from LB. The formula shows that if LB is used to police MCR, it will require a counter of very large size, which in turn causes a slower response and when the size of the counter is smaller, CLR will be higher. Thus, LB is appropriate to police only PCR.

The restriction of using the conventional traffic policing mechanism to police burst traffic source, is that the structure of mechanism is of static nature. Thus, it cannot adapt itself to the fluctuation of the incoming traffic. In order to solve this problem, the research work of [3] proposed to apply Artificial Intelligence (AI), to UPC. This work, proposed training two units of NNs to be able to predict the counting process of incoming cells in the next window, by using the counting process of traffic in the past, as the training sets. Each unit of NNs was trained by different training sets. One unit of NN was trained by the patterns of traffic that did not violate the contract, while the other unit was trained by all possible patterns of traffic (i.e., patterns of traffic that did or did not violate the contract). The difference of both NNs is used in the decision of cell discarding. The advantage of this mechanism is that it can police traffic in real time. However, its disadvantage is the problem in training NN. Not only does it require a long training time, but also unlimited training sets of inputs and a very large difference in training set patterns and thus make the selection of training set and training method have severe effects on the recognition of NN. The wrong selection may make NN unable to recognize the pattern of traffic at all. Moreover, NN must be repeatedly trained every time the characteristics of the traffic source (which are required to be controlled) are changed. This makes the NN approach complicated and inflexible.

Another tool, which was proposed to be applied to UPC, is Fuzzy Logic (FL). The research work of [4] proposed, for the first time, the usage of token LB, with FL controlling the number of excessive tokens which the system allows to have, besides the normal tokens produced with a constant rate equal to the negotiated MCR, to police traffic. The Cell Loss Priority (CLP) bit of the cell which receives the excessive token, will be set to "1" (low priority). Although this approach can decrease CLR compared with the conventional LB, the approach does not really control MCR of the traffic sources, because the controlled variable is the number of excessive tokens while the normal tokens are still produced with a rate equal to the negotiated MCR. Moreover, this work did not evaluate the performance of the mechanism, in terms of selectivity and responsiveness. The research works of [5] and [6], proposed a FL controller based on window mechanism with window limit, as its output. This approach makes the overload policing approach the ideal violation curve. However, the usage of window mechanism as the basic structure makes the responsiveness of the policing to the initialized overload much slower, than that of the conventional LB.

This chapter proposes new traffic policing using token LB mechanism, with fuzzy NN [7] controlling the token generation rate of token generator. The objective is to police MCR of the traffic in a way that it reduces CLR of the cells and improves the performance of overload policing when compared with the conventional LB mechanism [1] and its dynamic behavior must be better than that of the FL window mechanism [6].

7.2 Traffic Control in ATM Networks

ATM is one of the methods to transfer packet data. Its transferred data unit called "cell", has a fixed length of 53 bytes. Each cell consists of an information field and header field which specifies Virtual Channel (VC) in order to manage routing correctly. ATM is proposed to be a technology for Broadband Integrated Services Digital Networks (B-ISDN). It's advantage, is that it is a packet mode data transfer method, thus, network resources can be economized, as cells will only be created when there is information to be transferred. Moreover, circuit mode in ATM offers fast speed in data transfer, after the connection setup phase has been established.

In ATM networks, there is no exact time slot assignment to communication channels. The structure of the functions of data transfer is of the hierarchical type, that is, the transmission medium serves data transfer service to the Virtual Path (VP), and VP serves VC. In protocol aspect, the functions supported by transmission medium are in the physical layer and the functions supported by VP and VC, are in the ATM layer.

VC is a general term to describe one-way communication of transferring ATM cells. In the header of each cell, there are VC Identifier (VCI) and VP Identifier (VPI). In each VP, it is possible to have the same number of VCI. The values of VCI and VPI in the cell header, are changed according to the translation table, every time there is data across the switching node. VC link is capable of transferring one-way ATM cells between switching nodes. VCI of the adjacent two nodes and the corresponding VC link, form the VC Connection (VCC). The termination point of VCC is the point where the information field of the cell is exchanged between ATM layer and its user, i. e., ATM Adaptation Layer (AAL).

VP is a group of VC links. At VC switching node, the values of both of the VCI and VPI will be rewritten while in VP switching, only the value of VPI will be rewritten. In the same way, the connecting VP links form the VP Connection (VPC). The physical layer serves the ATM layer by transferring ATM cells which are loaded in the transmission path payload. An example of the physical layer supporting ATM technology, is the Synchronous Digital Hierarchy (SDH).

With the concept of transmission through VP and VC, more flexible network management can be achieved. However, a higher flexibility makes various aspects of resource management possible and thus creates a challenge in seeking the optimum scheme. The trend in congestion control in ATM networks, consists of the following:

1. Control of the subscriber access to the network. CAC protects the network from overload and ensures that the average service received by the subscriber, does not exceed the existing network resources.
2. Protection of the users' QoS from stochastic fluctuation of some other users' load and the ability to notify those users to adjust the transmission rate.
3. Ensuring that each VC or VP does not exceed traffic descriptor and allocated resources, such as bandwidth or memory.

Congestion control makes the utilization of the network resources efficient. This objective, can be achieved by two main mechanisms: CAC and UPC.

7.2.1 Call Admission Control: CAC

CAC is a set of actions consisting of a new connection setup decision, which is dependent on connection traffic characteristics expected to occur if approved to connect, requested QoS and network load at the time of decision. The connection traffic characteristics can be predicted from the source traffic descriptors. If the connection request is accepted, link bandwidth will be implicitly allocated. In ATM networks with high-speed transmission and long delay time, the reactive control mechanism cannot function efficiently. Thus, in this case, CAC, which is a preventive control mechanism, is one of the most important mechanisms in congestion control. In the case of Synchronous Transfer Mode (STM) networks, the bandwidth of each channel is a fixed value, because the time slot of each channel is cyclically allocated. Thus, the decision to admit the new connection or not, can be considered explicitly and conveniently. However, In ATM networks, where there is no explicit bandwidth allocation in each VC because of statistical multiplexing in each VC, the decision to admit the new connection or not is then dependent on the prediction of traffic characteristics of the new VC and the consideration of the effect on QoS of the existing connections.

Therefore, the network decides to admit the new connection, based on the assumption that the user will strictly behave according to the traffic descriptors. Thus, if not so, congestion will occur in the network and it is necessary to have a mechanism to protect the above event, i. e., UPC.

7.2.2 Usage Parameter Control: UPC

UPC or traffic policing, is the mechanism to monitor cell stream coming into the network at User-to-Network Interface (UNI), in order to ensure that there is no VC or VP violating the contract between user and network and penalize the violating traffic, with the following measures:

1. *Cell discarding.* UPC will discard the excessive traffic from VC or VP, before entering the network.
2. *Cell tagging.* UPC will mark cell of excessive traffic from VC or VP, by setting the CLP bit to "1". If congestion occurs at any point in the network, cells with

"1" CLP bits will be discarded immediately. Although this method is appropriate for the network that does not require very strict control and induces users to use the services, discarding marked cells after entering the network is unfair to all VCs, because the discarding is not proportional to traffic violation from each VC.

The required characteristics of UPC, are as follows:

1. *High selectivity.* Ideal UPC should detect and penalize, only traffic that violates the contract and should be transparent to the traffic that does not violate the contract.
2. *High responsiveness to the violation.* This characteristic, is in order to protect small buffers in the network from overflowing.
3. *Simple implementation and cost effectiveness.*
4. *Efficient network resource usage.*

Conventional mechanisms proposed to function as detection for UPC, can be categorized into two types of LB mechanisms and window mechanisms.

A. LB mechanisms can be sub-categorized as follows:

1. Conventional LB mechanism

This mechanism is an algorithm of traffic policing, with structure as shown in Fig. 7.1. The parameters that specify the performance of the mechanism, are the counter threshold and the leak rate. Every time the traffic source generates a cell, the counter will count up one step. At the same time, the counter will count down by a constant rate equal to leak rate. The cell that arrives when the counter counts up to counter threshold will be discarded.

Butto et al. [2] through analysis found an approximate formula of CLR of conventional LB mechanism (P_L) when on-off traffic source model is considered, as shown in Eq. (7.1).

$$P_L = \frac{b'}{b} \cdot \frac{a\lambda_1 - b'\lambda_2}{a\lambda_1 e^{-(\lambda_1/b' - \lambda_2/a)M} - b'\lambda_2} \tag{7.1}$$

Where b is peak bit rate of traffic source,
a is leak rate of LB (bps),
b' is net bit rate of the counter that increases in the burst duration (equal to $b - a$),
λ_1, λ_2 are the inverse values of mean burst duration and mean silent period, respectively,
M is counter threshold.

This mechanism performs efficiently when used to police PCR, or when $a \approx b$. However, when it is used to police MCR, CLR will be high. In solving this problem of high CLR, there are two approaches. The first approach increases leak rate by a factor called overdimension factor (C), while the second one increases the value of the counter threshold. The disadvantage of the first approach is that

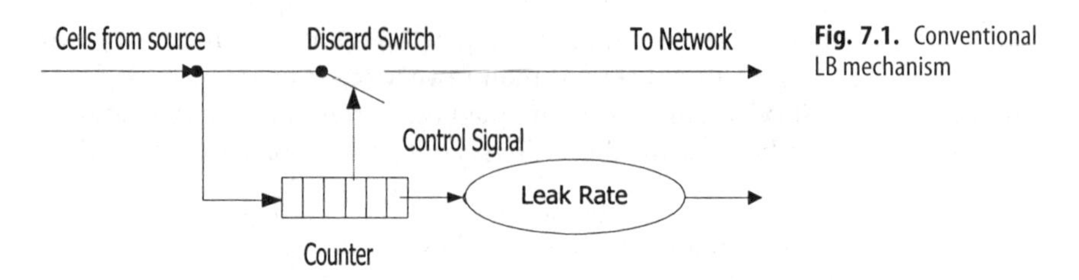

Fig. 7.1. Conventional LB mechanism

the bandwidth used must be increased, hense it is not an economical approach, while the disadvantage of the second approach, is that the responsiveness to the violation is slower.

2. Token LB mechanism, or buffered LB mechanism

To solve the problem of high CLR in unbuffered LB mechanism, in order to be able to utilize the LB mechanism in efficiently policing MCR, an input queue buffer, as shown in Fig. 7.2, is added to the LB mechanism. The traffic cell will be kept in the queue buffer until it receives a token from the token pool, then it is allowed to enter the network. The token is generated with a constant rate called token generation rate. Cell loss will occur when the queue buffer overflows. Thus, the parameters specifying the performance of the mechanism, are the size of the input queue buffer, the size of the token pool, and the token generation rate.

Although the queue buffer can decrease CLR, the mechanism responds to the violation slower than the conventional LB mechanism. If CLR is needed to be very much decreased, a very large size of buffer is needed. This, in turn, causes a delay time problem.

B. Window mechanisms

The principle of these mechanisms, is to limit the number of cells entering the network in each time period or window, so as not to exceed a specified value called window limit. Thus, two important parameters of the window mechanisms, are window size and window limit.

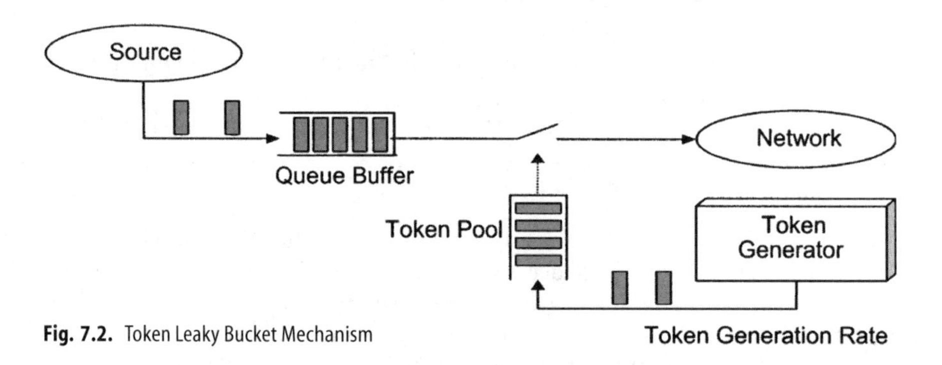

Fig. 7.2. Token Leaky Bucket Mechanism

1. Jumping window mechanism
This mechanism is the simplest window mechanism. The windows are laid in adjacent fashion, and of a fixed window size and a fixed window limit. The biggest disadvantage of this mechanism, is that the starting point of the window is fixed, thus, if the network subscriber knows where it is, they can reduce up to two times the window limit, by dividing data into two equal parts and sending each part in each window. The operation and the disadvantage of the jumping window mechanism, are shown in Fig. 7.3(a).

2. Moving window mechanism
This mechanism was developed from the jumping window mechanism, by fixing the starting point of each window at the arrival point of each cell. This may be considered as a window moving along the time axis, while the size and the limit of the window are fixed. The operation of the moving window mechanism, is shown in Fig. 7.3(b).

Although the moving window mechanism can solve the problem faced in the jumping window mechanism, in a certain time period, the former may need the maximum memory size equal to the window limit, especially, in evaluating long-term traffic characteristics, it is necessary that the window size is large. Thus, this mechanism may be considered as a mechanism which consumes too much memory source.

3. Triggered jumping window mechanism
This mechanism is similar to the jumping window mechanism. The difference is that the starting point of the next window, will occur when the previous window has already finished and the next cell has already arrived. The operation of this mechanism, is shown in Fig. 7.3(c).

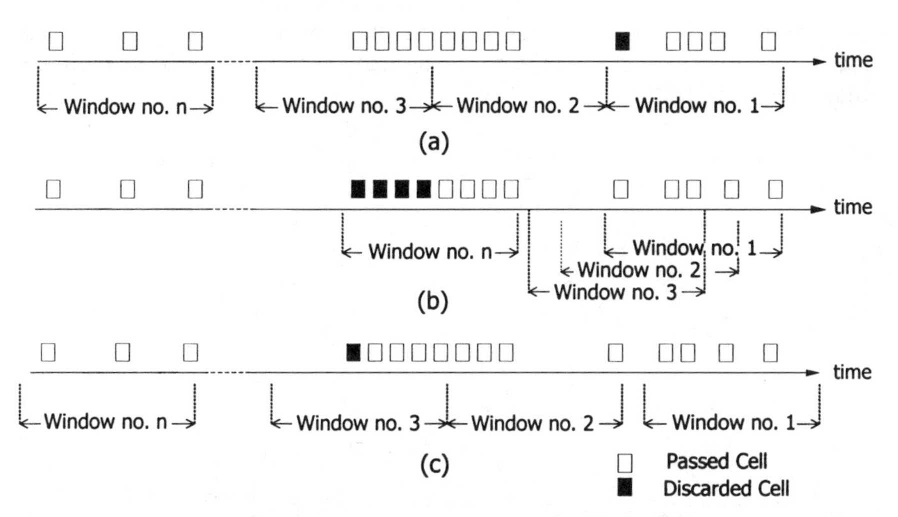

Fig. 7.3. The operation principles of various window mechanisms, when window size is equal to 4. (a) Jumping window (b) Moving window (c) Triggered jumping window

4. Exponentially Weighted Moving Average mechanism (EWMA)
EWMA mechanism is different from the other window mechanisms, in that it has unequal window limit in each window, while the window size is fixed. The window limit of the i-th window (N_i) can be calculated by Eq. (7.2).

$$N_i = \frac{N_{i-1} - \gamma S_{i-1}}{1-\gamma} \tag{7.2}$$

where γ is the parameter controlling the flexibility of the mechanism. When $\gamma = 0$, this mechanism is equivalent to the jumping window mechanism. S_i is the exponentially weighted sum of S_{i-1} and the number of cells arriving in the previous window (X_{i-1}). S_i can be calculated by Eq. (7.3).

$$S_{i-1} = (1-\gamma)X_{i-1} + \gamma S_{i-2} \tag{7.3}$$

7.2.3 Performance Evaluation of Traffic Policing Mechanism

From the required characteristics of traffic policing or UPC in section 7.2.2, in evaluating the capability of selectivity, the overload detection capability is compared. That is, CLR (due to traffic policing) should be approximately the same as the ideal violation probability, P_d, as in Eq. (7.4) and as shown in Fig. 7.4.

$$P_d = \frac{\sigma - 1}{\sigma} \tag{7.4}$$

Where σ is normalized long-term MCR.

One of the objectives of this chapter is to develop traffic policing, that polices MCR. Thus, the closer to the ideal violation probability curve can the traffic policing be designed to detect overload, the more economical the utilization of network resources will be. Moreover, traffic policing should not generate false alarm probability or CLR due to traffic policing at normal traffic load, higher than QoS of the traffic.

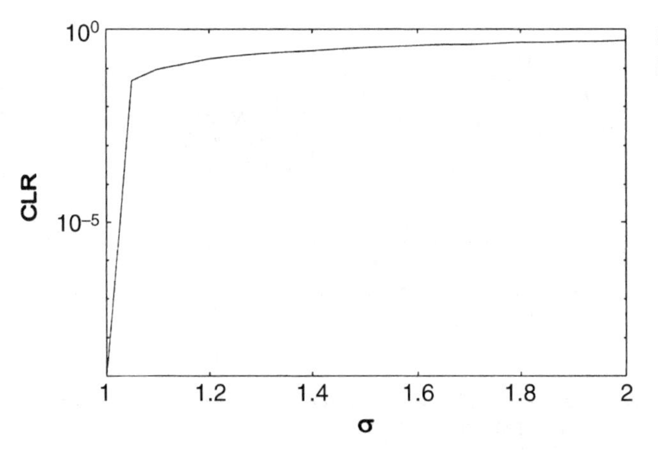

Fig. 7.4. Ideal violation probability curve

In evaluating responsiveness, a comparison is made between the number of cells that each mechanism allows to enter the network before the first cell is discarded, and the cell discarding rate after the first cell has been detected when the user transmits traffic with a fixed level of overload.

7.3 Traffic Source Model and Traffic Policing Mechanism

7.3.1 Traffic Source Model used in Simulation Test

Simulation test, uses the bursty source model or on-off source model, for packet voice traffic.

The traffic source is in one of two states: *burst state* and *silent state*, as shown in Fig. 7.5. In burst state, the traffic source transmits cells with a constant rate equal to the PCR. The number of cells in each burst (X) is a random variable with geometric distribution. Let p be the probability of no cells generated. Then, the probability density function and the mean value of $X(E(X))$, are as shown in Eqs. (7.5) and (7.6), respectively.

$$f(x;p)=\begin{cases} p(1-p)^x & ; x\in\{0,\ 1,\ \ldots\} \\ 0 & ; \textit{otherwise} \end{cases} \tag{7.5}$$

$$E(X)=\frac{1-p}{p} \tag{7.6}$$

The silent period (s) is a random variable with negative exponential distribution. Let β be the mean of the silent period. Then, the probability density function, is as shown in Eq. (7.7).

$$f(s;\beta)=\frac{1}{\beta}e^{-\frac{s}{\beta}} \tag{7.7}$$

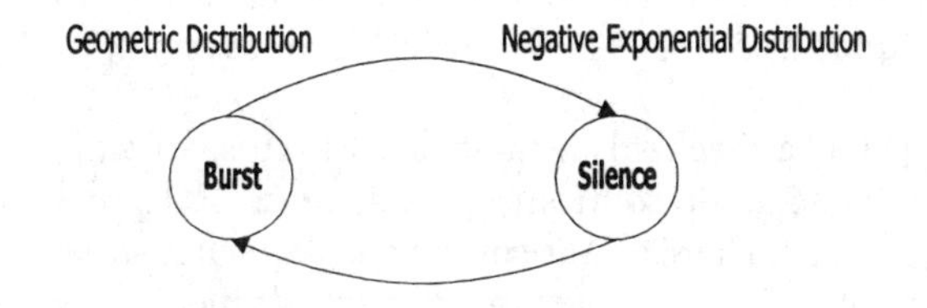

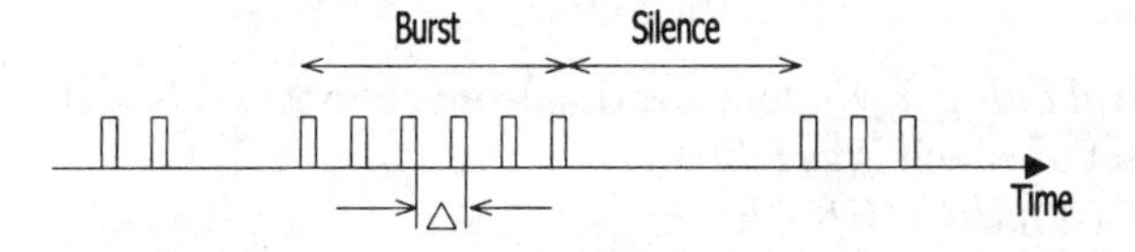

Fig. 7.5. On-off traffic source model

Table 7.1. Parameters of traffic source model used in simulation test

Parameters	Traffic Source Model 1	Traffic Source Model 2
Cell interarrival time: Δ	16 ms	12 ms
Average number of cells in each burst ($E(X)$)	5	29
Mean silence duration: $E(S)$	147.72 ms	650 ms
Peak Bit Rate	32 kbps	32 kbps
Mean Cell Rate: MCR	21.96 cells/s	29.06 cells/s

The reason for the selection of this model is due to its simplicity and flexibility, that is, various parameters, such as PCR, MCR and MBD, can be adjusted independently.

Due to the fact that performance of the proposed mechanism must be compared with those of the other two mechanisms [1, 6] as stated above, traffic source used in the simulation test, must also be according to those research works. Thus, there are two types of parameters in simulating packet voice traffic source, as shown in Table 7.1.

MCR of the traffic can be calculated according to Eq. (7.8).

$$MCR = \frac{E(x)}{E(x)\Delta + E(S)} \tag{7.8}$$

7.3.2 Structure of Traffic Policing Mechanism for Comparison

According to the objective, the performance of the proposed mechanism in this chapter, will be compared with those of conventional LB [1] and FL window [6] mechanisms. A detailed discussion and comparison of these two mechanisms, can be studied by referring to [1] and [6]. However, the overview of conventional LB mechanism is in section 7.2.2. Only type 1 of the traffic source in Table 7.1 is used in simulation test. The parameters of conventional LB mechanism, are as follows: leak rate of 31.24 cells per second or 1.42 times of negotiated MCR and the counter threshold of 45 cells. The following, is an overview of the FL window mechanism.

FL Window mechanism [6], is a mechanism the basic structure of which is the jumping window mechanism and FL controlling the increase or decrease of the next time window limit (ΔN_{i+1}). The structure of the mechanism is shown in Figure 7.6 and the structure of fuzzy controller for FL window mechanism, is shown in Fig. 7.7.

The inputs of the FL controller, consists of the following:

1. *Average number of arrival cells per window,* starting from connection initialization (A_{0i}) the fuzzy set of which is as follows:
   A_{0i} = {*Low (L), Medium (M), High (H)*}

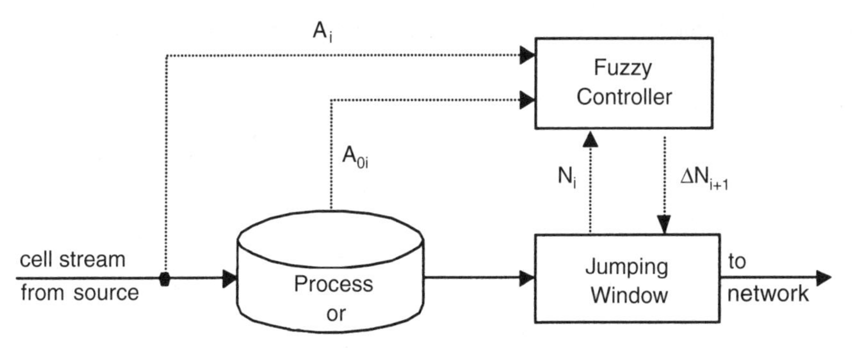

Fig. 7.6. Model of the operation of FL window mechanism

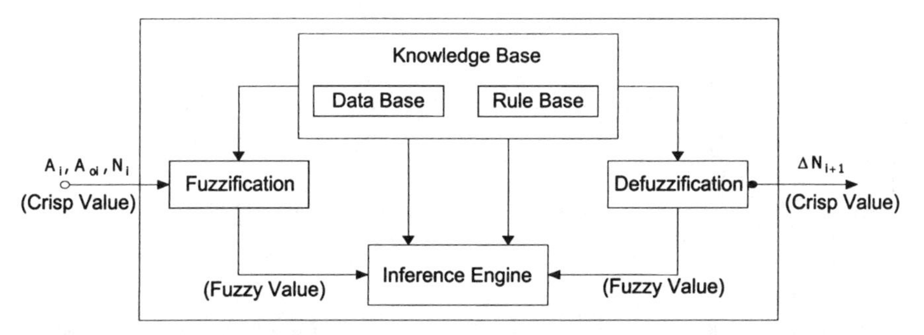

Fig. 7.7. Fuzzy controller for FL window mechanism, shown in Fig. 7.6

2. *The number of arrival cells in the i-th window (A_i)*, the fuzzy of which is as follows:
   $A_i = \{Low\ (L), Medium\ (M), High\ (H)\}$
3. *The limit of the i-th window (N_i)*, the fuzzy set of which is as follows:
   $N_i = \{Low\ (L), Medium\ (M), High\ (H)\}$

The output of the FL controller is the limit of the $(i+1)$-*th* window (ΔN_{i+1}), the fuzzy set of which is as follows:

ΔN_{i+1} = {*Positive Big (PB), Positive Medium (PM), Positive Small (PS), Zero Equal (ZE), Negative Small (NS), Negative Medium (NM), Negative Big (NB)*}

The membership function for fuzzy sets of inputs and output, are shown in Fig. 7.8(a)–7.8(c)

The values of the parameters of the FL window mechanism used in testing with type 1 of the traffic source in Table 7.1, are as follows: $N = 32$ cells, $N_1 = 3.5\ N$, $N_{i\_max} = 9\ N$, $MAX = 90$ cells and window size = 1,440 ms.

The values of the parameters of the FL window mechanism used in testing with type 2 of the traffic source in Table 7.1, are as follows: $N = 87$ cells, $N_1 = 3.5\ N$, $N_{i\_max} = 9\ N$, $MAX = 230$ cells and window size = 3,000 ms.

Fuzzy rule base used in the inference stage, is shown in Table 7.2.

Table 7.2. Fuzzy rule base used in the inference stage of FL controller in FL window mechanism

Rule #	A_{0i}	N_i	A_i	ΔN_{i+1}
1	L	H	L	PB
2	L	H	M	PS
3	L	H	H	ZE
4	M	M	L	PB
5	M	M	M	PS
6	M	M	H	ZE
7	M	H	L	PB
8	M	H	M	ZE
9	M	H	H	NB
10	H	L	L	PB
11	H	L	M	PM
12	H	L	H	PS
13	H	M	L	PB
14	H	M	M	PM
15	H	M	H	ZE
16	H	H	L	NS
17	H	H	M	NM
18	H	H	H	NB

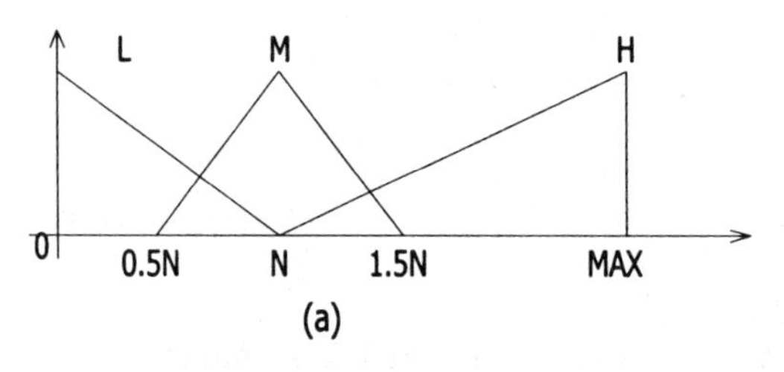

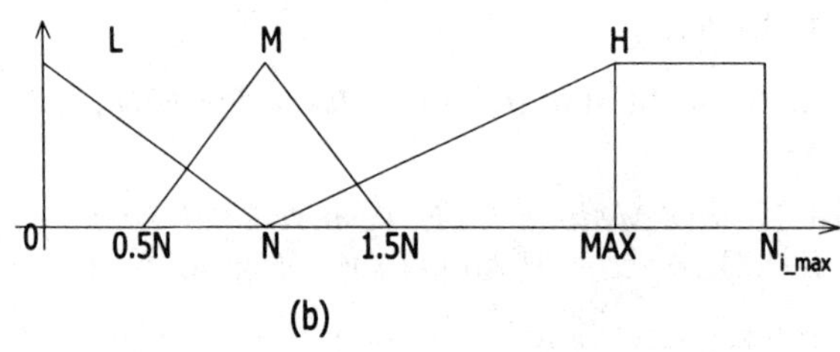

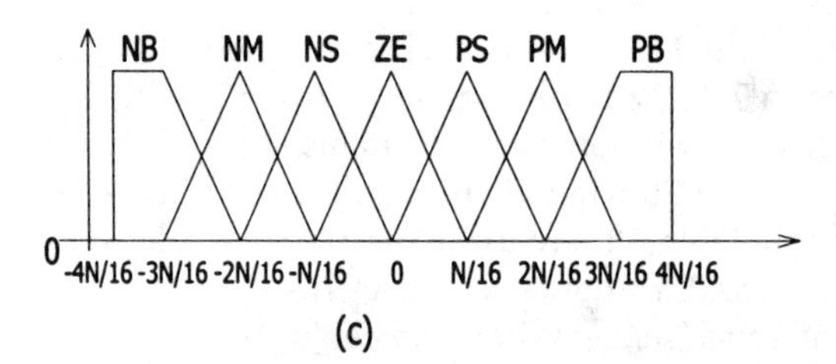

Fig. 7.8. Membership functions for fuzzy sets of FL window mechanism; (a) A_{0i}, A_i (b) N_i (c) ΔN_{i+1}

7.3.3 Structure of Traffic Policing Mechanism Using NFS

The structure of traffic policing mechanism using NFS, is shown in Fig. 7.9, and the structure of Neuro-Fuzzy System controller for FFN LB mechanism, is shown in Fig. 7.10. The reason of using token LB mechanism as the fundamental structure, is that token LB mechanism gives lower CLR than conventional LB mechanism, and faster responsiveness than window mechanism. Thus, when FL without NN is applied to this fundamental structure as its controller (called Fuzzy Logic Leaky Bucket (FLLB) mechanism), it is possible to develop traffic policing mechanism on this fundamental structure, to achieve a better performance than FL window and conventional LB mechanisms. Moreover, the parameters obtained from queue buffer and token pool of token LB mechanism, can be used as input traffic characteristics descriptor, without additional processors.

The operation of the mechanism starts from the input of the NFS, which is the descriptor of instantaneous traffic, sampled to measure the characteristics at a fixed appropriate period called sampling period. The measured inputs will be fed to the NFS controller in order to calculate the output, which is the token generation rate of the next period.

The design of the mechanism starts with assigning various parameters, i. e., the size of the input buffer *(B)*, the size of the token pool *(P)* and the sampling period (T_S), to correspond to the parameters of traffic from each type of traffic source, as shown in Table 7.3.

The reason as to why the queue buffer in case of testing with traffic source model 2 has the relative size with respect to E(X) of traffic source smaller than that of model 1, is that the buffer size relates directly to the traffic delay time. Thus, the assignment of a very large the buffer size is not good for the performance of mechanism. In practice, it is found that the buffer size and sampling period have strong effects on the assignment of membership function and

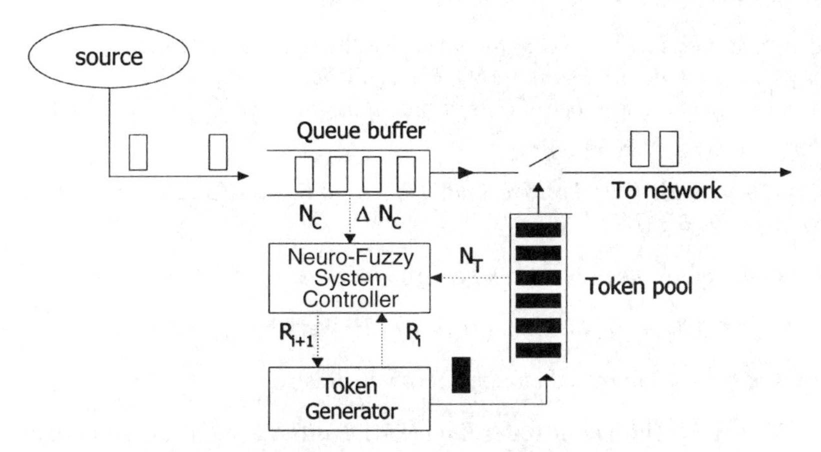

Fig. 7.9. Model of the operation of NFS LB mechanism

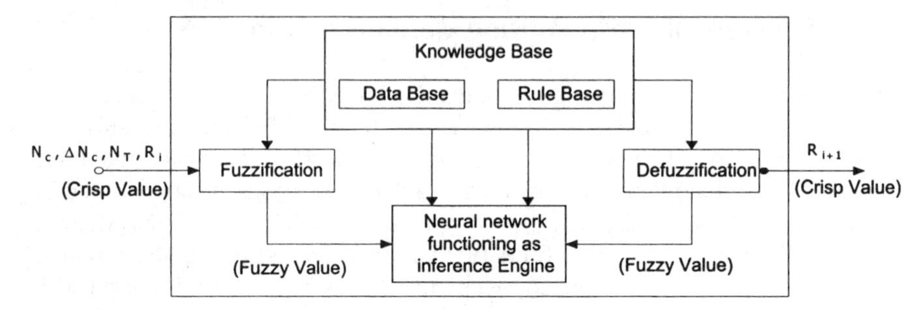

Fig. 7.10. Neuro-Fuzzy System controller for NFS LB mechanism, shown in Fig. 7.9

Table 7.3. Parameters of NFS LB mechanism

Parameters	Traffic Source Model 1	Traffic Source Model 2
B	$10 = 2E_1(X)$	$30 \approx E_2(X)$
P	500	3,000
T_S	230 ms $= E_1(X)\Delta + E_1(S)$	$1{,}000$ ms $= E_2(X)\Delta + E_2(S)$

fuzzy inference rule, afterwards. Only a small change to the values of both parameters can make a big difference in the results. Therefore, it is necessary to assign fixed values to both parameters, before finding the membership function and fuzzy rule base, which can be obtained by trial and error, and then the parameters are tuned until satisfactory results are obtained. In contrast, the size of the token pool can be appropriately adjusted afterwards, because under a certain membership function and fuzzy rule base, there may be many appropriate sizes of token pool and the test results are slightly different.

The inputs of the proposed NFS controller consist of:

1. *The number of cells in the input queue buffer*, the fuzzy set of which is N_c = {*Zero (Z), Low (L), Medium (M), High (H), Very High (VH)*},
2. *The changing of the number of cells in the input queue buffer*, the fuzzy set of which is

   ΔN_c = {*Positive Big (PB), Positive Small (PS), Zero Equal (ZE), Negative Small (NS), Negative Big (NB)*},

3. *The number of tokens in the token pool*, the fuzzy set of which is

   N_T = {*Zero (Z), Low (L), Medium (M), High (H), Very High (VH)*},

4. *The token generation rate*, the fuzzy set of which is

   R_i = {*Positive Big (PB), Positive Medium (PM), Positive Small (PS), Zero Equal (ZE), Negative Small (NS), Negative Medium (NM), Negative Big (NB)*}.

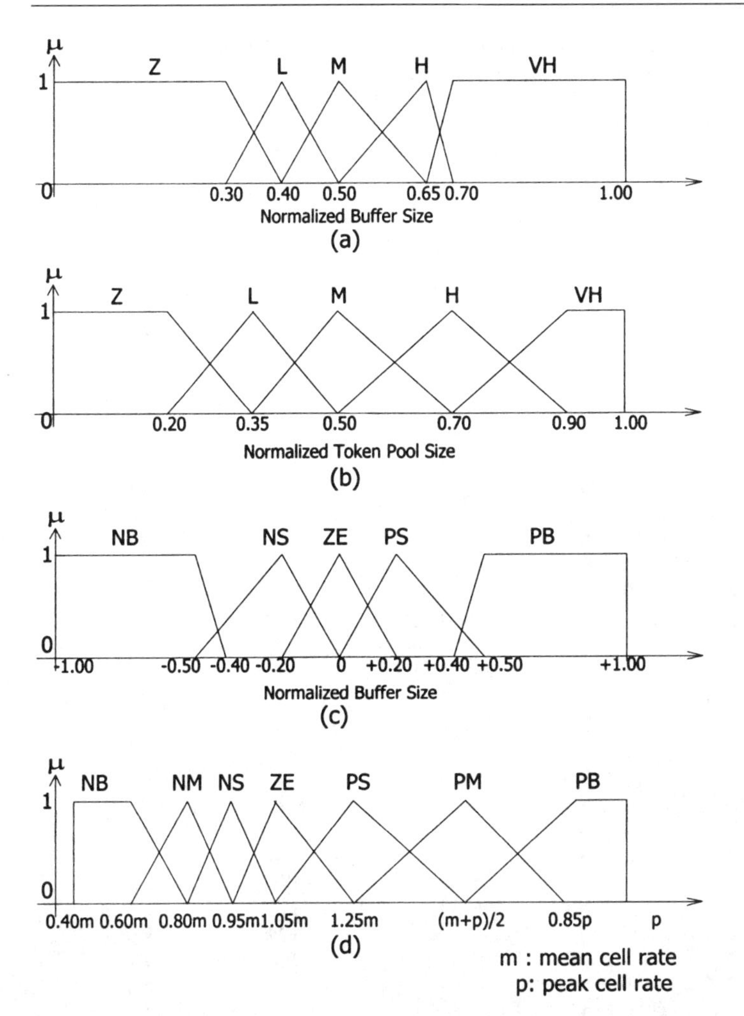

Fig. 7.11. Membership functions for fuzzy sets of NFS. (a) N_c (b) N_T (c) ΔN_c (d) R_i, R_{i+1}

The output of the proposed NFS controller is the token generation rate of the next window, the fuzzy set of which is

> R_{i+1} = {*Positive Big (PB), Positive Medium (PM), Positive Small (PS), Zero Equal (ZE), Negative Small (NS), Negative Medium (NM), Negative Big (NB)*}

The reason of designing the token generation rate to have the maximum number of fuzzy set members of 7 levels, is in order to make the adjustment of the token generation rate smooth and fine. The membership functions of inputs and output, are of a triangular shape as shown in Fig. 7.11(a)–(d) and the fuzzy rule base is shown in Table 7.4.

Table 7.4. Fuzzy rule base for inference of NFS controller in NFS LB mechanism

Rule #	N_c	ΔN_c	N_T	R_i	R_{i+1}
1	Z	NB	Z	NS	PB
2	Z	NB	Z	ZE	PM
3	Z	NB	Z	PS	PS
4	Z	NB	Z	PM	PS
5	Z	NB	Z	PB	PB
6	Z	NB	L	NS	PM
7	Z	NB	L	ZE	PS
8	Z	NB	L	PS	ZE
9	Z	NB	L	PM	ZE
10	Z	NB	L	PB	PS
11	Z	NB	M	ZE	PS
12	Z	NB	M	PS	ZE
13	Z	NB	M	PM	ZE
14	Z	NB	M	PB	ZE
15	Z	NB	H	PS	ZE
16	Z	NB	H	PM	NS
17	Z	NB	H	PB	NS
18	Z	NB	VH	PM	NM
19	Z	NB	VH	PB	NM
20	Z	NS	Z	NS	ZE
21	Z	NS	Z	ZE	PS
22	Z	NS	Z	PS	PS
23	Z	NS	Z	PM	ZE
24	Z	NS	Z	PB	NS
25	Z	NS	L	NS	PS
26	Z	NS	L	ZE	PS
27	Z	NS	L	PS	ZE
28	Z	NS	L	PM	PS
29	Z	NS	L	PB	PS
30	Z	NS	L	PS	NS
31	Z	NS	M	PM	ZE
32	Z	NS	M	PB	PS
33	Z	NS	H	PM	NS
34	Z	NS	H	PB	NS
35	Z	NS	VH	PM	NM
36	Z	NS	VH	PB	NM
37	Z	ZE	Z	NS	ZE
38	Z	ZE	Z	ZE	PS
39	Z	ZE	Z	PS	ZE
40	Z	ZE	Z	PM	PS
41	Z	ZE	L	NS	PM
42	Z	ZE	L	ZE	PS
43	Z	ZE	L	PS	ZE
44	Z	ZE	L	PM	PS
45	Z	ZE	L	PB	NS
46	Z	ZE	M	ZE	PM
47	Z	ZE	M	PS	NS
48	Z	ZE	M	PM	PS
49	Z	ZE	M	PB	ZE
50	Z	ZE	H	PS	ZE

Table 7.4 (continued)

Rule #	N_c	ΔN_c	N_T	R_i	R_{i+1}
51	Z	ZE	H	PM	ZE
52	Z	ZE	H	PB	NS
53	Z	ZE	VH	PM	NS
54	Z	ZE	VH	PB	NM
55	Z	PS	Z	NB	NM
56	Z	PS	Z	NM	NS
57	Z	PS	Z	NS	ZE
58	Z	PS	Z	ZE	PS
59	Z	PS	Z	PS	ZE
60	L	NB	Z	NS	PM
61	L	NB	Z	ZE	PS
62	L	NB	Z	PS	PS
63	L	NB	Z	PM	PS
64	L	NB	Z	PB	PM
65	L	NS	Z	NS	PS
66	L	NS	Z	ZE	ZE
67	L	NS	Z	PS	ZE
68	L	NS	Z	PM	PS
69	L	NS	Z	PB	PS
70	L	ZE	Z	NM	NS
71	L	ZE	Z	NS	ZE
72	L	ZE	Z	ZE	PS
73	L	ZE	Z	PS	ZE
74	L	ZE	Z	PM	ZE
75	L	ZE	Z	PB	NS
76	L	PS	Z	NB	NS
77	L	PS	Z	NM	ZE
78	L	PS	Z	NS	PS
79	L	PS	Z	ZE	PS
80	L	PS	Z	PS	ZE
81	L	PS	Z	PM	NS
82	L	PB	Z	NB	ZE
83	L	PB	Z	NM	PS
84	L	PB	Z	NS	PS
85	L	PB	Z	ZE	PS
86	L	PB	Z	PS	ZE
87	M	NB	Z	NS	PB
88	M	NB	Z	ZE	PM
89	M	NB	Z	PS	ZE
90	M	NB	Z	PM	PS
91	M	NS	Z	NS	PS
92	M	NS	Z	ZE	PS
93	M	NS	Z	PS	ZE
94	M	NS	Z	PM	ZE
95	M	NS	Z	PB	ZE
96	M	ZE	Z	NB	NS
97	M	ZE	Z	NM	ZE
98	M	ZE	Z	NS	ZE
99	M	ZE	Z	PS	ZE
100	M	ZE	Z	PM	PS

Table 7.4 (continued)

Rule #	N_c	ΔN_c	N_T	R_i	R_{i+1}
101	M	ZE	Z	PB	NS
102	M	PS	Z	NB	NM
103	M	PS	Z	NM	NS
104	M	PS	Z	NS	ZE
105	M	PS	Z	ZE	PS
106	M	PS	Z	PS	ZE
107	M	PS	Z	PM	ZE
108	M	PS	Z	PB	NS
109	M	PB	Z	NM	PS
110	M	PB	Z	NS	PS
111	M	PB	Z	ZE	PS
112	M	PB	Z	PS	NS
113	H	NB	Z	NS	PB
114	H	NB	Z	ZE	PM
115	H	NB	Z	PS	PM
116	H	NB	Z	PM	PS
117	H	NB	Z	PB	ZE
118	H	NS	Z	NS	PM
119	H	NS	Z	ZE	PS
120	H	NS	Z	PS	PS
121	H	NS	Z	PM	ZE
122	H	NS	Z	PB	NS
123	H	ZE	Z	NB	NM
124	H	ZE	Z	NM	NS
125	H	ZE	Z	NS	ZE
126	H	ZE	Z	ZE	PS
127	H	ZE	Z	PS	ZE
128	H	ZE	Z	PM	NS
129	H	ZE	Z	PB	NM
130	H	PS	Z	NB	NS
131	H	PS	Z	NM	ZE
132	H	PS	Z	NS	PS
133	H	PS	Z	ZE	PS
134	H	PS	Z	PM	NS
135	H	PS	Z	PB	NM
136	H	PB	Z	NB	ZE
137	H	PB	Z	NM	PS
138	H	PB	Z	NS	PS
139	H	PB	Z	ZE	NS
140	H	PB	Z	PS	NM
141	H	PB	Z	PM	NB
142	H	PB	Z	PB	NB
143	VH	NS	Z	PS	ZE
144	VH	NS	Z	PM	PS
145	VH	NS	Z	PB	ZE
146	VH	ZE	Z	NS	PS
147	VH	ZE	Z	PS	ZE
148	VH	ZE	Z	PM	ZE
149	VH	ZE	Z	PB	PS
150	VH	PS	Z	NB	ZE

Table 7.4 (continued)

Rule #	N_c	ΔN_c	N_T	R_i	R_{i+1}
151	VH	PS	Z	NM	PS
152	VH	PS	Z	NS	PS
153	VH	PS	Z	ZE	NS
154	VH	PS	Z	PS	ZE
155	VH	PS	Z	PM	NS
156	VH	PS	Z	PB	NM
157	VH	PB	Z	NB	NS
158	VH	PB	Z	NM	ZE
159	VH	PB	Z	ZE	PS
160	VH	PB	Z	PS	NS
161	VH	PB	Z	PM	NM
162	VH	PB	Z	PB	NB

The membership function is assigned by hand optimization and the decision rule to increase or decrease the token generation rate, in other words, the credit which is the origin of fuzzy rule base is summarized as follows:

1. If the behavior of the traffic source in the previous long-term period and in the previous short-term period violates the contract, the controller will largely decrease the credit (Negative Big).
2. If the behavior of the traffic source in the previous long-term period violates the contract and in the previous short-term period does not violate the contract, the controller will moderately decrease the credit (Negative Medium).
3. If the behavior of the traffic source in the previous long-term period does not violate the contract and in the previous short-term period violates the contract, the controller will slightly decrease the credit (Negative Small), or will not change the credit.
4. If the behavior of the traffic source in the previous long-term period does not violate the contract and in the previous short-term period also does not violate the contract, the controller will increase the credit, or will not change the credit.

The degree of adjustment in increasing or decreasing, depends on the degree of violation or not violation by trial and error fine tunings.

Fuzzy centroid method and piecewise linearity algorithm [8] are used in defuzzification as a quick solution.

7.3.4 General Problem Statement

The general problems to be studied and solved in this chapter, are as follows:

1. How to develop a novel MCR policing of a single traffic source in order to prevent the occurrence of congestion in ATM network and be able to utilize the network resource efficiently,

2. How to improve the performance of overload detection and decreasing CLR, caused by traffic policing in UPC when compared with LB mechanism, and improve the responsiveness to respond faster than FL window mechanism, and
3. How to decrease the complexity and complication of the inference engine of the fuzzy controller by feedforward neural network, to map the relation between the token generation rate of the traffic policing (input) and the variable showing the traffic characteristics (output), using a back propagation algorithm to train the neural network to achieve the output, as close to the output obtained from the conventional fuzzy controller as possible.

7.4 Performance of FLLB Policing Mechanism

To solve the problem of static behavior in conventional policing mechanisms, the FNN controller is utilized in order to adjust the token generation rate. The main structure of the FNN used is based on FL, and NN is applied as part of the FL structure to map the inputs and outputs of the inference engine. Therefore, to save time in the NN training process, the performance of the proposed policing when FL controller is used (called FLLB as stated in section 7.3.3), should be evaluated first. The parameter of token pool size should then be tuned until the best performance is achieved, before the training process of NN is made.

7.4.1 Effects of Token Pool Size on Policing Performance

Since the system parameters consisting of input queue buffer size and sampling period are tuned to be harmonious for a set of membership functions and fuzzy rules, these parameters are fixed and cannot be changed. On the contrary, more than one values of the token pool size that give the required performance in the acceptable range can be found, because the token pool is only used for excessive token storage, and is not involved in the CLR of the system directly.

When testing with the first source model ($E(x) = 5$ cells, $E(S) = 147.72$ ms), the CLR caused by the policing with different sizes of token pool (P) which are the member of the set $\{500, 750, 1000, 1250\}$, is measured. The traffic load fed is varied from $\sigma \in \{1.00, 1.05, 1.10, \ldots, 2.00\}$. The traffic load, which is the normalized value of actual MCR of the source with the negotiated MCR, can be varied by two methods, varying $E(X)$ while fixing $E(S)$, and varying $E(S)$ while fixing $E(X)$. The overload detection performances of the proposed policing (FLLB or Fuzzy LB), the conventional LB and the FL window (Fuzzy Window), are shown in Fig. 7.12 and Fig. 7.13, when varying the traffic load by varying $E(X)$ and by varying $E(S)$ of the source, respectively. To give an obvious view of performance comparison, the Root of Sum Square Error (RSSE) is calculated from the difference between the CLR caused by the policing (CLR_m) and the ideal violation

Table 7.5. Overload detection comparison among different types of policing when tested with source model type 1

Policing	Conventional LB	FL Window	FLLB			
			$P = 500$	$P = 750$	$P = 1{,}000$	$P = 1{,}250$
RSSE Vary $E(X)$	1.023511	0.161944	0.008563	0.009743	0.010369	0.010221
Vary $E(S)$	1.145818	0.207957	0.007548	0.008612	0.009030	0.010270

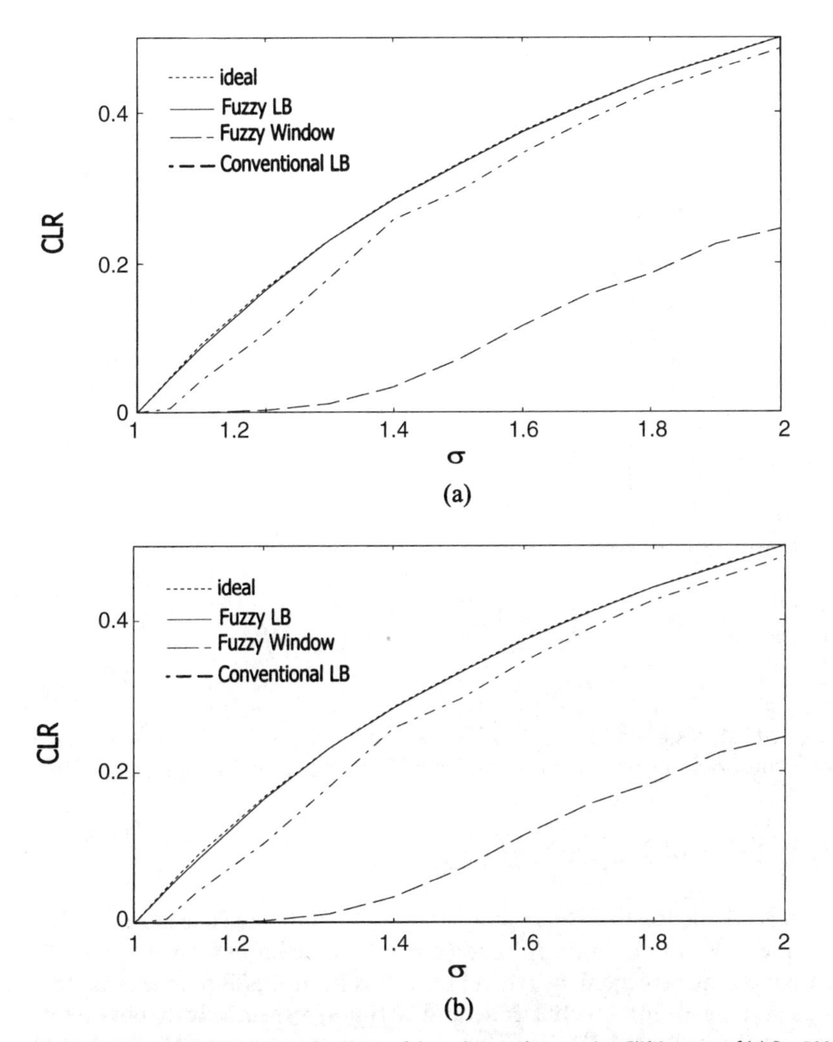

Fig. 7.12. Overload detection performance of the policing when varying $E(X)$ in cases of (a) $P = 500$, (b) $P = 750$, (c) P = 1,000 and (d) $P = 1{,}250$

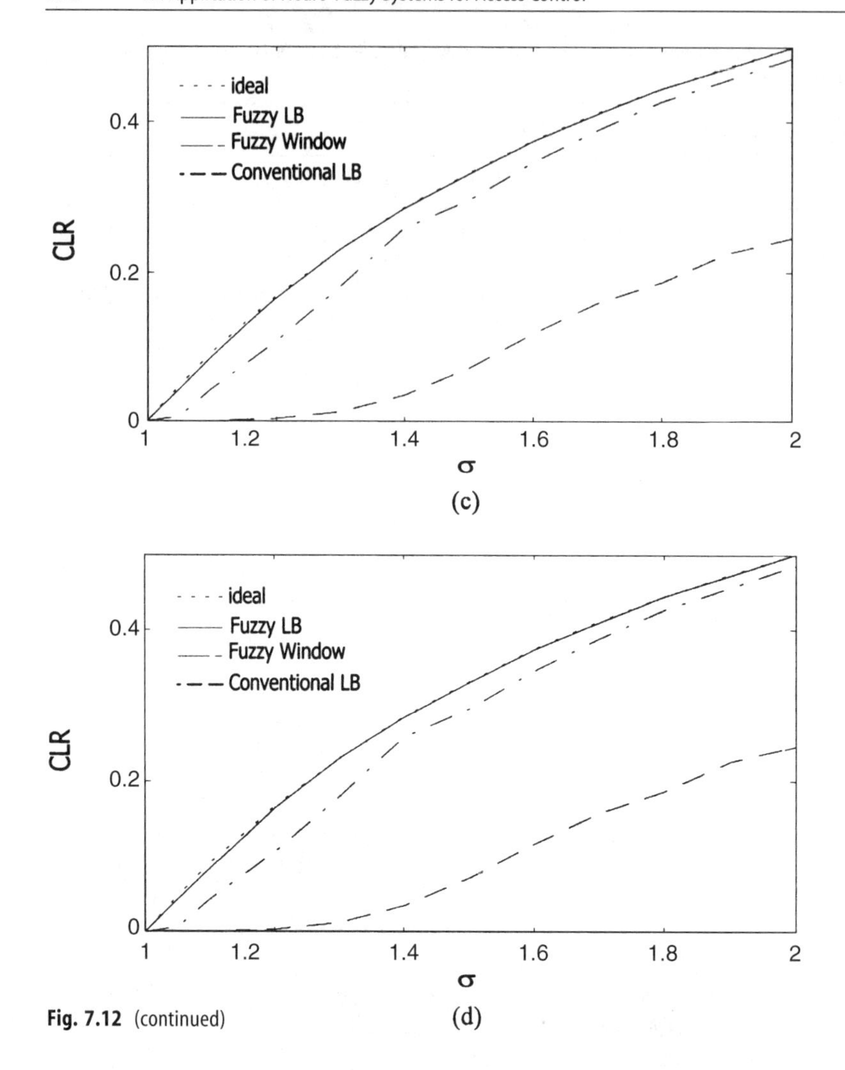

Fig. 7.12 (continued)

probability (P_d) as shown in Eq. (7.9). The less the RSSE is, the closer to the ideal behavior the policing performance is. The RSSEs obtained are given in Table 7.5.

$$RSSE = \sqrt{\sum_{\sigma} (P_d - CLR_m)^2} \tag{7.9}$$

From Table 7.5, the least RSSE is found at $P = 500$, while the false alarm probability or the CLR at normal traffic load ($\sigma = 1.0$) is the lowest, when $P = 1{,}250$. Compared to other mechanisms, the FLLB yields lower RSSE than the conventional LB and the FL window 100 times and 20 times, respectively. To obtain this greatly better overload detection performance, the proposed policing has to trade off with the higher false alarm probability that is in the order of 10^{-7}, com-

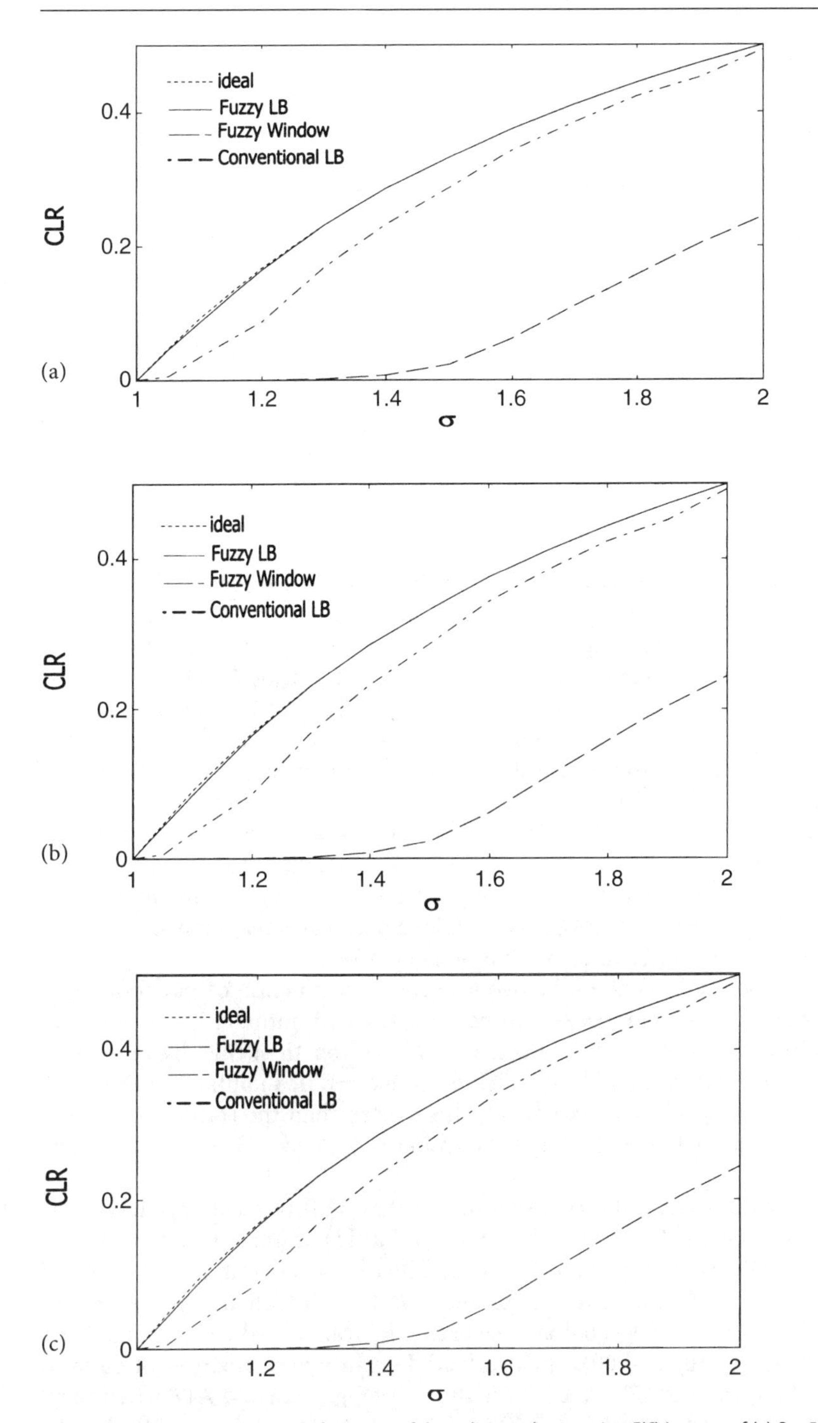

Fig. 7.13. Overload detection performance of the policing when varying $E(S)$ in cases of (a) $P = 500$, (b) $P = 750$, (c) $P = 1{,}000$ and (d) $P = 1{,}250$

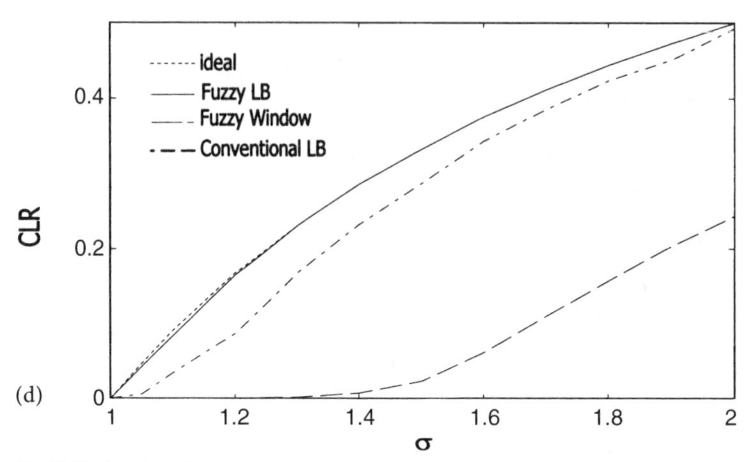

Fig. 7.13 (continued)

pared to those of FL window which is in the order of 10^{-8}. However, the false alarm of the FLLB, is still lower than those of the conventional LB that is in the order of 10^{-5}, and lower than the QoS of packet voice that is in the order of 10^{-3}.

On the responsiveness aspect, the evaluation, in terms of the number of cells entering the network from the violating source before the policing discards the first cell, is performed. The performance comparison is conducted at traffic load $\sigma = 1.50$. The responsiveness results, are shown in Figs. 7.14, 7.15 and Table 7.6.

From Figs. 7.14, 7.15 and Table 7.6, increasing token pool size degrades the initial responsiveness behavior of the proposed policing. However, the long-term dynamic performances show only a slight difference. Compared to the other two mechanisms, the proposed FLLB yields, on average, a much better dynamic behavior in both initial and long-term periods.

Another aspect that cannot be overlooked, is the reaction of the policing to different kinds of violations. As can be seen, the conventional LB responds to the violation by decreasing $E(S)$ considerably slower than that, by increasing $E(X)$ since the traffic, the $E(S)$ of which is reduced while maintaining $E(X)$, still has the frequency of recovery periods more often than the traffic, the $E(S)$ of which is increased. But when the FL controller is applied, this difference can be reduced.

In order to evaluate the performance of the FLLB toward a model of real packet voice source, the source which has the traffic parameters, $E(X) = 29$ cells and $E(S) = 650$ ms at the peak bit rate of 32 kbps, is considered. The CLR caused by the policing with different sizes of token pool (P), which are the member of the set {3000, 3600, 4200, 4800} is measured. The traffic load fed is varied from $\sigma \in \{1.00, 1.05, 1.10, \ldots, 2.00\}$. The overload detection performances of the proposed policing and the FL window, are shown in Figs. 7.16 and 7.17, when varying the traffic load by varying $E(X)$ and $E(S)$ of the source, respectively. The RSSEs obtained are given in Table 7.7.

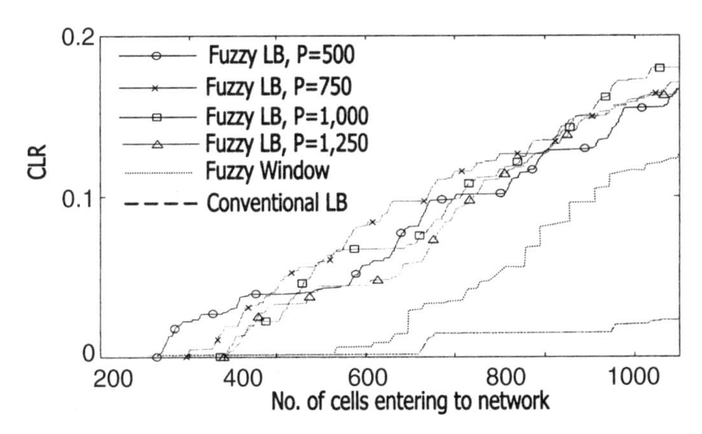

Fig. 7.14. Responsiveness behavior comparison of policing, tested with source model 1, the load of which is varied by varying $E(X)$

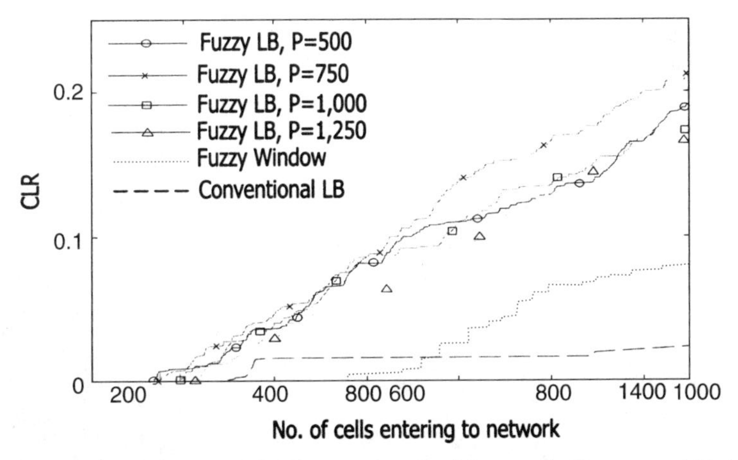

Fig. 7.15. Responsiveness behavior comparison of policing, tested with source model 1, the load of which is varied by varying $E(S)$

Table 7.6. Initial responsiveness behavior comparison among different types of policing, when tested with source model 1

Number of cells entering network before first rejection	Conventional LB	FL Window	FLLB			
			$P = 500$	$P = 750$	$P = 1{,}000$	$P = 1{,}250$
Vary $E(X)$	309	731	328	394	472	482
Vary $E(S)$	496	757	334	346	392	424

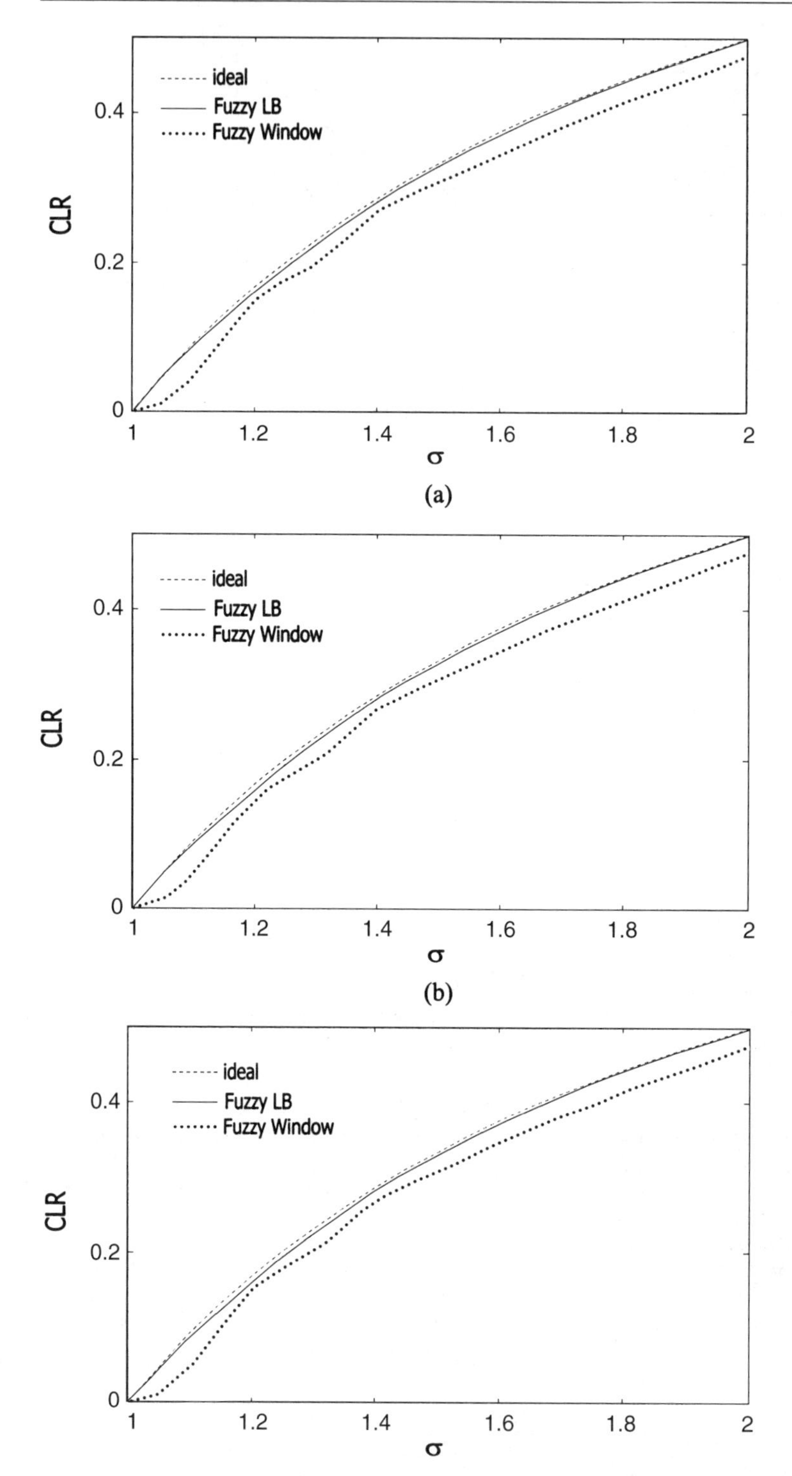

Fig. 7.16. Overload detection performance of the policing, when varying $E(X)$ in cases of (a) $P = 3{,}000$, (b) $P = 3{,}600$, (c) $P = 4{,}200$ and (d) $P = 4{,}800$

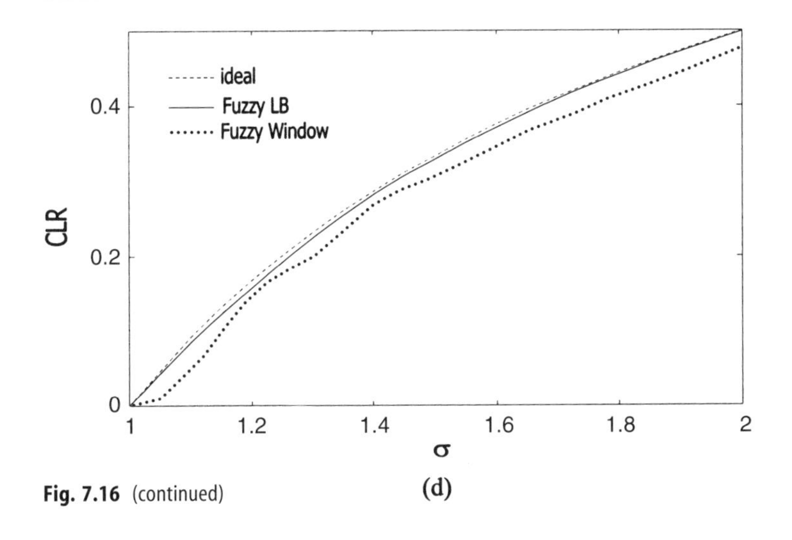

Fig. 7.16 (continued) **(d)**

Table 7.7. Overload detection comparison among different types of policing when tested with source model type 2

Policing	FL Window	FLLB			
		$P = 3{,}000$	$P = 3{,}600$	$P = 4{,}200$	$P = 4{,}800$
RSSE Vary $E(X)$	0.130790	0.019124	0.020493	0.020825	0.023615
Vary $E(S)$	0.181735	0.067424	0.069386	0.071187	0.072953

Table 7.8. Initial responsiveness behavior comparison among different types of policing when tested with source model 2

Number of cells entering network before first rejection	FL Window	FLLB			
		$P = 3{,}000$	$P = 3{,}600$	$P = 4{,}200$	$P = 4{,}800$
Vary $E(X)$	1,244	768	845	838	1,301
Vary $E(S)$	1,967	943	962	1,097	1,134

The proposed policing still gives the least RSSE when using the smallest token pool size, 3,000 tokens. When compared to the FL window, the FLLB gives better overload detection performance in every size of token pool.

On the responsiveness aspect, with the same method as testing with source model type 1, the results are shown in Figs. 7.18, 7.19 and Table 7.8.

From Table 7.8, the FLLB takes action faster than the FL window except when using token pool size of 4,800. An increase in the token pool size tends to yield slower initial rejection time.

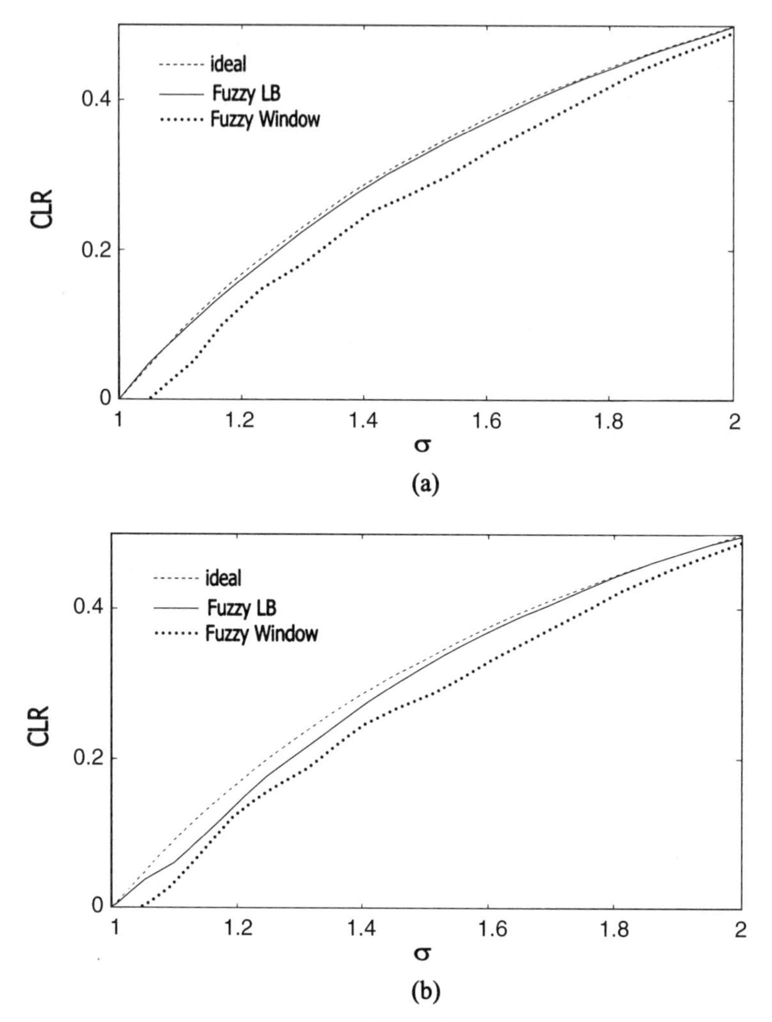

Fig. 7.17. Overload detection performance of the policing when varying $E(S)$ in cases of (a) $P = 3{,}000$, (b) $P = 3{,}600$, (c) $P = 4{,}200$ and (d) $P = 4{,}800$

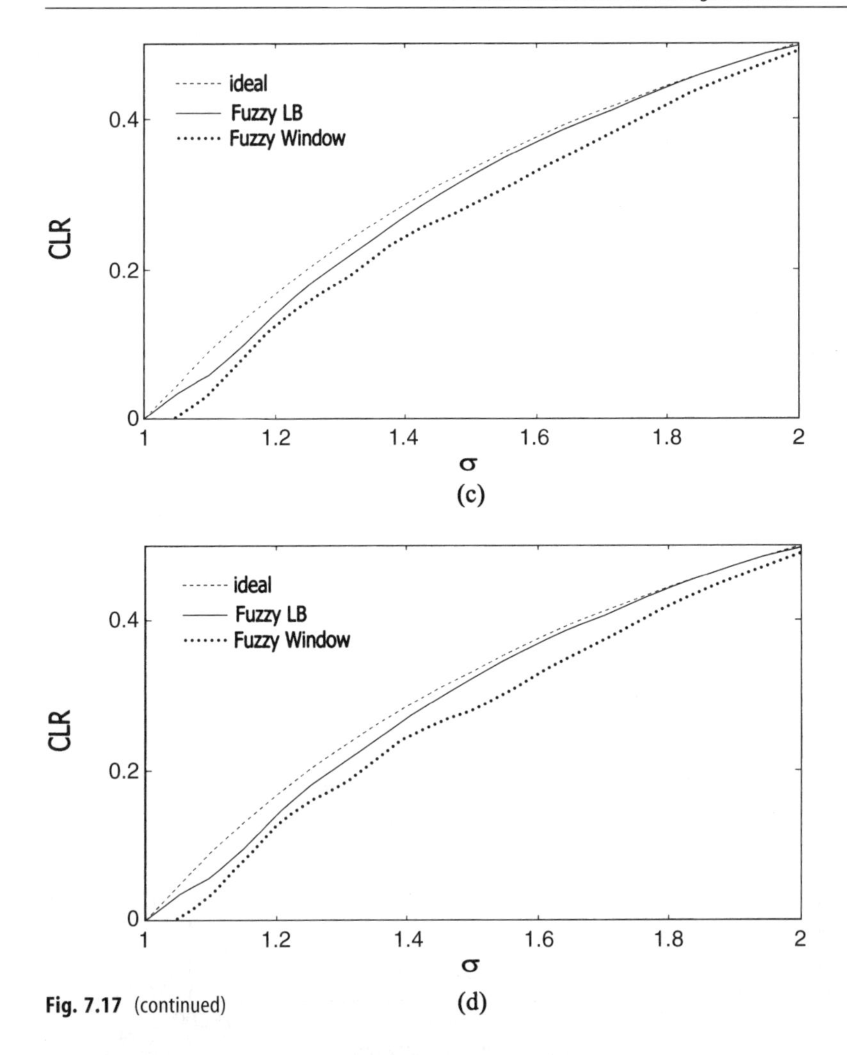

Fig. 7.17 (continued)

7.5 Performance of NFS LB Policing Mechanism

Since the FL controller has 4 inputs and up to 162 fuzzy rules, it is proposed to use NN to map the input and output of inference engine in FL, in order to reduce complication and processing time. This composite structure is called "Neuro-Fuzzy System" (NFS). The NN is trained with a supervised method, the input and target output are fed to the network until the target Sum of Square Error (SSE) calculated from the Eq. (7.10), is not over 10^{-5}.

$$SSE = \sum_{p=1}^{P} \sum_{i=1}^{N} (t_{ip} - o_{ip})^2 \tag{7.10}$$

Where t and o represent the target output and the real output, respectively.

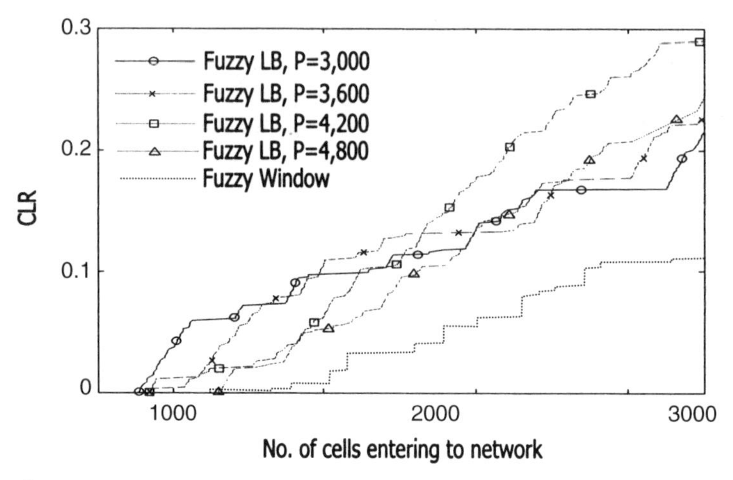

Fig. 7.18. Responsiveness behavior comparison of policing tested with source model 2 the load of which is varied by varying $E(X)$

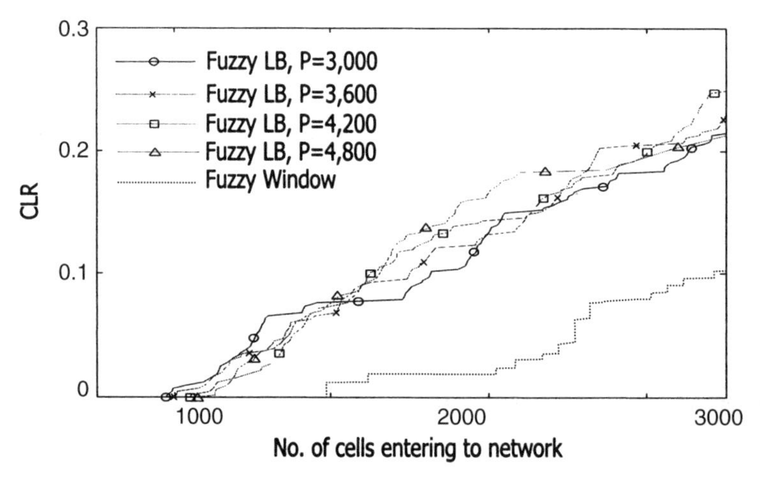

Fig. 7.19. Responsiveness behavior comparison of policing tested with source model 2 the load of which is varied by varying $E(S)$

7.5.1 NN Structure

Multilayer perceptron with single hidden layer, is chosen to perform the mapping function. The number of input nodes and output nodes in input layer and output layer, are equal to 22 and 7 nodes, respectively, as shown in Fig. 7.20. However, the suitable number of hidden nodes in hidden layer cannot be determined in advance, because it depends on the training data. Therefore, it is necessary to have a test to find the suitable number of hidden nodes of all policing performance tests, in advance.

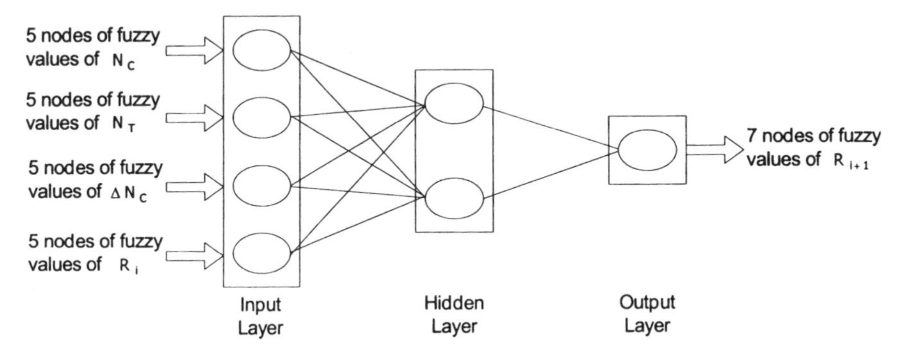

Fig. 7.20. Neural network functioning as inference engine for Neuro-Fuzzy controller, shown in Fig. 7.10

22 input nodes, consist of:

1) 5 nodes of fuzzy values of the number of cells in input queue buffer (N_c)
2) 5 nodes of fuzzy values of the number of tokens in token pool (N_T)
3) 5 nodes of fuzzy values of the difference between the number of cells in the input queue buffer, from the previous sampling period to the present one (ΔN_c)
4) 7 nodes of fuzzy values of token generation rate in the present sampling period (R_i)
   7 output nodes are fuzzy values of token generation rate in the next sampling period (R_{i+1})

The activation function from input layer to hidden layer is *Bipolar Sigmoid Function* as shown in Eq. (7.11).

$$f_{ih}(x) = \frac{2}{1+\exp(-x)} - 1 \ ; \ f'_{ih}(x) = \frac{1}{2}\,[1+f_{ih}(x)]\,[1-f_{ih}(x)] \tag{7.11}$$

and the activation function from hidden layer to output layer is *Binary Sigmoid Function,* as shown in Eq. (7.12).

$$f_{ho}(x) = \frac{1}{1+\exp(-x)} \ ; \ f'_{ho}(x) = f_{ho}(x)[1-f_{ho}(x)] \tag{7.12}$$

7.5.2 Simulation Results when Tested with Source Model 1

In order to have the most strict policing, the token pool size of 500 is chosen to control traffic from the source model 1, since it gives the least RSSE and the fastest rejection time while the cell discard ratio never exceeds the ideal violation probability. The complete training input consists of $N_C \in \{0, 1, \ldots, 10\}$, $\Delta N_T \in \{0, 1, \ldots, 500\}$, $\Delta N_C \in \{-10, -9, \ldots, +10\}$ and $R_i \in (8,80, 62,50)$. It is noticeable that the input R_i is a continuous value that gives unlimited members in the input set. Moreover, there are too many combinations between inputs N_C, N_T, ΔN_C

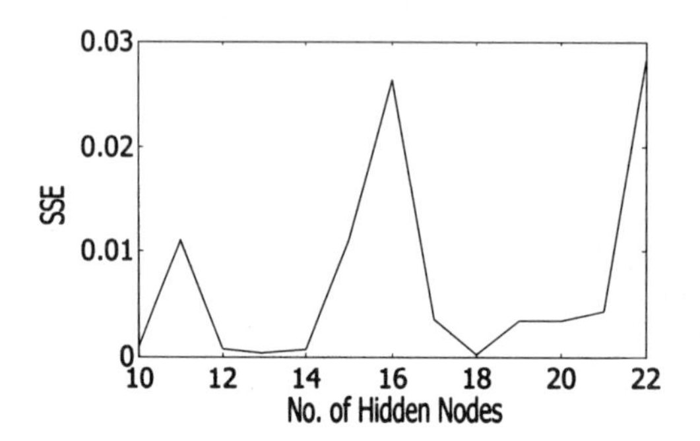

Fig. 7.21. Finding the number of hidden nodes to be used when tested with source model 1

and R_i. Therefore, only some of all inputs will be chosen to be training sets which are $N_C \in \{0, 1, \dots, 10\}$, $N_T \in \{0, 10, \dots, 500\}$, $\Delta N_T \in \{-10, -9, \dots, +10\}$ and $R_i \in \{10.0, 12.0, \dots, 62.0\}$. These inputs have to be fuzzified into fuzzy values before they are fed into the NN. The training algorithm is Marquardt-Lavenberg Backpropagation. The test to find the appropriate number of hidden nodes is conducted, and the result is shown in Fig. 7.21.

From Fig. 7.21, the number of hidden nodes to be used is 18 nodes. After training the NN which has the selected number of hidden nodes with the selected input until the SSE is quite steady, the obtained weights are used in NFS controller, and the overload detection of the policing is evaluated by comparison to those of the FL controller. The results are shown in Figs. 7.22 and 7.23.

From Figs. 7.22 and 7.23, errors obtained when using NFS controller instead of FL controller are 5.14% and 5.11% for policing the $E(X)$-varying source and the $E(S)$-varying source, respectively. Although the errors are not high, they are still unacceptable, because the CLRs caused by the NFS controller are higher than those caused by the FL controller, and tend to exceed the ideal violation probability. To solve the problem, the number of combinations of inputs required for training is reduced, by discarding the middle input in three adjacent inputs, that give the same value of output. Since more than one parameters have

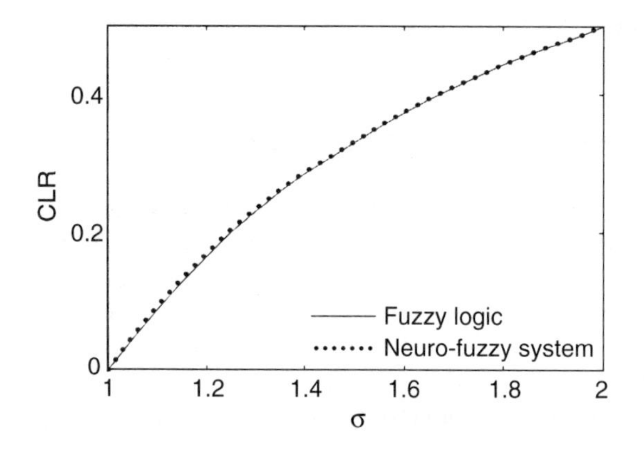

Fig. 7.22. Overload detection performance comparison between NFS (Neuro-Fuzzy System) controller and FL (Fuzzy logic) controller, when varying $E(X)$ of the source

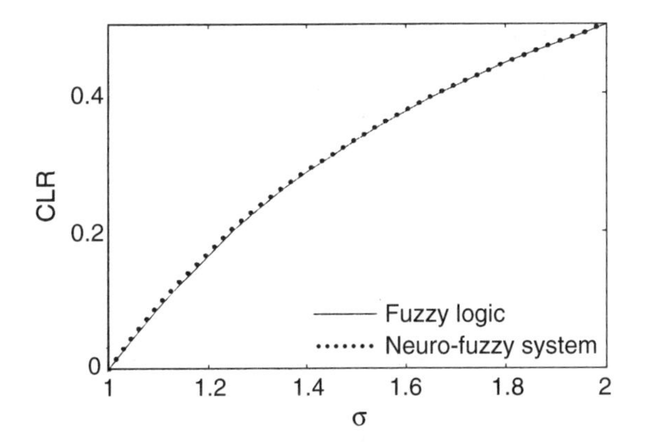

Fig. 7.23. Overload detection performance comparison between NFS (Neuro-Fuzzy System) controller and FL (Fuzzy logic) controller, when varying $E(S)$ of the source

been chosen to be the inputs, the size of input set, 18,117 combinations, is reduced by 3 methods:

1) by reducing the successive inputs N_T that give repeated outputs to 11,714 combinations,
2) by reducing the successive inputs R_i that give repeated outputs to 13,691 combinations and
3) by reducing the successive inputs N_T, R_i that give repeated outputs to 9,605 combinations.

The number of hidden nodes test is conducted for each method, and the results are shown in Fig. 7.24.

From Fig. 7.24, it can be seen that the most suitable number of hidden nodes for method 1, 2 and 3, are 12, 12, and 11 nodes, respectively. After training NN with inputs selected by the above three methods, the overload detection performance of policing which uses NFS is compared with those of policing which uses FL, as shown in Figs. 7.25, 7.26 and 7.27.

From Figs. 7.25–7.27, the SSEs between NFS controller and FL controller when reducing input by method 1, 2 and 3, are shown in Table 7.9.

Input reduction can help decrease error, and the trained NN gives closer outputs to the targets, especially in the case of method 1. Nevertheless, reducing too many inputs can make training performance worse, as in the case of method 3. From the results, it can be concluded that giving sufficient training inputs is necessary, on the other hand, training NN with as many inputs as possible, does not always give the best result.

The trained NN with method 1 is tested for overload detection performance compared with the conventional LB and the FL window, and the results are shown in Figs. 7.28, 7.29 and Table 7.10.

From Table 7.10, the NFS LB still gives the best overload detection performance with the least RSSE. It trades off with a higher false alarm probability than the FL window, in the order of 10^1. When compared with the policing using FL, the policing with NFS yields lower false alarm probability, while giving a slightly higher value of RSSE. This means that the NFS LB (with method 1), be-

Table 7.9. Error of overload detection performance between the policing with NFS controller and the policing with FL controller, when reducing input with different methods

Input reduction method	(i)	(ii)	(iii)
Error when varying $E(X)$ (%)	0.80	1.97	7.11
Error when varying $E(S)$ (%)	0.94	1.87	7.37

Table 7.10. Selectivity performance comparison between the NFS LB (with method 1) and other mechanisms when tested with source model 1

Mechanisms		Conventional LB	FL Window	NFS LB
RSSE	Vary $E(X)$	1.023511	0.161944	0.011550
	Vary $E(S)$	1.145818	0.207957	0.011744
False Alarm Probability		2.72×10^{-5}	1.20×10^{-8}	5.22×10^{-7}

haves less strictly than others, not only in the overload case but also in the normal load case. Using the same method as in the above section, the responsiveness behavior of the policing, is compared with those of other mechanisms. The results are shown in Figs. 7.30 and 7.31.

It can be simply described that the error of NN training, causes less strict behavior of the policing. Having the less strict policing, does not mean that all of the policing performance will be worse. In the process of policing development, there will always be conflict between the requirements needed, such as false alarm and responsiveness, therefore, in this case, when the policing shows the worse performance in overload detection, it gives the lower false alarm probability. However, it must be noted that this error cannot be avoided. The more suitable the method of training that is conducted, the less error will appear.

When tested with source model 2, from the result in the above section, the policing with the token pool size of 3,000 tokens, yields the most strict behavior of policing, but it is noticeable that the CLR caused by the policing exceeds the ideal violation probability, at low-level overload traffic. This means that the policing violates the QoS of users, therefore the token pool size of 3,600 tokens, which gives the second most strict behavior, is chosen instead. The complete training input consists of $N_C \in \{0, 1, \ldots, 30\}$, $N_T \in \{0, 1, \ldots, 3600\}$, $\Delta N_C \in \{-30, -29, \ldots, +30\}$ and $R_i \in (11.60, 83.30)$. The size of input set to be trained (which has 50,688 combinations) will be reduced by the same method, as when tested with source model 1 as follows:

1) Reducing the successive inputs N_T that give repeated outputs to 37,694 combinations,
2) Reducing the successive inputs R_i that give repeated outputs to 40,470 combinations and
3) Reducing the successive inputs N_T, R_i that give repeated outputs to 28,511 combinations.

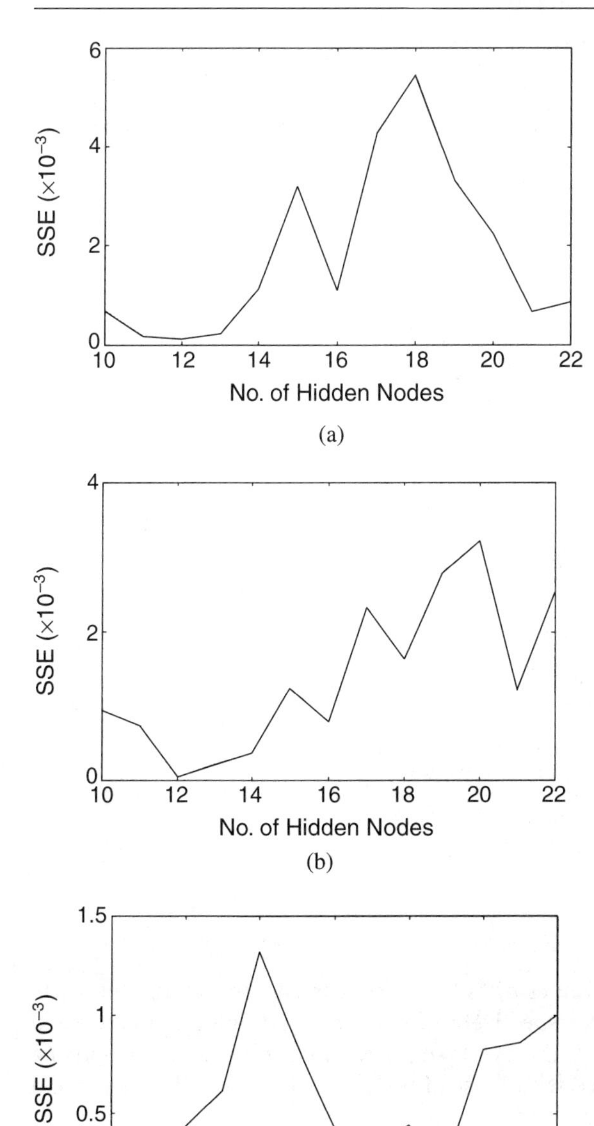

Fig. 7.24. Finding the number of hidden nodes to be used after reducing the size of input set by (a) method 1, (b) method 2, and (c) method 3

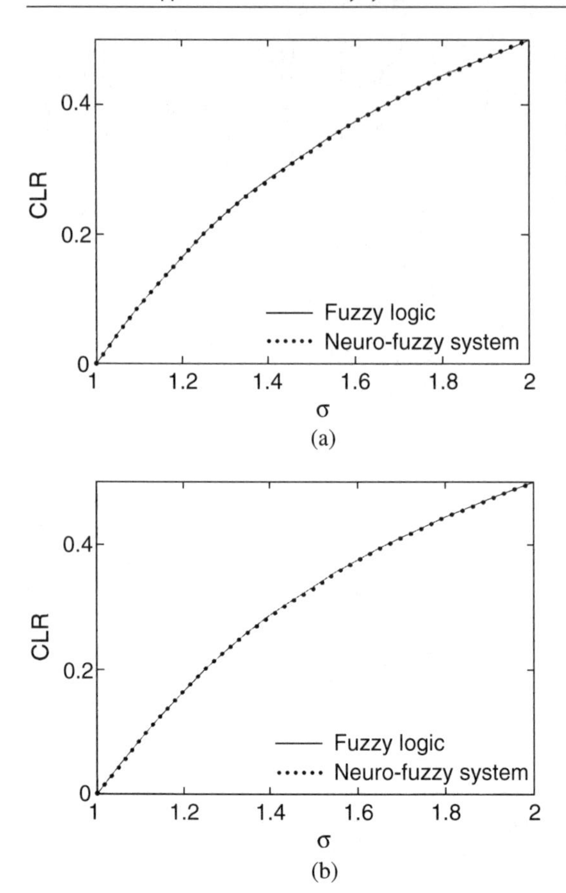

Fig. 7.25. Overload detection performance comparison between NFS controller trained with inputs selected by method 1, and FL controller when (a) varying $E(X)$ and (b) varying $E(S)$, of the source

From the test result, the number of hidden nodes most suited to method (i)–(iii) are 14, 16 and 16 nodes, respectively. After training NN with inputs selected by the above three methods, the overload detection performance of policing which uses NFS, is compared with those of policing which uses FL, as shown in Figs. 7.32, 7.33 and 7.34.

From Figs. 7.32, 7.33 and 7.34, the SSEs between NFS controller and FL controller, when reducing input by method 1, 2 and 3, are shown in Table 7.11.

Table 7.11. Error of overload detection performance between the policing with NFS controller and the policing with FL controller, when reducing input with different methods

Input reduction method	1	2	3
SSE when varying $E(X)$ (%)	29.10	39.95	66.42
SSE when varying $E(S)$ (%)	37.64	47.91	74.11

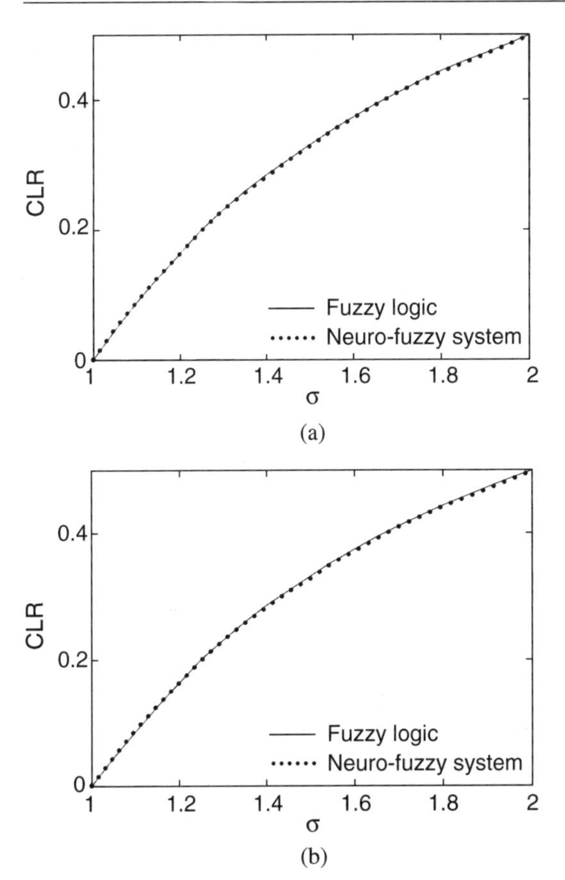

Fig. 7.26. Overload detection performance comparison between NFS controller trained with inputs selected by method 2, and FL controller when (a) varying *E*(*X*) and (b) varying *E*(*S*), of the source

From the results, in this case, the same method (as in the case of source model 1 to solve the problem of NN training) cannot be used. The solution is to use more than one NN called the parallel-structure NN. It helps reduce the number of inputs to be trained for a NN, while the data resolution is not affected. The structure of parallel NN with n subgroups of NNs used, is shown in Fig. 7.35 [9]. Each of the subgroups of NNs has structure similar to that shown in Fig. 7.20.

In the simulation, the input set is divided into 2, 3, 4 and 5 groups, according to the input parameter R_i as follows:

1) 2 groups, $R_{i1} \in \{11.0, 13.0, \ldots, 45.0\}$ and $R_{i2} \in \{47.0, 49.0, \ldots, 83.0\}$
2) 3 groups, $R_{i1} \in \{11.0, 13.0, \ldots, 33.0\}$, $R_{i2} \in \{35.0, 37.0, \ldots, 57.0\}$ and $R_{i3} \in \{59.0, 61.0, \ldots, 83.0\}$
3) 4 groups, $R_{i1} \in \{11.0, 13.0, \ldots, 27.0\}$, $R_{i2} \in \{29.0, 31.0, \ldots, 45.0\}$, $R_{i3} \in \{47.0, 49.0, \ldots, 63.0\}$ and $R_{i4} \in \{65.0, 67.0 \ldots, 83.0\}$
4) 5 groups, $R_{i1} \in \{11.0, 13.0, \ldots, 23.0\}$, $R_{i2} \in \{25.0, 27.0, \ldots, 37.0\}$, $R_{i3} \in \{39.0, 41.0, \ldots, 51.0\}$, $R_{i4} \in \{53.0, 55.0, \ldots, 67.0\}$ and $R_{i5} \in \{69.0, 71.0, \ldots, 83.0\}$

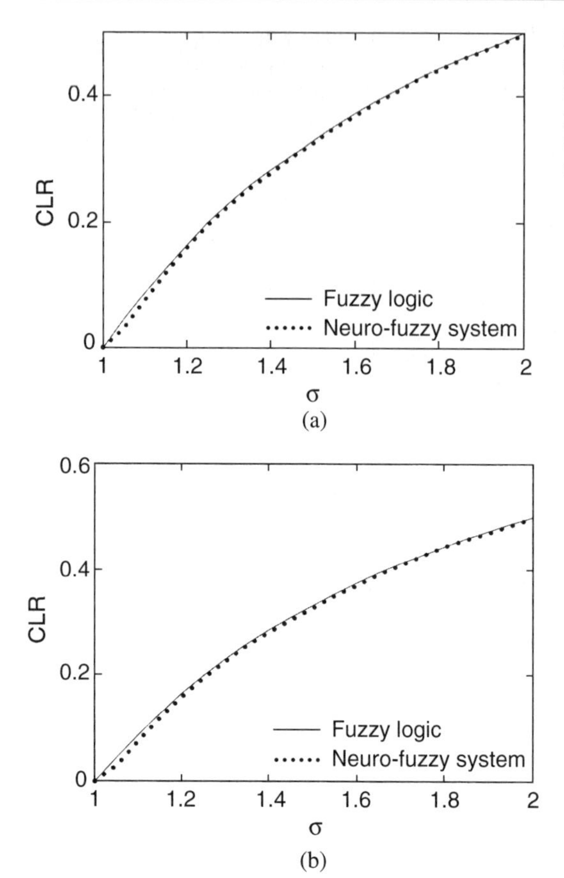

Fig. 7.27. Overload detection performance comparison between NFS controller trained with inputs selected by method 3, and FL controller when (a) varying $E(X)$ and (b) varying $E(S)$ of the source

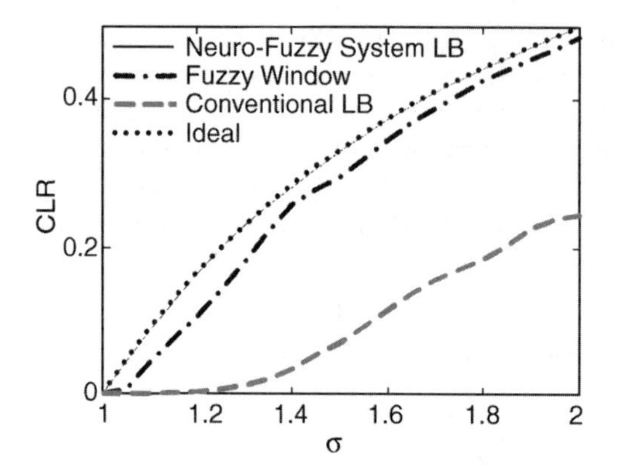

Fig. 7.28. Overload detection performance comparison between the NFS LB and other mechanisms, when varying $E(X)$ of the source

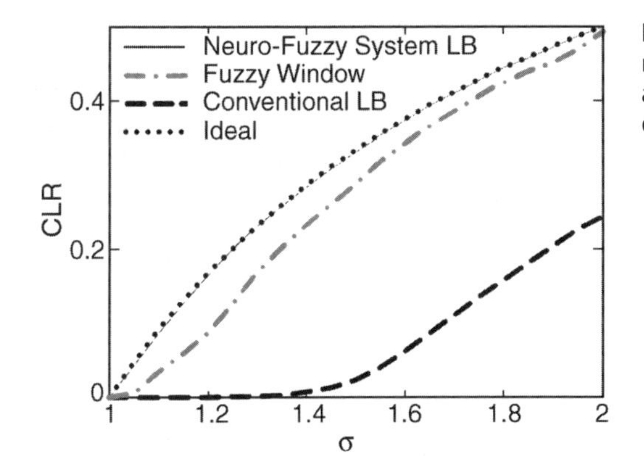

Fig. 7.29. Overload detection performance comparison between the NFS LB and other mechanisms, when varying $E(S)$ of the source

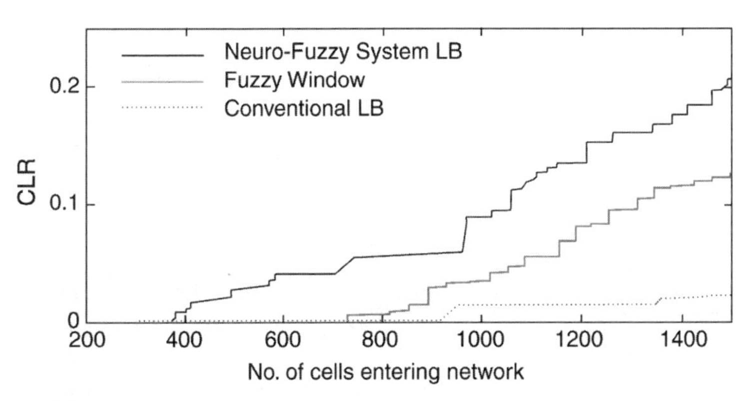

Fig. 7.30. Responsiveness behavior comparison between NFS controller and FL controller, when varying $E(X)$ of the source

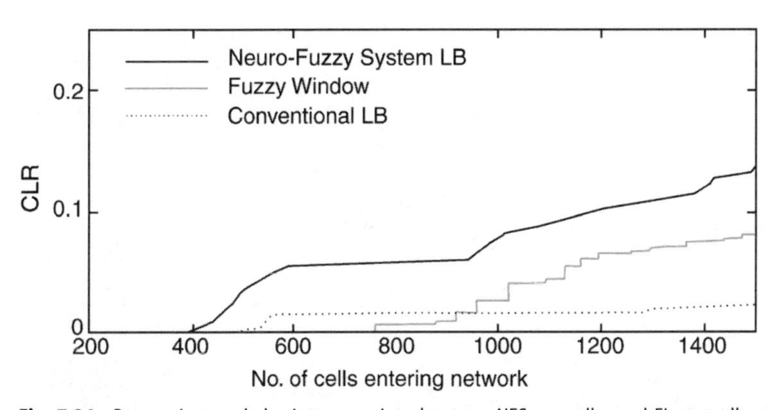

Fig. 7.31. Responsiveness behavior comparison between NFS controller and FL controller, when varying $E(S)$ of the source

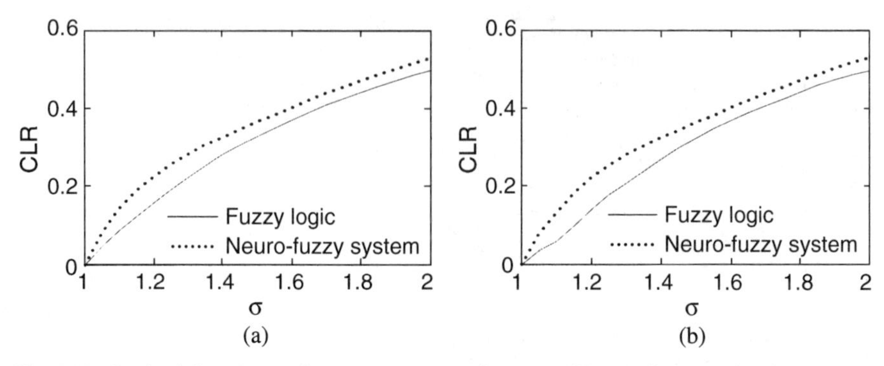

Fig. 7.32. Overload detection performance comparison between NFS controller trained with inputs selected by method 1, and FL controller when (a) varying $E(X)$ and (b) varying $E(S)$ of the source

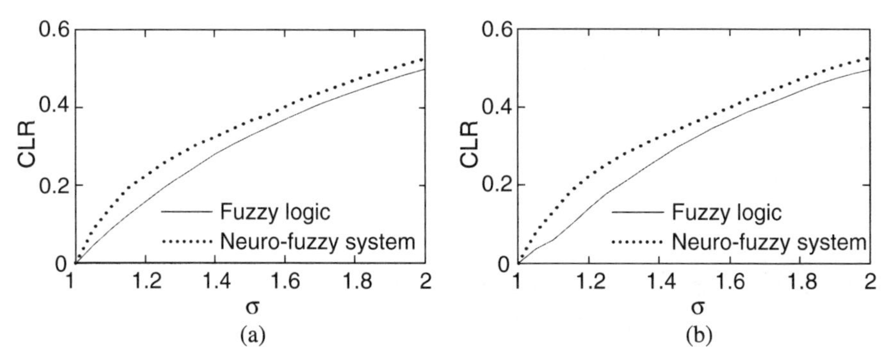

Fig. 7.33. Overload detection performance comparison between NFS controller trained with inputs selected by method 2, and FL controller when (a) varying $E(X)$ and (b) varying $E(S)$ of the source

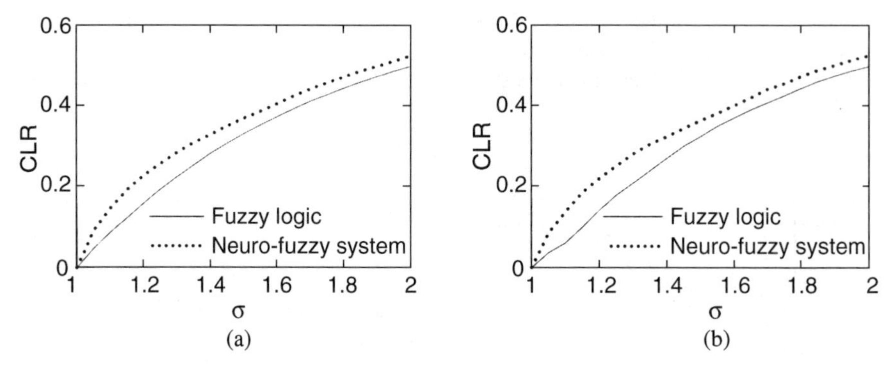

Fig. 7.34. Overload detection performance comparison between NFS controller trained with inputs selected by method 3, and FL controller when (a) varying $E(X)$ and (b) varying $E(S)$ of the source

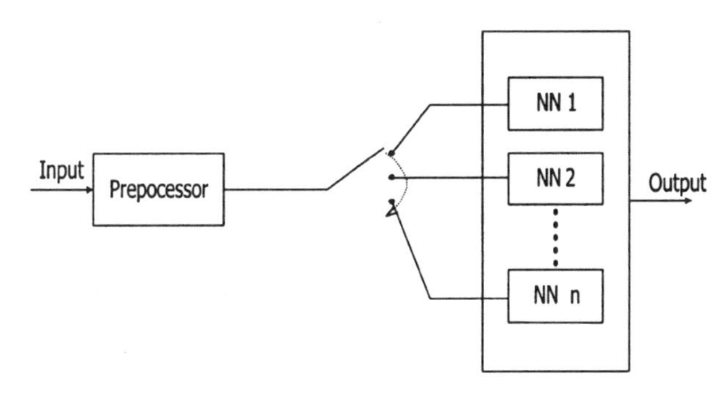

Fig. 7.35. Parallel NN model with n subgroups of NNs

The parallel NN is trained with the number of hidden nodes that have been proven to be most suitable for each NN, then the policing with the trained parallel NN is tested for the overload detection performance, compared with the policing with FL. The results are shown in Figs. 7.36–7.39, and the error of overload detection performance is shown in Table 7.12.

From the results, it is found that in this case, dividing the input set into subgroups, helps increase the training efficiency. However, it cannot be concluded that the more subgroups divided, the more efficient the training is. Since, it is noticeable that dividing inputs into 5 subgroups does not give any better performance than dividing into 3 or 4 subgroups. Therefore, another important factor for NN training efficiency, apart from the amount of inputs, is the pattern of inputs to be trained.

The trained NN is tested for the overload detection performance compared with the FL window, and the results are shown in Figs. 7.40 and 7.41.

The proposed scheme still shows better overload detection performance compared with FL window, but this improvement unavoidably results in a degradation in false alarm of the policing, as shown in Table 7.13.

On the responsiveness aspect, the policing is tested with the same method, as when tested with source model 1. The results are shown in Figs. 7.42 and 7.43.

From Figs. 7.42 and 7.43, the proposed policing can react to the violation faster than the FL window, and, in long-term rejection behavior, the NFS LB also shows much better performance. This improved strict behavior has some costs and one of those is the degradation of policing false alarm. However, for the QoS of the packet voice traffic which guarantees CLR of 10^{-3}, this false alarm probability of the NFS is still acceptable.

Table 7.12. Error of overload detection performance between the policing with NFS controller and the policing with FL controller when divided the input set to be trained into subgroups

No. of subgroups		2	3	4	5
Error	vary $E(X)$	10.64	2.41	1.79	8.69
(%)	vary $E(S)$	11.54	2.44	1.83	8.34

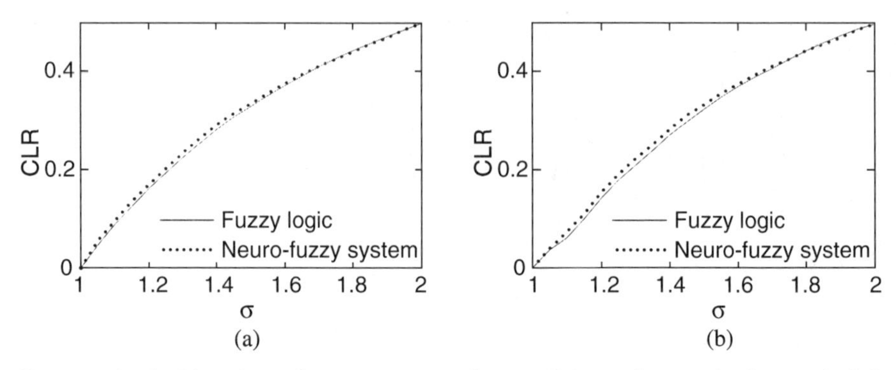

Fig. 7.36. Overload detection performance comparison between NFS controller trained with inputs divided into 2 sets and FL controller when (a) varying $E(X)$ and (b) varying $E(S)$ of the source

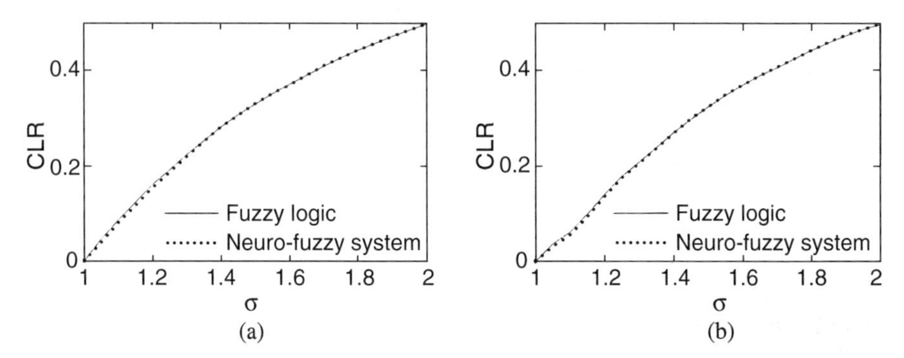

Fig. 7.37. Overload detection performance comparison between NFS controller trained with inputs divided into 3 sets and FL controller when (a) varying $E(X)$ and (b) varying $E(S)$ of the source

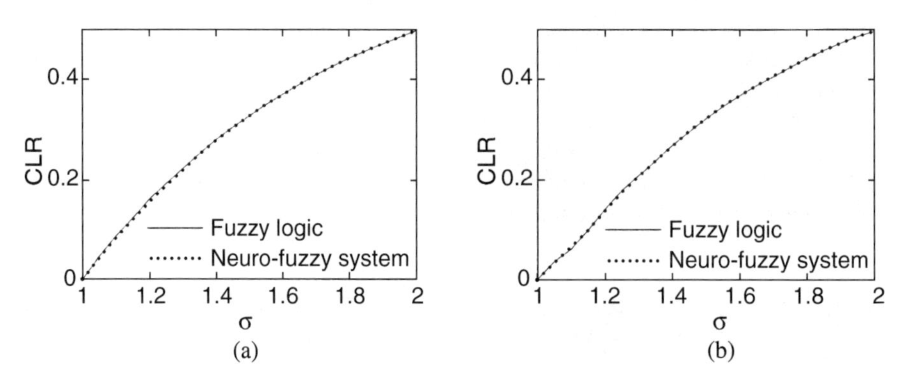

Fig. 7.38. Overload detection performance comparison between NFS controller trained with inputs divided into 4 sets and FL controller when (a) varying $E(X)$ and (b) varying $E(S)$ of the source

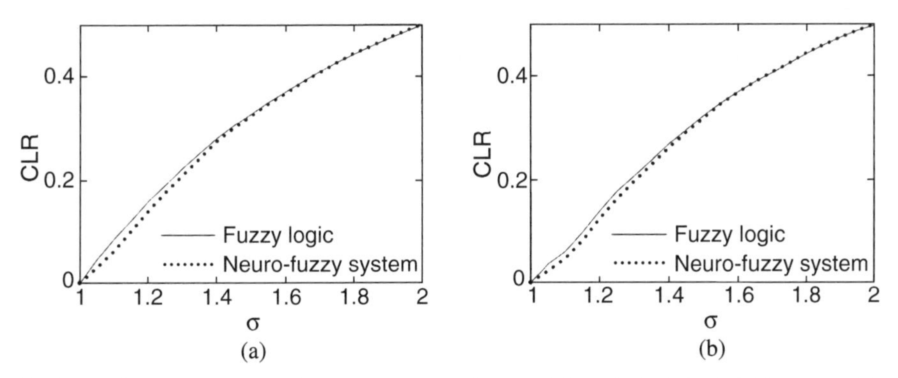

Fig. 7.39. Overload detection performance comparison between NFS controller trained with inputs divided into 5 sets and FL controller when (a) varying $E(X)$ and (b) varying $E(S)$ of the source

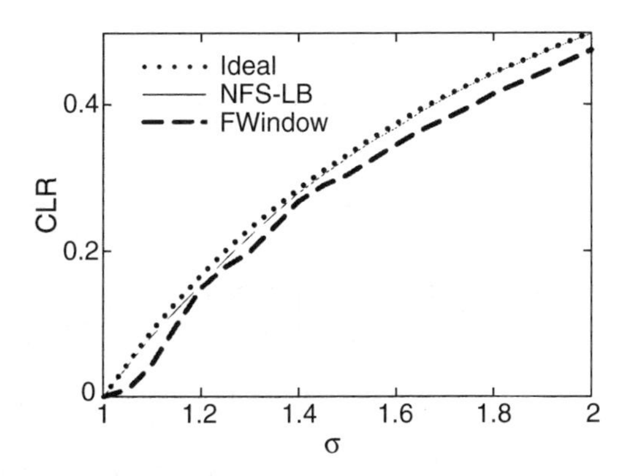

Fig. 7.40. Overload detection performance comparison between the NFS LB and the FL window (FWindow) when varying $E(X)$ of the source

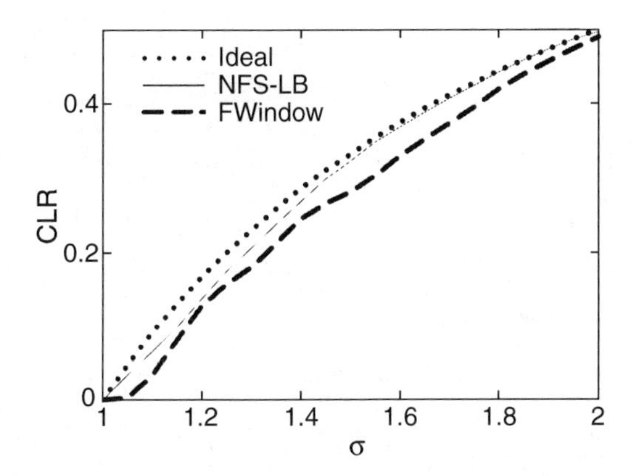

Fig. 7.41. Overload detection performance comparison between the NFS LB and the FL window (FWindow) when varying $E(S)$ of the source

Table 7.13. Selectivity comparison when tested with source model 2

Mechanisms		FL Window	NFS LB
RSSE	vary $E(X)$	0.130790	0.027591
	vary $E(S)$	0.181735	0.070226
False Alarm Probability		1.14×10^{-8}	3.69×10^{-7}

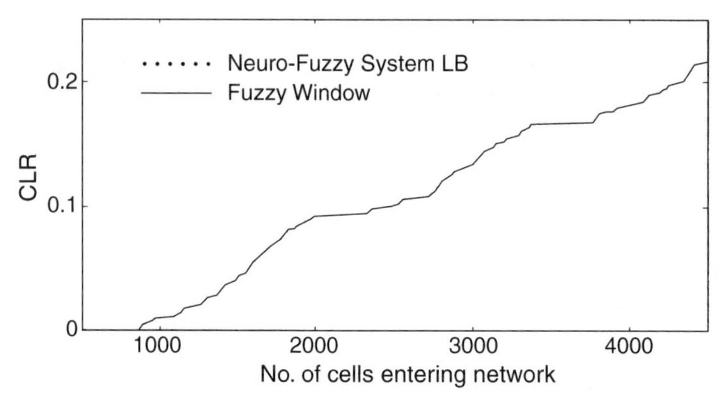

Fig. 7.42. Responsiveness behavior comparison between NFS controller with parallel NN and FL controller when varying $E(X)$ of the source

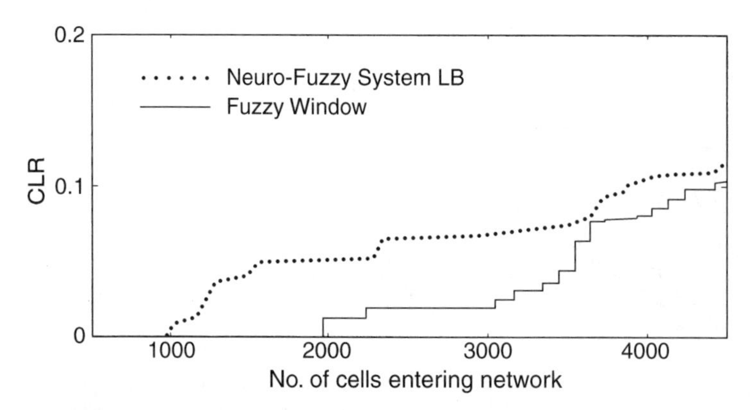

Fig. 7.43. Responsiveness behavior comparison between NFS controller with parallel NN and FL controller when varying $E(S)$ of the source

Table 7.14. Comparison of processing time and token generation rate of FL and NFS controllers

Controller	Traffic Source Model 1	Traffic Source Model 2
Fuzzy Logic	0.0980 sec.	0.0977 sec.
Neuro-Fuzzy System	0.0553 sec.	0.0543 sec.
Difference of results obtained from the above two controllers	2.17%	1.85%

7.5.3 Comparison of Processing Time of FL and NFS Controllers

The objective of bringing NN to this application, is in order to reduce the processing time in the fuzzy inference stage of FL. However, training NN takes a long time, thus, it is necessary to test the cost and benefit of the application of NN.

The test was performed by measuring the processing time of NFS and FL controllers for 100 times of processing and then calculating the mean of the processing time. Comparison of the token generation rate, obtained from both controllers was also performed. The results are shown in Table 7.14.

It can be seen form Table 7.14, that the application of NFS can save processing time by approximately 44% with respect to that of FL, while the token generation rates obtained from both controllers are very slightly different.

It is also found that the emergence of the preprocessor in parallel NN structure very slightly increases the processing time compared with the other processing times in NFS controller, the overall processing time of which depends on the number of NN nodes.

7.6 Evaluation of Simulation Results

In the simulation test of the two types of packet voice traffic sources with different burst duration, it is found that the proposed traffic policing mechanism based on NFS controller is efficient in overload detection and more responsive than the conventional LB and FL window mechanisms. Except in some cases, such as, when testing with traffic violating MCR by increasing burst duration ($E(S)$) in traffic source model 1, where the mechanism performs the initial responsiveness a little later than the conventional LB. However, cell discard rate of the proposed mechanism against the violating traffic in long-term, increases faster than the other mechanisms. Moreover, in testing with two types of violation, increasing the burst duration and decreasing the silent period, the proposed mechanism responds to both types similarly while the conventional LB responds to the case of decreasing silent period much slower than the case of increasing burst duration.

In spite of FL being used to help the decision, the limitation of the mechanism in exactly identifying a little overload state from normal state, cannot be avoided. Thus, in designing the mechanism to perform strict behavior to the violation makes the false alarm probability or CLR at normal load higher than that of FL window mechanism for 1 order (10 times) but still lower than that of the conventional LB for 2 orders (100 times). Thus, the proposed NFS LB mechanism is appropriate to the ATM network that has the risk of being congested and requires a high level of strictness in traffic policing. The structure of the mechanism has an input buffer, in order to decrease CLR and be used to evaluate input traffic characteristics from the number and the change of the number of cells in the buffer. However, the following problem is time delay. It is found in the simulation test, that the proposed traffic policing mechanism causes time delay to the traffic less than 1 ms and the delay can be reduced by increasing the size of the token pool, which is traded off with lower responsiveness and overload detection capabilities.

In training NN to perform mapping input to output of inference engine, it is found that training set selection and the number of training sets have effects on the performance of NN learning, and what training method gives the least error cannot be specified exactly. Too little training sets cannot force NN to learn efficiently, and too many training sets cannot train NN to know all of the global data. The solution to this problem is to apply the parallel NN and allocate appropriate training sets to each of the NN in that parallel structure. Training NN until the error is low enough, makes the NN function approximately the same as inference engine with no impact on any efficiency aspects of the policing.

The advantages and disadvantages of the NFS LB policing mechanism, are summarized below:

A. Advantages:

1) Improves the efficiency of overload detection and responsiveness, when compared with the conventional LB and FL window mechanisms.
2) Reduces false alarm probability when compared with the conventional LB.
3) Using NN instead of inference engine, makes the processing time faster.

B. Disadvantages:

1) In designing and testing in this chapter, two types of packet voice traffic source model are used. Although their burst duration is different, their burstiness is the same. Thus, if the proposed mechanism is to be applied to another type of traffic, such as, image or video, the burstiness of which is higher, FL and NFS controllers must be redesigned to suit that type of traffic, but they can be designed by the same concept and procedure, as stated in this chapter.
2) Training NN takes a long time and it is necessary to test with many training methods to find the most appropriate one but after it is found, the processing time will be faster.

This chapter proposes a novel traffic policing mechanism by applying NFS to police traffic of single source of packet voice only. However, in the real ATM network, there are other types of traffic and the designed mechanism in this chapter cannot be utilized directly. The structure of mechanism must be redesigned, for example, buffer size, token pool size, membership function and fuzzy rules, to suit that type of traffic. However, the results show the efficiency and the possibility of applying NFS for the objective of bursty traffic policing, which can also be applied to the other types of traffic, such as image and video.

References

[1] Rathgeb E P (1991) Modeling and Performance Comparison of Policing Mechanisms for ATM Networks, IEEE J. Select. Areas in Commun., Vol. 9, No. 3, pp. 325–334.

[2] Butto M, Cavallero E, Tonietti A (1991) Effectiveness of the Leaky Bucket Policing Mechanism in ATM Networks, IEEE J. Select. Areas Commun., Vol. 9, No. 3, pp. 335–342.

[3] Tarraf A A, Habib I W, Saadawi T N (1994) A Novel Neural Network Traffic Enforcement Mechanism for ATM Networks, IEEE J. Select. Areas Commun., Vol. 12, No. 6, pp. 1088–1096.

[4] Ndousse T D (1994) Fuzzy Neural Control of Voice Cells in ATM Networks, IEEE J. Select. Areas Commun., Vol. 12, No. 9, pp. 1488–1494.

[5] Catania V, Ficili G, Palazzo S, Panno D (1995) A Fuzzy Expert System for Usage Parameter Control in ATM Networks, IEEE GLOBECOM '95, Vol. 2, pp. 1338–1342.

[6] Catania V, Ficili G, Palazzo S, Panno D (1996) A Comparative Analysis of Fuzzy versus Conventional Policing Mechanism for ATM Networks, IEEE/ACM Trans. Networking, Vol. 4, pp. 449–459.

[7] Kwok A, McLeod R (1996) ATM Congestion Control using a Fuzzy Neural Network, Canadian Conference on Electrical and Computer Engineering, Vol. 2, pp. 814–817.

[8] Runkler T A, Glesner Z E (1994) Efficient Algorithms for High Resolution Fuzzy Controllers with Piecewise Linearities. Proc. 3rd IEEE Conf. Fuzzy System, Vol. 1, pp. 189–191.

[9] Laongmal J, Benjapolakul W (1999) Aggregate Bandwidth Allocation of Heterogeneous Sources in ATM Networks, IEEE Symposium on Intelligent Signal Processing and Communication Systems, pp. 215–218.

8 Appendix A. Overview of Neural Networks

Peter Stavroulakis

A.1 Introduction

Artificial neural systems can be considered simplified mathematical models of brain-like systems and they function as parallel distributed computing networks. However, in contrast to conventional computers, which are programmed to perform specific tasks, most neural networks must be taught, or trained.

Neural Networks can learn new associations, new functional dependencies and new patterns. Although computers outperform both biological and artificial neural systems for tasks based on precise and fast arithmetic operations, artificial neural systems represent the promising new generation of information processing networks.

The study of brain-style computation has its roots over 50 years ago in the work of McCulloch and Pitts (1943) [1] and slightly later in Hebb's famous Organization of Behavior (1949) [2]. The early work in artificial intelligence was torn between those who believed that intelligent systems could be best built on computers modelled after brains, and those like Minsky [3] who believed that intelligence was fundamentally symbol processing of the kind readily modelled on the von Neumann computer.

For a variety of reasons, the symbol-processing approach became the dominant theme in Artificial Intelligence in the 1970s. However, the 1980s showed a rebirth in interest in neural computing: In 1982 Hopfield [4] provided the mathematical foundation for understanding the dynamics of an important class of networks, while in 1984 Kohonen developed unsupervised learning networks for feature mapping into regular arrays of neurons. In 1986, Rumelhart and McClelland [5] introduced the back propagation learning algorithm for complex, multilayer networks. In 1986–87, many neural networks research programs were initiated. The list of applications that can be solved by neural networks has expanded from small test-size examples, to large practical tasks. Very-large-scale integrated neural network chips have been fabricated. In the long term, we can expect to use artificial neural systems in applications involving vision, speech, decision making, and reasoning, but also as signal processors such as filters, detectors, and quality control systems.

Artificial neural systems, or neural networks, are physical cellular systems that can acquire, store, and utilize experimental knowledge. The knowledge is in the form of stable states or mappings embedded in networks that can be recalled, in response to the presentation of cues. The basic processing elements of neural networks are called artificial neurons, or simply neurons or nodes.

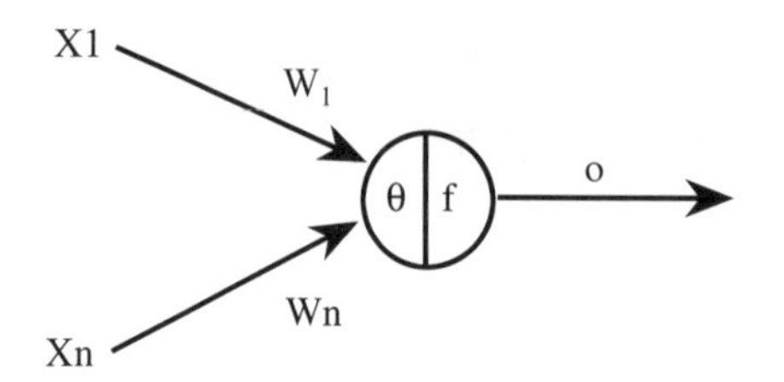

Fig. A.1. A processing element with single output connection

Each processing unit, illustrated in Fig. A.1, is characterized by an activity level (representing the state of polarization of a neuron), an output value (representing the firing rate of the neuron), a set of input connections, (representing synapses on the cell and its dendrite), a bias value (representing an internal resting level of the neuron), and a set of output connections (representing a neuron's axonal projections).

Each of these aspects of the unit, are represented mathematically by real numbers. Thus, each connection has an associated weight (synaptic strength), which determines the effect of the incoming input on the activation level of the unit. The weights may be positive (excitatory), or negative (inhibitory).

The signal flow of neuron inputs, x_j, is considered to be unidirectional as indicated by arrows, as is a neuron's output signal flow. The neuron output signal is given by the following relationship:

$$o = f(\langle w, x \rangle) = f(w^T x) = f\left(\sum_{j=1}^{n} w_j x_j \right) \tag{A.1}$$

where $w = (w_1, \ldots, w_n)^T \in \boldsymbol{R}^n$ is the weight vector. The function $f(w^T x)$ is often referred to as an activation (or transfer function). Its domain is the set of activation values, *net*, of the neuron model, so this function is often used as $f(net)$. The variable net is defined as a scalar product of the weight and input vectors, and in the simplest case the output value o is computed as:

$$o = f(net) = \begin{cases} 1 & \text{if } w^T x \geq \theta \\ 0 & \text{otherwise} \end{cases} \tag{A.2}$$

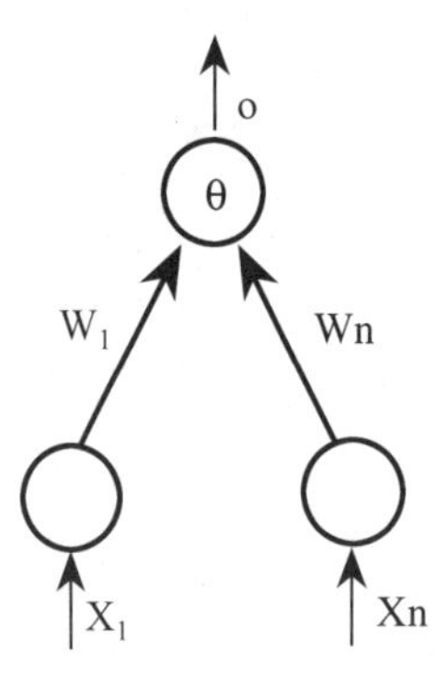

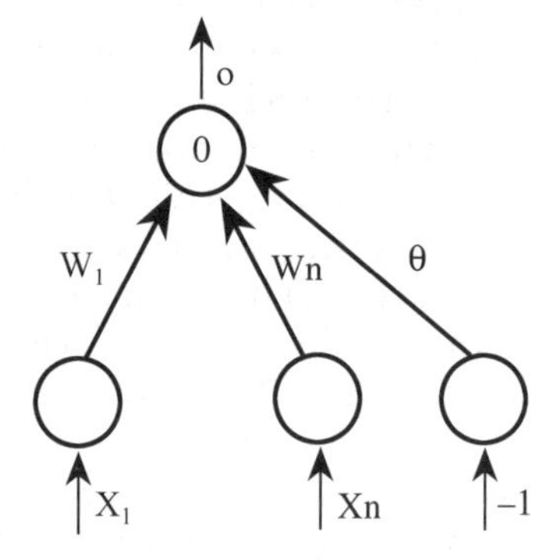

Fig. A.2. Removal of the threshold

where θ is called threshold-level and this type of node is called a linear threshold unit. The removal of the threshold from the network, is very easily done by increasing the dimension of input patterns. Really, the identity:

$$w_1x_1 + \ldots + w_nx_n > \theta \Leftrightarrow w_1x_1 + \ldots + w_nx_n - 1 \times \theta > 0 \tag{A.3}$$

means that, by adding an extra neuron to the input layer with fixed input value -1 and weight θ the value of the threshold becomes zero. That is why in the following we suppose that the thresholds are always equal to zero.

A.2 Learning by Neural Networks

The problem of learning by neural networks, is simply the problem of finding a set of connection strengths (weights) which allow the network to carry out the desired computation. The network is provided with a set of example input/output pairs (a training set) and must modify its connections, in order to approximate the function from which the input/output pairs have been drawn. The networks are then tested for their ability to generalize. The error correction learning procedure is simple enough in concept. The procedure is as follows: During training, an input is placed into the network and flows through it, generating a set of values on the output units. Then, the actual output is compared with the desired target, and a match is computed. If the output and target match, no change is made to the net. However, if the output differs from the target, a change must be made to some of the connections.

A.2.1 Multilayer, Feedforward Network Structure

A neural network with a layered, feed forward structure and error gradient-based training algorithm, will be discussed in this section. Although a single-layer network of this type, known as the 'perceptron,' has existed since the late 1960s by Minsky/Papert, it did not experience widespread application, due to its limited classification ability and the lack of a training algorithm for the multi-layer case. Furthermore, the training procedure evolved from the early work of Widrow [6] in single-element, nonlinear adaptive systems such as ADALINE. The *feedforward network* is composed of a hierarchy of processing units, organized in a series of two or more mutually exclusive sets of neurons or layers. The first, or input layer, serves as a holding site for the values applied to the network. The last, or output layer, is the point at which the final state of the network is read. Between these two extremes, lie zero or more layers of hidden units. Links, or weights, connect each unit in one layer, to those only in the next-higher layer. There is an implied directionality in these connections, in that the output of a unit, scaled by the value of a connecting weight, is fed forward to provide a portion of the activation for the units in the next-higher layer. Figure A.3, illustrates the typical feed forward network. The network as shown, consists of a layer of d

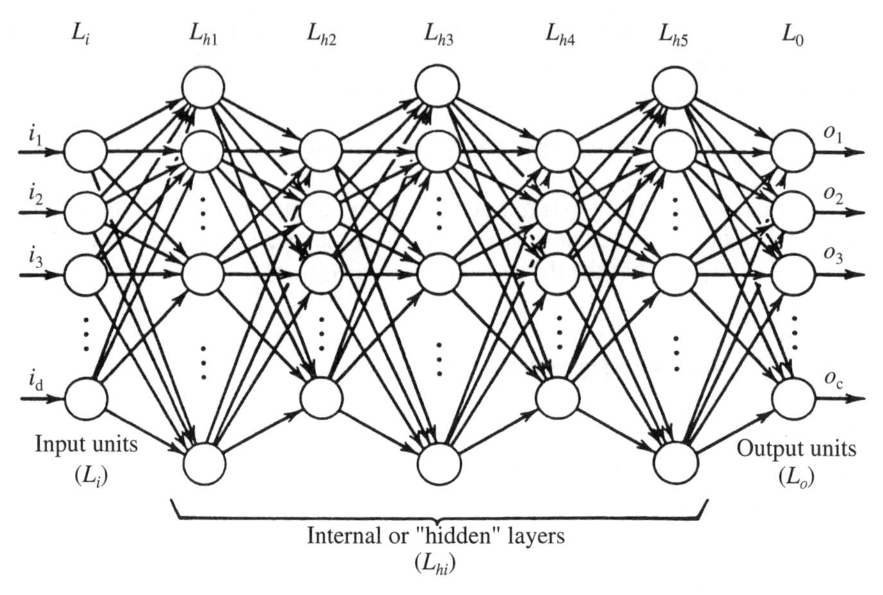

Fig. A.3. Structure of a multiple-layer feedforward network

input units (L_i), a layer of c output units (L_o), and a variable number (5 in this example) of internal or 'hidden' layers (Lh_i) of units. Observe the *feedforward* structure, where the inputs are directly connected to units in L_i only and the outputs of layer L_k units are connected to units in layer L_{k+1} only.

The role of the input layer is somewhat fictitious, in that input layer units are only used to 'hold' input values and distribute these values to all units in the next layer. Thus, the input layer units do not implement a separate mapping or conversion of the input data, and their weights are insignificant. The feedforward network must have the ability to learn pattern mappings. The network may be made to function as a pattern associator through training. Training is accomplished by presenting the patterns to be classified to the network and determining its output. The actual output of the network is compared with a 'target' and an error measure is calculated. The error measure is then propagated backward through the network and used to determine weight changes within the network. This process is repeated, until the network reaches a desired state of response. Although this is an idealized description of training, it does not imply that an arbitrary network will converge to the desired response.

A.2.2 Training the Feedforward Network: The Delta Rule (DR) and the Generalized Delta Rule (GDR) (Back-Propagation)

The GDR is *a product learning rule* for a feedforward, multiple-layer structured neural network, that uses gradient descent to achieve training or *learning by er-*

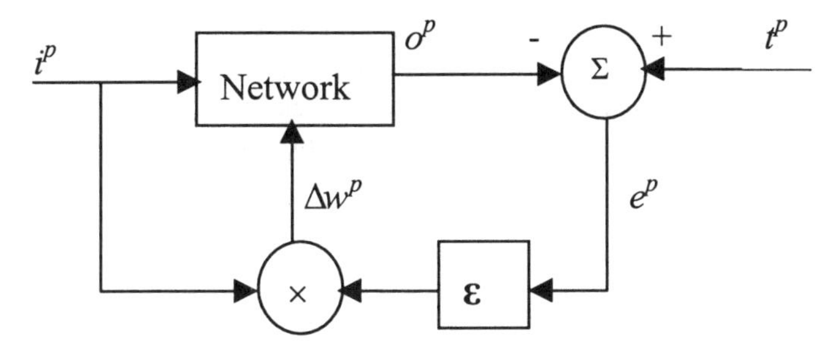

Fig. A.4. Structure of iterative training strategy, using pth element of H

ror correction. Network weights are adjusted to minimize an error, based on a measure of the difference between desired and actual feedforward network output. Desired input output behavior, is given by the training set. The GDR is one instance of a training algorithm for a NN-based pattern associator, with the structure shown in Fig A.4.

The basic Operation of the Generalized Delta Rule (GDR), is summarized in the following steps:

1) Apply input (stimulus) vector to network.
2) 'Feed forward' or propagate input pattern, to determine all unit outputs
3) Compare unit outputs in output layer with desired pattern response
4) Compute and propagate error measure backward (starting at output layer) through the network
5) Minimize error at each stage, through unit weight adjustments.

A.2.3 The Hopfield Approach to Neural Computing

Hopfield [2], characterized a neural computational paradigm, for using a neural net as an autoassociative memory. In the Hopfield network, every neuron is allowed to be connected to all other neurons, although the value of w_{ij} varies (it may also be 0 to indicate no unit interconnection). To avoid false reinforcement of a neuron state, the constraint $w_{ii} = 0$ is also employed. The w_{ij} values, therefore, play a fundamental role in the structure of the network. In general, a Hopfield network has significant interconnection (i.e., practical networks seldom have sparse W matrices, where $W = [w_{ij}]$.

A.2.4 Unsupervised Classification Learning

Unsupervised classification learning is based on clustering of input data. No *a priori* knowledge is assumed to be available regarding an input's membership in a particular class. Rather, gradually detected characteristics and a history of training will be used to assist the network in defining classes and possible boundaries between them. Clustering is understood to be the grouping of similar objects and separating of dissimilar ones. The Kohonen's network will be discussed, which classifies input vectors into one of the specified number of m categories, according to the clusters detected in the training set $\{x^1, \ldots, x^k\}$

The learning algorithm treats the set of m weight vectors as variable vectors, that need to be learned. Prior to the learning, the normalization of all (randomly chosen) weight vectors is required. The weight adjustment criterion for this mode of training, is the selection of w_r such that:

$$\left\| x - w_r \right\| = \min_{i=1,\ldots,m} \left\| x - w_i \right\| \tag{A.4}$$

The index r denotes the *winning* neuron number corresponding to the vector w_r, which is the closest approximation of the current input x.

The Kohonen's learning algorithm can be summarized in the following three steps:

Step 1: $w_r := w_r + \eta(x - w_r)$, $o_r = 1$, (r is the winner neuron)
Step 2: $w_r := w_r / \|w_r\|$ (normalization)
Step 3: $w_i := w_i$, $o_i := 0$, $i \neq rv$ (losers are unaffected)

It should be noted that from the identity,

$$w_r := w_r + \eta(x - w_r) = (1 - \eta)w_r + \eta x \tag{A.5}$$

it follows that the updated weight vector is a convex linear combination of the old weight and the pattern vectors, as shown in Fig A.5.

In the end of the training process, the final weight vectors point to the centers of gravity of classes. The network will only be trainable, if classes/clusters of patterns are linearly separable from other classes by hyperplanes passing through the origin. To ensure reparability of clusters with *a priori* unknown number of training clusters, the unsupervised training can be performed with an excessive number of neurons, which provides a certain reparability safety

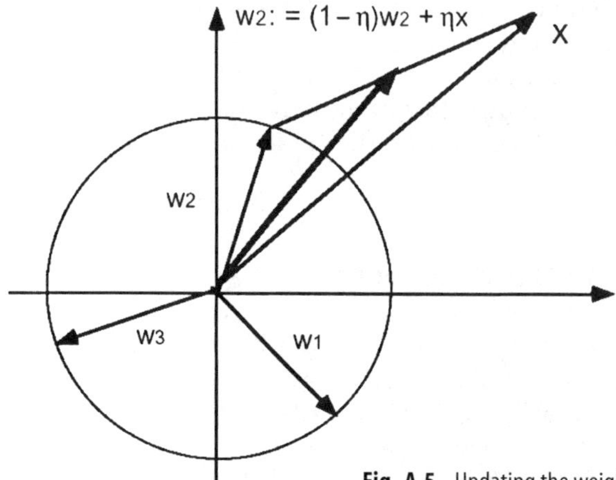

Fig. A.5. Updating the weight of the winner neuron

margin. During the training, some neurons are not likely to develop their weights, and if their weights change chaotically, they will not be considered as indicative of clusters. Therefore, such weights can be omitted during the recall phase, since their output does not provide any essential clustering information. The weights of the remaining neurons should settle at values that are indicative of clusters.

A.3 Examples of Neural Network Structures for PR Applications

The best way to show how the above theoretical foundation works in practice, is to apply it to a real world problem. This application will help us understand how this approach can be combined with fuzzy logic, in order to create a very powerful neurofuzzy system.

A.3.1 Neural Network Structure

The connectivity of a neural network determines its structure. Groups of neurons could be locally interconnected to form 'clusters' that are only loosely, weakly, or indirectly connected to other clusters. Alternately, neurons could be organized in groups or layers that are (directionally) connected to other layers. Thus, neural implementation of PR approaches, requires an initial assessment of neural network architectures. Possibilities include the following:

1) Designing an application-dependent network structure that performs some desired computation.
2) Selecting a 'commonly used' pre-existing structure, for which training algorithms are available. Examples are the feed forward and Hopfield networks.
3) Adapting a structure in item 2 to suit a specific application. This includes using semantics or other information to give meaning to the behaviour of units or groups of units.

Several different 'generic' neural network structures, indicated in item 2, are useful in a class of PR problems. Examples are:

The Pattern Associator (PA). Feed forward networks exemplify this neural implementation. A sample feed forward network is shown in Fig. A.6(a).

The Content-Addressable or Associative Memory Model (CAM or AM). This neural network structure, best exemplified by the Hopfield model which was described in A.2.3, is another attempt to build a pattern recognition system with useful pattern association properties. A sample structure is shown in Fig. A.6(b).

Self-Organizing Networks. These networks exemplify neural implementations of unsupervised learning, in the sense that they typically cluster, or self-organize input patterns into classes or clusters based on some form of similarity. Although these network structures are only examples, they seem to be receiving the vast amount of attention. Figure A.6(c) and Fig. A.6(d), give a more 'generic' viewpoint of these structures.

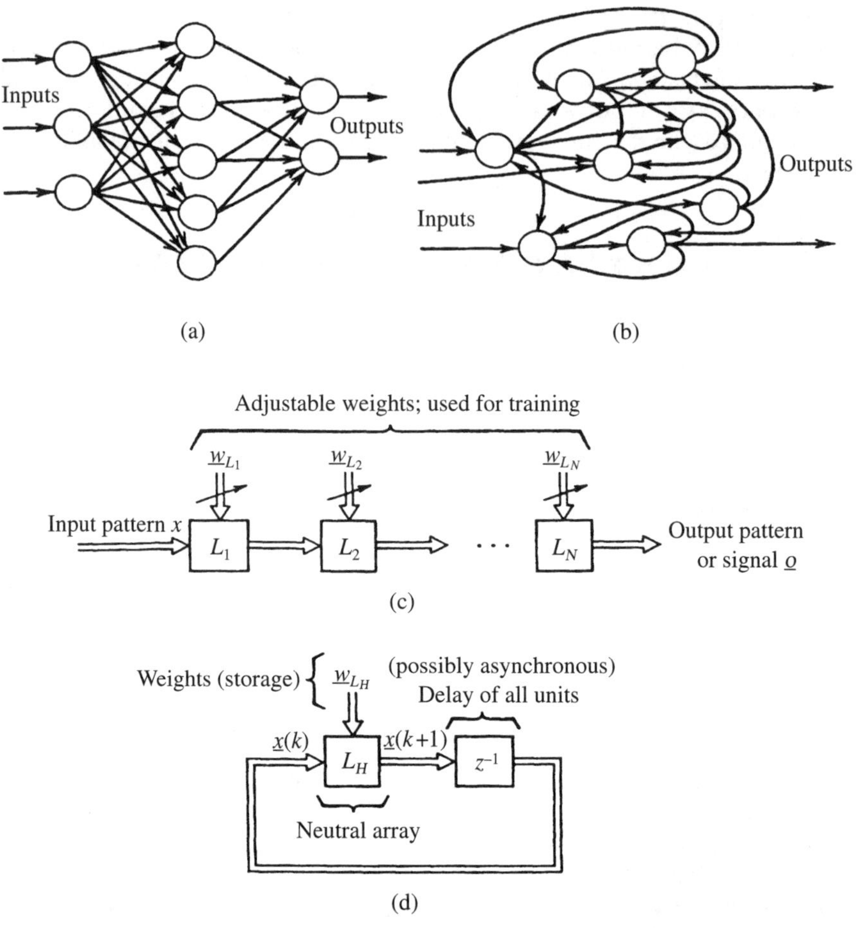

Fig. A.6. Overview of a neural network structure. (a) Sample feedforward neural network structure. (b) Sample Hopfield-like neural network structure. (c) 'Generic' or block diagram view of 1(a) – a multilayer network. d) 'Generic' or block diagram view of 1(c) – a recurrent or feedback network

Feedback interconnections and Network Stability. The feedback structure of a recurrent network shown in Fig. A.6(b) and Fig. A.6(d), suggests that network temporal dynamics, that change over time, should be considered. In many instances, the resulting system, due to the non-linear nature of unit activation output characteristics and the weight adjustment strategies, is a highly non-linear dynamic system. This raises concerns over overall network stability, including the possibility of network oscillation, instability, or lack of convergence to a stable state. The stability of non-linear systems is often difficult to ascertain.

A.3.2 Learning in Neural Networks

Learning in neural network PR may be either supervised or unsupervised. An example of supervised learning in a feedforward network, is the generalized delta rule, described in A.2.2. An example of supervised learning in a recurrent structure is the Hopfield (CAM) approach. Unsupervised learning in a feedforward (nonrecurrent) network is exemplified by the Kohonen self-organizing network, whereas the ART approach exemplifies unsupervised learning with a recurrent network structure.

A.3.3 Reasons to Adapt a Neural Computational Architecture

Generally, the achievement of a computational paradigm based on a neural network, is attractive for the following reasons:

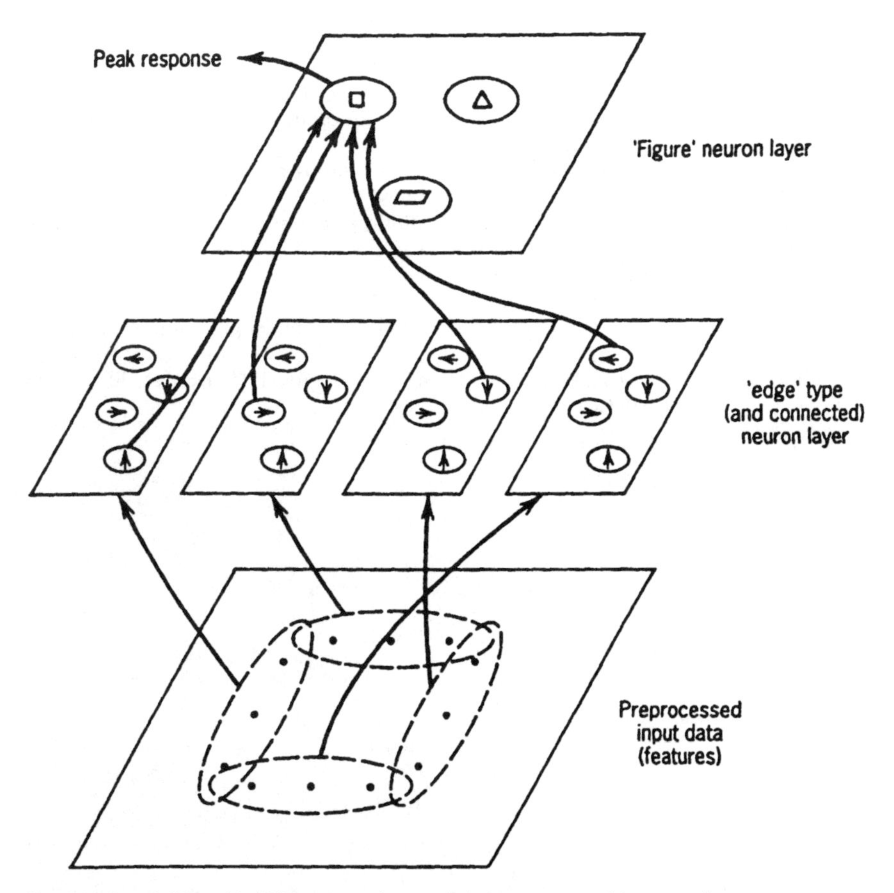

Fig. A.7. Sample hierarchical NN structure that parallels data structure-vision example

1) The local computation is simple, due to the local activation characteristic of the neural unit.
2) Computations proceed inherently in parallel. The possibility of a time delay for network convergence must, however, be taken into account.
3) In many instances, the structure of the problem is reflected directly on the structure of the network. This has been referred to as the isomorphism hypotheses and is depicted in Fig. A.7. An example of the structural isomorphism, is found in the implementation of problems in syntactic pattern recognition, where the structure of word formation via productions is reflected on the structure of the neural network used for recognition.
4) The VLSI or optical implementation of neural network (neglecting the difficulty of massive interconnections) is straightforward, due to the simplicity and regularity of the network.
5) Emulation of the biological computing paradigm, is probably desirable. What is not simple is the mapping of an arbitrary PR problem into a neural network solution.

References

[1] McCulloch W S and Pitts W (1943) A logical calculus of ideas immanent in nervous activity. Bull. Math. Biophys. 5:115–133
[2] Hebb D O (1949) The Organization of behavior: a neuropsychological theory. New York. John Wiley
[3] Minsky M L (1967) Computation: finite and infinte machines. Englewood Cliffs, NJ. Prentice Hall
[4] Hopfield J J (1982) Neural networks and physical systems with emergent collective computational abilities, Proc. Nat. Acad. Sci. USA 79:2554–2558
[5] McClelland J L and Rumelhart D E (1986) Parallel distributed processing, Cambridge, MA, MIT Press
[6] Widrow B, Hoff M E (1960) Adaptive switching circuits, Proc. 1960 IRE West, Elect. Show Conv. Rec., Part 4, 96–104, New York

8

Appendix B. Overview of Fuzzy Logic Systems

Peter Stavroulakis

B.1 Introduction

The logic to infer a crisp outcome from fuzzy input values, is fuzzy logic. A generalized logic that includes not only crisp values but all possible values between 1 and 0 and there is some degree of fuzziness about the exact value in {1,0}. In order to express the distribution of truth of a variable, membership functions are employed. Theoretically, a fuzzy set F of a universe of discourse $X = \{x\}$ is defined as a mapping, $\mu_F(x): X \rightarrow [0, \mathrm{a}]$, by which each x is assigned a number in the range $[0, \mathrm{a}]$, indicating the extent to which x has the attribute F. In Fig. B.1, the membership functions of runners' speeds are illustrated by using the fuzzy variables *slow, moderate* and *fast.*

B.2 Overview of Fuzzy Logic

In Boolean logic, the function of Boolean operators (or gates) AND, OR, and INVERT, is well known. For instance, by 'gating' the value of two variables using an AND, we get $11 \rightarrow 1, 10 \rightarrow 0, 01 \rightarrow 0$, or $00 \rightarrow 0$. In fuzzy logic the values are not crisp, and their fuzziness exhibits a distribution described by the membership function. The problem of 'gating' two fuzzy variables has been addressed by various fuzzy logics and here the min-max logic will be considered. In simple terms, if 'union' is considered (equivalent to OR), the outcome is equal to the input variable with the greatest value, $\max(x_1, x_2, \ldots, x_n)$. That is, if $A = 0.5, B = 0.7$, and $C = A$ OR B, then $C = \max(0.5, 0.7) = 0.1$. If we consider 'intersection' (equivalent to AND), the outcome is equal to the least value of the input variables, $\min(x_1, x_2, \ldots, x_n)$. In this case, if $C = A$ AND B, then $C = \min(0.5, 0.7) =$

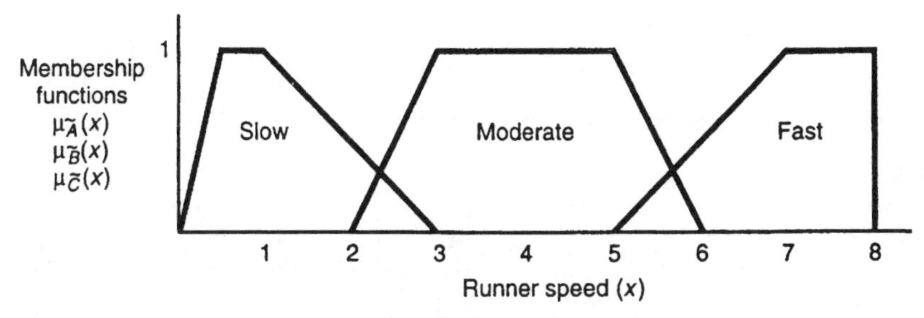

Fig. B.1. Membership functions (normalized) for Slow, Moderate and Fast

0.5. If we consider 'complement' (equivalent to NOT), then the outcome is the complement of 1, or $\overline{x} = 1 - x$. If $C = \overline{B}$, then, $C = 1 - 0.1 = 0.3$.

B.2.1 Fuzzy Rule Generation

In most fuzzy problems, the rules are generated based on past experience. Concerning problems that deal with fuzzy engines or fuzzy control, one should know all possible input-output relationships even in fuzzy terms. The input-output relationships, or rules, are then easily expressed with **if ... then** statements, such as:

If A_1 and/or B_1, then H_{11}, else.
If A_2 and/or B_1, then H_{21}, else.
If A_1 and/or B_2, then H_{12}, else.
If A_2 and/or B_2, then H_{22}.

Here 'and/or' signifies logical union or intersection, the A's and B's are fuzzified inputs, and the H's are actions for each rule.

The case where rules are expressed by a single input variable-if A_1 then H_1, if A_2 then H_2, ... , if A_n then H_n-represents a simple translation (or transformation) of input variables to the output. However, the most common fuzzy logic problems involve more than one variable. The set of if ... then rules with two input variables, is tabulated in Table B.1

Table B.1. Fuzzy Rule Tabulation

A_1	H_{11}	H_{12}
A_2	H_{21}	H_{22}
	B_1	B_2

The *if ... then* rule becomes more difficult to tabulate if the fuzzy statements are more involved (i.e., have many variables), such as if A and B and C or D, then H. Tabulation is greatly simplified, if one follows a statement decomposition process. For example, consider the original problem statements of the form:

If A_i and B_j and C_k, then H_{ijk}

This statement is decomposed as:

If A_i and B_j, then H_{ij}.
If H_{ij} and C_k, then H_{ijk}.

H_{ij} is an intermediate variable. This process is illustrated in Table B.2

Table B.2. Decomposition Process of Three Variables

A_1	H_{11}	H_{12}	C_1	H_{111}	H_{121}	H_{211}	H_{221}
A_2	H_{21}	H_{22}	C_2	H_{112}	H_{122}	H_{212}	H_{222}
	B_1	B_2		H_{11}	H_{12}	H_{21}	H_{22}

B.2.2 Defuzzification of Fuzzy Logic

The defuzzification process is an important step. Based on this step, the output action may or may not be successful. It is not uncommon to have, based on the rules and membership functions, two (or more) answers to a question; see for example Fig. B.1, where μ_{slow} (2.5) = 0.25 and $\mu_{moderate}$. (2.5) = 0.5. In general, defuzzification is the process where the membership functions are sampled to find the grade of membership; then the grade of membership(s) is used in the fuzzy logic equation(s) and an outcome region is defined. From this, the output is deduced. Several techniques have been developed to produce an output. The three most commonly used, are:

maximizer, by which the maximum output is selected,
weighted average, which averages weighted possible outputs,
centroid (and its variations), which finds the output's centre of mass.

In summary, the key steps for solving a fuzzy problem are as follows:

1) Define the fuzzy problem in detail.
2 Identify all important variables and their ranges
3) Determine membership profiles for each variable range
4) Determine rules (propositional statements), including action needed.
5) Select the defuzzification methodology
6) Test the system for correct answers; if needed, go back to step 3.

B.3 Examples

In the example that follows, we shall show how a fuzzy logic system can be applied to a real world problem. This example is chosen in order to help us point out the areas which can be improved, if we apply a neural network to enhance performance.

B.3.1 Fuzzy Pattern Recognition

Much of the information that we have to deal with in real life, is in the form of complex patterns. Pattern recognition involves the search for structure in these complex patterns. Many recognition schemes have been developed and among these, fuzzy set theory has long been considered a suitable framework for pattern recognition, especially classification procedures, because of the inherent fuzziness involved in the definition of a class or a cluster. Indeed, fuzzy set theory has introduced several new methods of pattern recognition which have led to successful realizations in various areas including speech recognition, intelligent robots, image processing, character recognition, scene analysis, recognition of geometric objects, signal classification, and medical applications. Major references include [1, 2, 4]

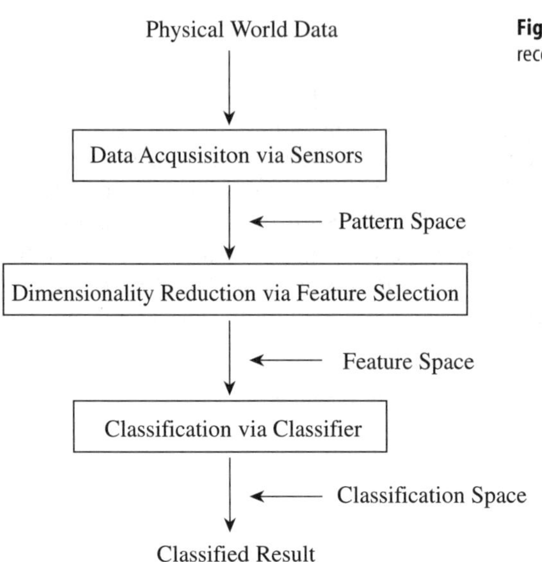

Fig. B.2. General scheme of pattern recognition

The concept of fuzzy set theory can be introduced into the pattern recognition process in Fig. B.2, to cope with uncertainty in several different ways. Two of them include; (i) fuzziness involving the feature space, and (ii) fuzziness involving the classification space. It is understood that most of the information gathered in the recognition processes of a human being, is of a non-numerical type. Even if numerical data are available, the process is worked out by the human mind at a level of non-numerical labels. This indicates that the classification is performed not on the basis of a mass of numbers, but by elicitation relationships between the classes and linguistic labels attached to the object for recognition. These linguistic labels can be represented by fuzzy sets specified in appropriate spaces. Thus, there may be, for example, a classification rule like, 'If an object is heavy and small and it moves fast, then it belongs to class ω_i.' A second important way of introducing fuzziness is regarding class assignment ('labelling') in the classification space. Unlike 'hard' labelling in which an object is classified as belonging to only one class crisply, 'fuzzy' labelling allows an object to be identified as belonging to different classes, to different degrees. That is, the boundaries of the classes are vague. For example, we can say, 'If an object is black and cubic, then it possibly belongs to class ω_i.' Obviously, the above two approaches can be merged into a classification rule like, 'If an object is heavy and small and it moves fast, then it most possibly belongs to class ω_i'. In the following sections, some fuzzy pattern recognition approaches will be introduced based on fuzzy relations and fuzzy clustering.

B.3.1.1 Classification Methods Based on Fuzzy Relations

In this section, we consider pattern classification performed in a fuzzy environment, assuming both features represented by fuzzy labels and class assignment are fuzzy, such as 'If the object is light and large and it moves fast, then it possibly belongs to class ω_i'.

Let X denote the pattern from the environment observed via the data acquisition process. The pattern is described in linguistic terms, rather than in numerical fashion. Therefore, the ith feature of X, X_i, is represented as a fuzzy set with membership function μ_{Xi}. We further assume that there exists finite number of classes $W = \{\omega_1, \omega_2, \ldots, \omega_c\}$. The goal is to find a fuzzy set Ω with membership function μ_Ω such that the value $\mu_\Omega(\omega_i)$, $i = 1, 2, \ldots, c$, denotes the grade of membership of the pattern X in the ith class.

To solve the above problem, we firstly need to define the feature space. We assume that the feature space χ consists of n coordinates, that is,

$$\chi = \chi_1 \times \chi_2 \times \cdots \times \chi_n \tag{B.1}$$

where each χ_i, i =1, 2, ..., n, is a linguistic variable consisting of a term set,

$$T(\chi_i) = \{\chi_{i1}, \chi_{i2}, \cdots \chi_{in_i}\} \tag{B.2}$$

where each term χ_{ij_i}, in a term set is a fuzzy set with membership function μ_{ij_i}. These fuzzy sets form a partition of the feature space χ_1 and χ_2, respectively. Sometimes, it is additionally assumed that these fuzzy sets satisfy the orthogonal condition, that is, that their membership degrees sum up to 1 at each element of the universe of discourse.

After the feature space is defined, the next step is to transform the input pattern X into this feature space. One convenient way to accomplish such a transformation is to use the possibility measure, which indicates the degree of matching of two fuzzy sets,

$$\prod(X_i \mid \chi_{ij_i}) = \sup_{x \in X_i} [\min(\mu_{X_i}(x), \mu_{\chi_{ij_i}}(x))] \tag{B.3}$$

for all $i = 1, 2, \ldots, n$ and $j_i = 1, 2, \ldots, n$. The higher the value of the possibility measure, the better the fuzzy set X_i fits the label expressed by the fuzzy set χ_{ij_i}. By such a transformation, the pattern X is represented by the Cartesian product of the (row} vectors,

$$\chi = \chi^{(1)} \times \chi^{(2)} \times \cdots \times \chi^{(n)} \tag{B.4}$$

where each vector is derived from Eq. (B.3) and is defined as:

$$\chi^{(i)} \triangleq [\prod(X_i \mid \chi_{i1}), \prod(X_i \mid \chi_{i2}), \ldots, \prod(X_i \mid \chi_{in_i})]. \tag{B5}$$

Hence, χ is viewed as a fuzzy relation defined in the Cartesian product of the reference fuzzy sets, that is, $\chi : \chi \to [0, 1]$.

The next step is to perform classification, that is, to transform the fuzzy set (fuzzy relation) χ in the feature space to a fuzzy set Ω in the classification space W. The wanted classifier should be a relation R (χ, W}, that will relate χ and Ω.

Such a transformation can be conveniently expressed by a fuzzy relation equation as:

$$\Omega = \chi \circ R \tag{B.6}$$

where the max-min composition $\circ$ can be replaced by any max-t or min-s composition. In the following derivation, we shall use the only max-min composition for clarity:

$$\mu_\Omega(\omega_j) = \max_{x \in \chi} \min\left[\mu_\chi(x), \mu_R(x, \omega_j)\right] \tag{B.7}$$

for $j = 1, 2, \ldots, c$. The above equation indicates that the degree of membership $\mu_\Omega(\omega_j)$ in the given class ω_j results from the intersection of the features of the pattern and the fuzzy relation R. The more similar χ is to the fuzzy relation R, the higher the value of $\mu_\Omega(\omega_j)$ will be.

For the design of the classifier, a set of training patterns is required that will satisfy the fuzzy relation Eq. (B.6):

$$(\chi_1, \Omega_1), (\chi_2, \Omega_2), \ldots, (\chi_L, \Omega_L) \tag{B.8}$$

then the fuzzy relation of the classifier is given by:

$$R = \bigcap_{i=1}^{L} (\chi_i \overset{a}{\leftrightarrow} \Omega_i) \tag{B.9}$$

the system of equations in (B.6), may not always produce solutions. In such cases, approximate solutions of the system of equations must be pursued. Details on obtaining approximate solutions, can be found in [3]

B.3.1.2 Classification Methods Based on Fuzzy Clustering

Clustering, deals essentially with the task of splitting a set of patterns into a number of more-or-less homogeneous classes (clusters) with respect to a suitable similarity measure, such that the patterns belonging to any one of the clusters are similar and the patterns of different clusters are as dissimilar as possible. The similarity measure used has an important effect on the clustering results since it indicates which mathematical properties of the data set (for example, distance, connectivity, and intensity) should be used and the way they should be used in order to identify the clusters. In non-fuzzy 'hard' clustering, the boundary of different clusters is crisp, such that one pattern is assigned to exactly one cluster. On the contrary, fuzzy clustering provides partitioning results with additional information supplied by the cluster membership values, indicating different degrees of belongingness.

The fuzzy clustering problem can be formulated by considering a finite set of elements $X = \{x_1, x_2, \ldots, x_n\}$, as being elements of the p-dimensional Euclidean space $\Re^p$. The problem is to perform a partition of this collection of elements into c fuzzy sets with respect to a given criterion, where c is a given number of clusters. The criterion is usually to optimize an objective function that acts as a performance index of clustering. The end result of fuzzy clustering can be expressed by a partition matrix U such that:

$$U = [u_{ij}]_{i=1\ldots c,\, j=1\ldots n} \tag{B.10}$$

where u_{ij} is a numerical value in [0,1] and expresses the degree to which the element x_j belongs to the ith cluster. However, there are two additional constraints on the value of u_{ij}. First a total membership of the element $x_j \in X$ in all classes is equal to 1; that js:

$$\sum_{i=1}^{c} u_{ij} = 1 \qquad \text{for all } j = 1, 2, \ldots, n \tag{B.11}$$

Second, every constructed cluster is nonempty and different from the entire set that is:

$$0 < \sum_{j=1}^{n} u_{ij} < n \qquad \text{for all } i = 1, 2, \ldots, c \tag{B.12}$$

A general form of the objective function is:

$$J(u_{ij}, v_k) = \sum_{i=1}^{c} \sum_{j=1}^{n} \sum_{k=1}^{c} g[w(x_j), u_{ij}]\, d(x_j, v_k) \tag{B.13}$$

where $w(x_j)$ is the a priori weight for each x_j, $g[w(x_j), u_{ij}]$ influencing the degree of fuzziness of the partition matrix, and $d(x_j, v_\kappa)$ is the degree of dissimilarity between the data x_j and the supplemental element v_κ, which can be considered the central vector of the kth cluster. (i) $d(x_i, v_\kappa) > 0$.

Fuzzy clustering can now be precisely formulated as an optimization problem:

$$\text{Minimize } j(u_{ij}, v_k), \qquad i,k = 1, 2, \ldots, c \;\; j = 1, 2, \ldots, \tag{B.14}$$

Subject to Eqs. (B.11) and (B.12).

One of the widely used clustering methods based on Eq. (B.14), is the fuzzy c-means (FCM) algorithm developed by Bezdek [4]. This objective function of the FCM algorithm takes the form of:

$$J(u_{ij}, v_i) = \sum_{i=1}^{c} \sum_{j=1}^{n} (u_{ij})^m \, \| x_j - v_i \|^2, \qquad m > 1 \tag{B.15}$$

where m is called the exponential weight which influences the degree of fuzziness of the membership (partition) matrix. To solve this minimization problem, the objective function in Eq. (B.15) is differentiated with respect to v_j and to u_{ij} and the conditions of Eq. (B.11) are applied obtaining:

$$v_i = \frac{1}{\sum_{j=1}^{n} (u_{ij})^m} \sum_{j=1}^{n} (u_{ij})^m x_j, \qquad i = 1, 2, \ldots c \tag{B.16}$$

$$u_{ij} = \frac{(1/\|x_j - v_i\|^2)^{1/(m-1)}}{\sum_{k=1}^{c}(1/\|x_j - v_i\|^2)^{1/(m-1)}}, \quad i = 1, 2, \ldots, c \quad j = 1, 2, \ldots, n \tag{B.17}$$

The system described by Eqs. (B.16) and (B.17) cannot be solved analytically. However, the FCM algorithm provides an iterative approach to approximating the minimum of the objective function starting from a given position. This algorithm is summarized below:

Algorithm FCM: Fuzzy *c*-Means Algorithm
Step 1: Select a number of clusters c ($2 \le c \le n$) and exponential weight $m(1 < m < \infty)$, Choose an initial partition matrix $U^{(0)}$ and a termination criterion ε. Set the iteration index l to 0.
Step 2: Calculate the fuzzy cluster centers $\{v_i^{(l)} \mid i = 1, 2, \ldots, c\}$ by using $U^{(l)}$ and Eq. (B.16).
Step 3: Calculate the new partition matrix $U^{(l+1)}$ by using $\{v_i^{(l)} \mid i = 1, 2, \ldots, c\}$ and Eq. (B.17).
Step 4: Calculate $\Delta = \|U^{(l+1)} - U^{(l)}\| = \max_{i,j} |u^{(l+1)} - u^{(l)}|$. If $\Delta > \varepsilon$, then set $l = l + 1$ and go to step 2. If $\Delta \le \varepsilon$, then stop.

References

[1] Kandel A (1982) Fuzzy techniques in Pattern Recognition, New York, John Wiley
[2] Pal S K, Dutta Majumder D K (1986) Fuzzy Mathematical Approach to Pattern Recognition, New York, John Wiley
[3] Pedrycz W (1991) Processing in relational structures: Fuzzy relational equations, Fuzzy Sets Syst., 40:77–106
[4] Bezdek J C (1981) Pattern Recognition with Fuzzy Objective Function Algorithms, New York, Plenum Press

8 Appendix C. Examples of Fuzzy-Neural and Neuro-Fuzzy Integration

Peter Stavroulakis

C.1 Fuzzy-Neural Classification

C.1.1 Introduction

This chapter will deal with various practical examples of how the combined fuzzyneural or neurofuzzy techniques can be used to study classification, clustering, image processing, speech recognition and fault diagnosis. Fuzzy neural models for pattern recognition (as a classification problem) will be explored in the following section and shown how this tool can help overcome endogenous difficulties.

Pattern recognition may be characterized as an information reduction, information mapping, or information labelling process. An abstract view of the PR classification/description problem is shown in Fig. C.1. We postulate a mapping between class-membership space, C, and pattern space, P. This mapping is done via a relation, G_i, for each class, and may be probabilistic. Each class, w_i, generates a subset of 'patterns' in pattern space, where the ith pattern is denoted p_i. Note, however, that these subspaces overlap, to allow patterns from different classes to share attributes. Another relation, M, maps patterns from subspaces of P into observations or measured patterns or features, denoted m_i. Using this concept, the characterization of many PR problems is simply that, given measurement m_i, we desire a method to identify and invert mappings M and G_i for all i. Unfortunately, in practice, these mappings are not functions. Even if they were, they are seldom 1:1, or invertible. For example, Fig. C.1 shows that identical measurements or observations may result from different p_i, which in turn correspond to different underlying classes. This suggests a potential problem with ambiguity. Nevertheless, it seems reasonable to attempt to model and understand these processes, in the hope that this leads to better classification/description techniques.

The abstract formulation of Fig. C.1 is well suited for modelling in both StatPR and SyntPR. We may view the realized patterns in Fig. C.1 as basic 'world data', which is then measured. Thus, another important aspect of Fig. C.1 concerns M. This mapping reflects our choice of measurement system. Measurement system design is an important aspect of PR system design, in the sense that good 'features' or 'primitives' (to be derived subsequently) probably require good, or at least adequate, measurements. It is unlikely that erroneous or incomplete measurements will facilitate good PR system performance. Finally,

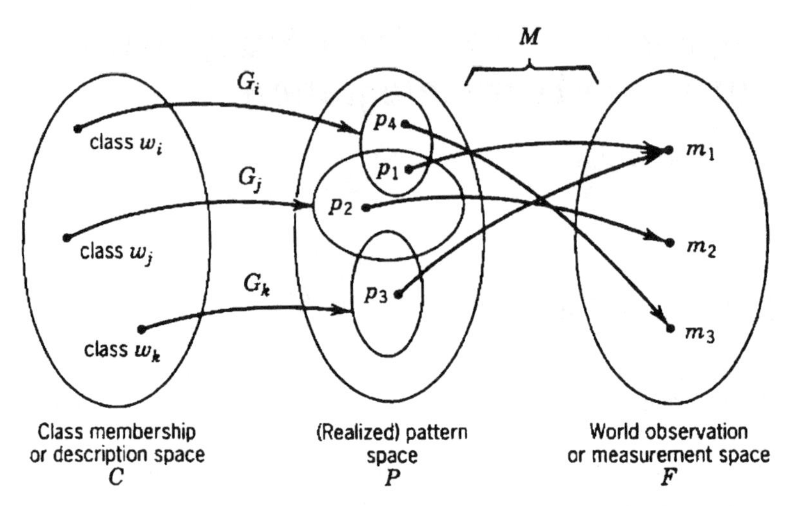

Fig. C.1. Mappings in an abstract representation of pattern generation/classification/interpretation systems

note that patterns that are generated by the same class (p_4 and p_1, from w_i, for example) and 'close' in pattern space P, do not necessarily yield measurements (m_l and m_3 in this case) that are also 'close'. This is significant when 'clustering' of measurement (or feature or primitive) data is used for measuring pattern similarity.

C.1.1.1 Patterns and Feature Extraction

Patterns

It comes as little surprise that much of the information that surrounds us manifests itself in the form of patterns. The ease with which humans classify and describe patterns often leads to the incorrect assumption that this capability is easy to automate. Sometimes the similarity in a set of patterns is immediately apparent, whereas in other instances it is not. Recognizing characters is an example of the former; economic forecasting illustrates the latter.

Pattern recognition, naturally, is based on patterns. A pattern can be as basic as a set of measurements or observations, perhaps represented in vector or matrix notation. Thus, the mapping M, from P to F in Fig. C.1, is an identity mapping. The use of measurements already presupposes some pre-processing and instrumentation system complexity. These measurements could be entities such as blood pressure, age, number of wheels, a 2-D line drawing, and the like. Furthermore, patterns may be converted from one representation to another.

The measurement, as shown in Fig. C.2, may be a two-dimensional image, a drawing, a waveform, a set of measurements, a temporal or spatial history (sequence) of events, the state of a system, the arrangement of a set of objects, and so forth.

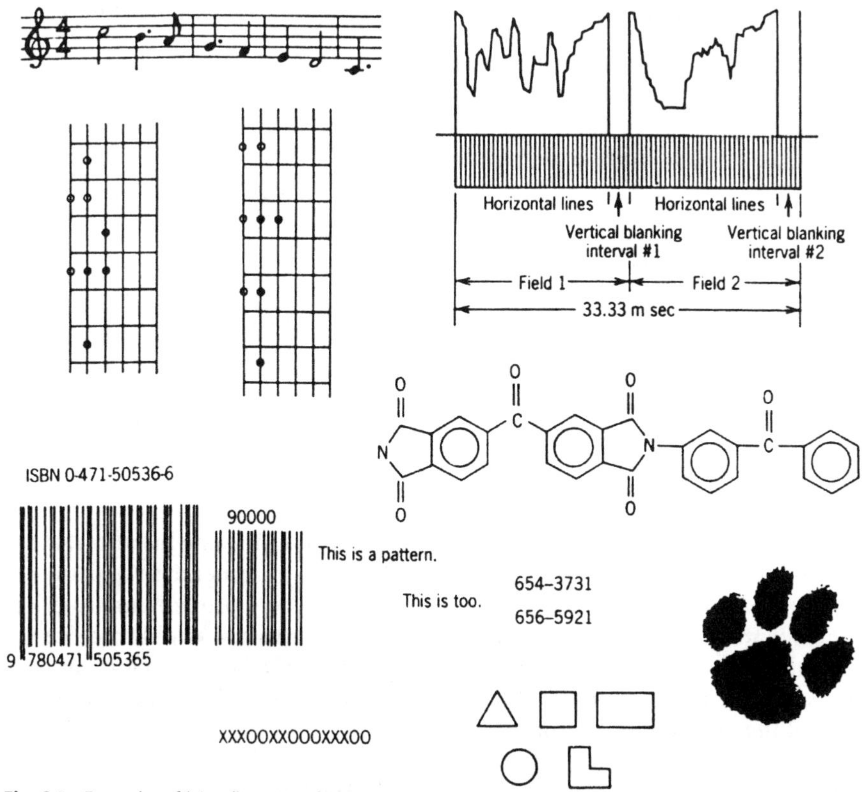

Fig. C.2. Examples of (visual) patterns/measurements

Features

Broadly speaking, features are any extractable measurement used. An example of low-level features are signal intensities. Features may be symbolic, numerical or both. An example of a symbolic feature is colour; an example of a numerical feature is weight (measured in pounds). Features may also result from applying a feature extraction algorithm or operator to the input data. Additionally, features may be higher-level entities: for example, geometric descriptors of either an image region or a 3D object appearing in the image. For instance, aspect ratio and Euler number are higher-level geometric features. Note that: (1) significant computational effort may be required in feature extraction; and (2) the extracted features may contain errors or 'noise'. Features may be represented by continuous, discrete, or discrete-binary variables. Binary features may be used to represent the presence or absence of a particular feature.

The related problems of feature selection and feature extraction must be addressed at the outset of any PR system design. The key is to choose and to extract features that (1) are computationally feasible; (2) lead to 'good' PR system success; and (3) reduce the problem data (e.g., raw measurements) into a manageable amount of information, without discarding valuable (or vital) information.

Feature Selection
Feature selection is the process of choosing input to the PR system and involves judgment. It is important that the extracted features be relevant to the PR task at hand. In some cases, there are mathematical tools that help in feature selection. In other cases, simulation may aid in the choice of appropriate features. Clearly, restrictions on measurement systems may limit the set of possible features. Furthermore, features need not be low-level data. However, the level at which features are extracted determines the amount of necessary pre-processing and may also influence the amount of error that is introduced into the extracted features. For example, in medical applications, the features may be symptoms and the classes might be states of health, including w_1 = 'healthy', w_2 = 'has the flu', and so forth.

C.1.1.2 Pattern Recognition Approaches

Pattern recognition applications come in many forms. In some instances, there is an underlying and quantifiable statistical basis for the generation of patterns. In other instances, the underlying structure of the pattern provides information fundamental for PR. In others, still, neither of the above cases hold true, but we are able to develop and 'train' an architecture to correctly associate input patterns with desired responses. Thus, a given problem may allow one or more of these different solution approaches.

The Statistical Pattern Recognition Approach (StatPR)
Statistical (or 'decision-theoretic') PR assumes, as its name implies, a statistical basis for classification of algorithms. A set of characteristic measurements, denoted features, is extracted from the input data and is used to assign each feature vector to one of c classes. Features are assumed to be generated by a state of nature, and therefore the underlying model is of a state of nature or class-conditioned set of probabilities, and/or probability density functions.

The Syntactic Pattern Recognition Approach (SyntPR)
Many times, the significant information in a pattern is not merely in the presence or absence, or the numerical values, of a set of features. Rather, the interrelationships or interconnections of features yield important structural information, which facilitates structural description or classification. This is the basis of syntactic (or structural) pattern recognition. However, in using SyntPR approaches, we must be able to quantify and extract structural information and to assess structural similarity of patterns. One syntactic approach is to relate the structure of patterns with the syntax of a formally defined language, in order to capitalize on the vast body of knowledge related to pattern (sentence) generation and analysis (parsing).

Typically, SyntPR approaches formulate hierarchical descriptions of complex patterns built up from simpler sub patterns. At the lowest level, primitive elements or 'building blocks' are extracted from the input data. One discrepancy or distinguishing characteristic of SyntPR involves the choice of primitives. Primitives must be sub patterns or building blocks, whereas features are any measurements. Musical patterns are one of the best examples in which structural information is paramount. At the risk of being overly simplistic we observe that all (Western) music basically consists of the same elemental features – an octave is subdivided into distinct tones, and we use about 6 octaves. Therefore, the feature set for all types of music is about 72 distinct tones, which are common to all musical classes. It is the temporal arrangement and structure of these notes, that defines the music.

The Neural Pattern Recognition Approach (NeurPR)

Modern digital computers do not emulate the computational paradigm of biological systems. The alternative of neural computing emerged from attempts to draw on knowledge of how biological neural systems store and manipulate information. This leads to a class of artificial neural systems termed neural networks and involves an amalgamation of research in many diverse fields, such as psychology, neuroscience, cognitive science, and systems theory, and has recently received considerable renewed worldwide attention. Neural networks are a relatively new computational paradigm. It is probably safe to say that the advantages, disadvantages, applications, and relationships to traditional computing are not fully understood. Expectations (some might say 'hype') for this area are high. The notion that artificial neural networks can solve all problems in automated reasoning, or even all PR problems, is probably unrealistic.

Incorporation of fuzzy logic to Pattern Recognition

Consider the case of a decision-theoretic approach to pattern classification. With conventional probabilistic and deterministic classifiers [1, 2], the features characterizing the input vectors are quantitative (i.e., numerical) in nature. Vectors having imprecise or incomplete specification are usually either ignored, or discarded from the design and test sets. Impreciseness (or ambiguity) in such data may arise from various sources. For example, instrumental error or noise corruption in the experiment may lead to partially reliable information available on a feature measurement. Again, in some cases the expense incurred in extracting a precise, exact value of a feature may be high, or it may be difficult to decide on the most relevant features to be extracted. For these reasons, it may be convenient to use linguistic variables and hedges (e.g., small, medium, high, very, more or less, etc.), in order to describe the feature information. In such cases, it is not appropriate to give an exact numerical representation to uncertain feature data. Rather, it is reasonable to represent uncertain feature information by fuzzy subsets.

Again, uncertainty in classification or clustering of patterns may arise from the overlapping nature of the various classes; for example, the amorphous forms of sulfur do not have clear temperature boundaries, they really do overlap. In conventional classification techniques, it is usually assumed that a pattern can belong to one and only one class, which is not necessarily realistic physically, and certainly not mathematically. Thus, feature vectors (and the objects they represent) can and should be allowed to have degrees of membership in more than one class.

Similarly, consider the problem of determining the boundary or shape of a class from its sampled points or prototypes. Conventional approaches attempt to estimate an exact shape for the area in question by determining a boundary that contains (i.e., passes through) some or all of the sample points. However, this property is not necessarily desirable for boundaries in real images. For example, it may be necessary to extend the boundaries to represent obscured portions not represented in the sampled points. Extended portions should have a lower membership in the boundary for such a class, than the portions explicitly highlighted by the sample points. The size of extended regions should decrease with an increase in the number of sample points. This leads us to define multi-valued or fuzzy (with continuum grade of belonging) shapes and boundaries of certain classes.

C.1.2 Uncertainties with Two-Class Fuzzy-Neural Classification Boundaries

In fuzzy classification, the original sharp decision boundaries are usually replaced by fuzzy decision boundaries described by proper membership functions called fuzzy boundary membership functions. A two-class classification problem will be considered with a membership function $\mu_{C1}(\boldsymbol{x})$ that represents the membership of $\boldsymbol{x}$ in class 1, where $\boldsymbol{x}$ is a point with n attribute dimensions. For illustration, the membership function $\mu_{C1}(x)$ in the case of $n = 1$ is depicted as the solid curve in Fig. C.3. When n is greater than 1, $\mu_{C1}(x)$ is a surface or hyper surface. To accommodate the fuzzy boundary membership function to a sharp classification boundary, let

$$\begin{cases} x \in C_1 & \text{when } \mu_{C_1}(x) \geq 0.5 \\ x \in C_2 & \text{otherwise} \end{cases} \tag{C.1}$$

Hence, the values of x that satisfy $\mu_{C1}(x) = 0.5$ define the sharp classification boundary (see Fig. C.3). These points are the crossover points. In order to make Eq. (C.1) meaningful, the membership function $\mu_{C1}(\boldsymbol{x})$ must be ensured to be monotonic. Hence, it is assumed that all membership functions considered in this section, are monotonic.

A more general expression of fuzzy complementation needs to be adopted, the λ complement, that is:

$$\mu^{\lambda}_{\bar{C}_i}(\boldsymbol{x}) = \frac{1-\mu_{C_i}(\boldsymbol{x})}{1+\mu_{C_i}(\boldsymbol{x})}, \qquad \lambda > -1 \tag{C.2}$$

where λ is a parameter. Based on existing misclassifications, a fuzzy boundary membership function $\mu_{C1}(\boldsymbol{x})$ can be supplemented with another membership

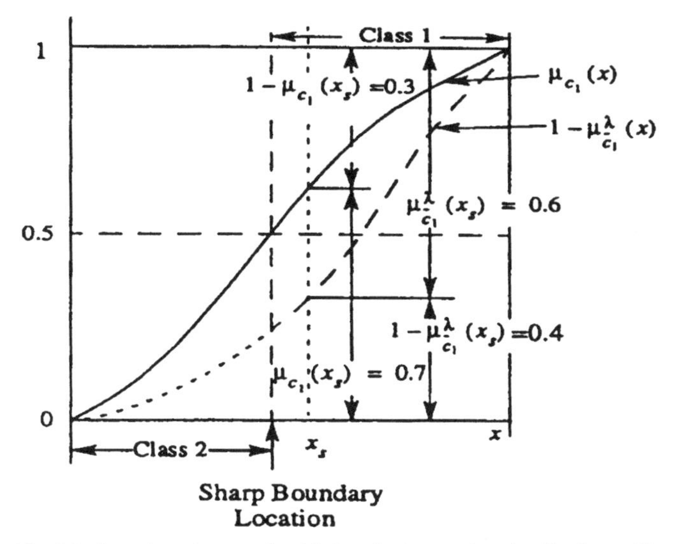

Fig. C.3. Fuzzy boundary membership function in two-class classification problems

function $\mu^{\lambda}_{\bar{C}_i}(x)$ which is a λ complement of $\mu_{C1}(x)$. This provides more information about the uncertainty caused by misclassification. Two fuzzy boundary membership functions [i.e. $\lambda_C(x)$ and $\mu^{\lambda}_{\bar{C}_i}(x)$] provide better uncertainty information, than a single fuzzy boundary membership function in the two-class classification problem, when both factors of potential misclassification and existing misclassification are taken into account. Since a back propagation network (BP net) can learn to function as a membership function, a neural network model would be able to provide more complete information about uncertainty, in terms of membership functions. Such a neural network model, called the neural fuzzy membership model (NFMM), has been proposed by Archer and Wang [3]. The NFMM consists of three individual BP nets, each of which corresponds to one of the membership functions shown in Fig. C.4. An algorithm for implementing the NFMM, is described in the following section.

C.1.2.1 Algorithm NFMM: Neural Fuzzy Membership Model [3]

Step 1: Based on given training samples, train a BP net to find a monotonic membership function that determines a sharp classification boundary according to (C.1) and denote this BP net as NN_0. This can be done by first finding the sharp boundary using a BP net and then determining the proper fuzzy boundary membership function according to the distance of each pattern point from this sharp boundary. Note that the points on the sharp boundary should form the crossover points of this fuzzy boundary membership function.

Step 2: Find misclassification sets Mc_1 and Mc_2 such that $x_m \in Mc_1$ if misclassified point x_m, is in the C_1, region (i.e., $x_m \in C_2$ but is misclassified to be in C_1)

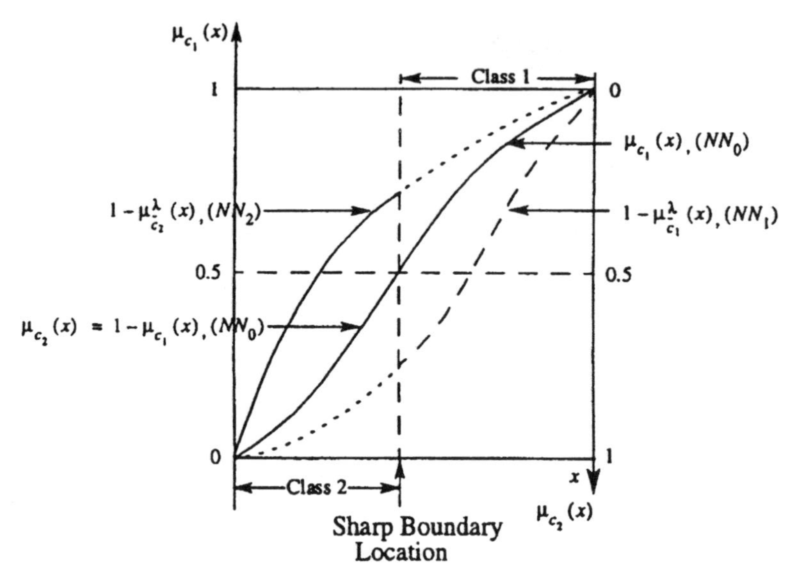

Fig. C.4. Fuzzy boundary membership function implemented in conjunction with neural classification

and $x_m \in Mc_2$ if misclassified point x_m is in the C_2 region. If Mc_1 and Mc_2 are empty, this means that there is no information available regarding the fuzzy nature of the given problem and the sharp boundary must be accepted; otherwise go on to the next step to develop the fuzzy boundary.

Step 3: Based on the ratio of the number of misclassifications needed to correct classification, determine λ subjectively, so that $-1 < \lambda < 0$.

Step 4: For each $x_m \in Mc_1$ or Mc_2, compute the corresponding outputs of the neural network NN_0, denoted by $y(x_m.)$

Step 5: Normalize the membership value for these misclassified points, so that the λ complement (C.2) is applicable:

$$y'(x_m) = \frac{y(x_m) - y_{\min}}{y_{\max} - y_{\min}} \tag{C.3}$$

where $y_{\max}$ and $y_{\min}$, are the extreme output values of NN_0.

Step 6: For each misclassified point x_m assign:

$$\begin{cases} y'_{C_1}(x_m) = y'(x_m) & \text{if } x_m \in M_{C_1} \\ y'_{C_2}(x_m) = 1 - y'(x_m) & \text{if } x_m \in M_{C_2} \end{cases} \tag{C.4}$$

Step 7: Calculate λ-complement values for the misclassified points:

$$y^{\lambda'}_{\bar{C}_2}(x_m) = \frac{1 - y'_{C_2}(x_m)}{1 + \lambda y'_{C_2}(x_m)} \tag{C.5}$$

$$y^{\lambda'}_{\bar{C}_1}(x_m) = 1 - \frac{1 - y'_{C_1}(x_m)}{1 + \lambda y'_{C_1}(x_m)} \tag{C.6}$$

Step 8: Demoralize $y^{\lambda'}_{\bar{C}_1}(x_m)$ *and* $y^{\lambda'}_{\bar{C}_2}(x_m)$ for neural network learning purposes:

$$y^{\lambda}_{\bar{C}_1}(x_m) = y^{\lambda'}_{\bar{C}_1}(x_m)[y_{\max} - y_{\min}] + y_{\min} \tag{C.7}$$

Step 9: Train the BP net NN_1 (with the same topology as NN_0 and the same extreme output values y_{max} and y_{min}) with the sample set Mc_1, such that each sample point has the λ-complement value $y^{\lambda}_{\bar{C}_1}(x_m)$.

Step 10: Repeat step 9 for the BP net NN_2 and it is trained with the sample set Mc_2.

C.1.3 Multilayer Fuzzy-Neural Classification Networks

A general concept of using multilayer fuzzy neural networks as pattern classifiers, is to create fuzzy subsets of the pattern space in the hidden layer and then aggregate the subsets to form a final decision in the output layer. In this section, a fuzzy neural network classifier that creates classes by aggregating several smaller fuzzy sets into a single fuzzy set class, will be introduced. This classifier is called the *fuzzy min-max classification neural network* and was proposed by Simpson [4]. It is constructed using hyperbox fuzzy sets, each of which is an n-dimensional box defined by a min point and a max point with a corresponding membership function. The hyper box fuzzy sets are aggregated to form a single fuzzy set class. Learning in this network, is performed by determining proper min-max points or, equivalently, by properly placing and adjusting hyper boxes in the pattern space.

A fuzzy min-max classification neural network has a three-layer structure, as shown in Fig. C.5(a). The input layer has n nodes, one for each of the n dimen-

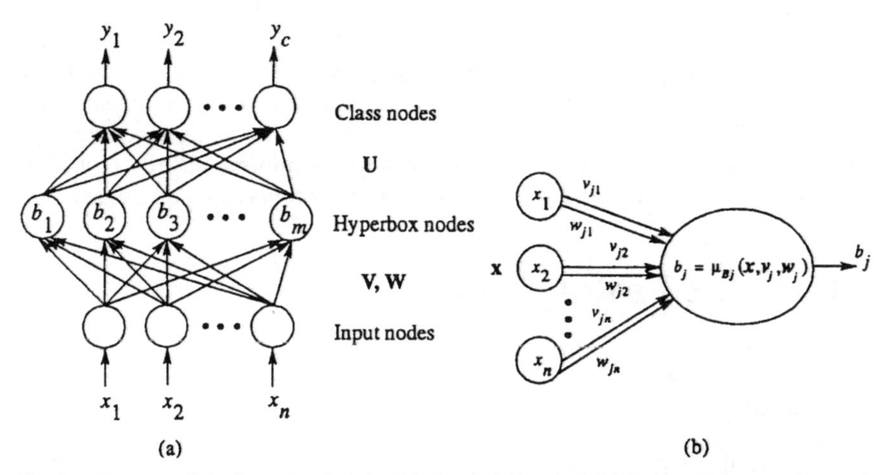

Fig. C.5. Structure of the fuzzy min-max classification neural network. (a) The three-layer neural network that implements the fuzzy min-max classifier. (b) The hyperbox node b_j for hyperbox B_j

sions of the input pattern $\boldsymbol{x}$. The input node just passes input data into the hidden layer. Each hidden node, called a hyper box node, represents a hyper box fuzzy set, where the hyper box node transfer function is the hyper box membership function defined by:

$$\mu_{B_j}(\boldsymbol{x}) = \frac{1}{n}\sum_{i=1}^{n} [1 - f(x_i - w_{ij}, r) - f(v_{ij} - x_p r)] \tag{C.8}$$

where $f(.)$ is the two-parameter ramp threshold function:

$$f(z,r) = \begin{cases} 1 & \text{if } zr > 1 \\ zr & \text{if } 0 \leq zr \leq 1 \\ 0 & \text{if } zr < 0 \end{cases} \tag{C.9}$$

the hyperbox node is illustrated in Fig. C.5 (b)

In summary, the fuzzy min-max classification-learning algorithm is a three-step process:

Step 1 (Expansion): Identify the hyperbox that can expand and expand it. If an expandable hyperbox cannot be found, add a new hyperbox for that class.

Step 2 ***(Overlapping Test):*** Determine if any overlapping exists between hyperboxes from different classes.

Step 3 ***(Contraction):*** If overlapping between hyperboxes that represent different classes exists, eliminate it, by minimally adjusting each of the hyperboxes.

C.2 Fuzzy-Neural Clustering

C.2.1 Fuzzy Competitive Learning for Fuzzy Clustering

Competitive learning, is one of the major clustering techniques in neural networks. By generalizing the conventional competitive learning rules, a fuzzy competitive learning rule based on the fuzzy *c*-means (FCM) formula, has been proposed by Jou [5]. Since competitive learning is designed for crisp clustering, it is conceivable that fuzzy competitive learning is suitable for fuzzy clustering. The fuzzy competitive learning rule is derived by minimizing the objective function J in Eq. (C.10) and applying the FCM formulas in Eqs. (C.11) and (C.12).

$$J(u_{ij}, v_k) = \sum_{i=1}^{c}\sum_{j=1}^{n}\sum_{k=1}^{c} g[w(x_j), u_{ij}]\, d(x_j, v_k) \tag{C.10}$$

$$v_i = \frac{1}{\sum_{j=1}^{n}(uij)^m}\sum_{j=1}^{n}(u_{ij})^m x_j, \qquad i = 1, 2, \ldots, c \tag{C.11}$$

$$u_{ij} = \frac{(1/\|x_j - v_i\|^2)^{1/(m-1)}}{\sum_{k=1}^{c}(1/\|x_j - v_k\|^2)^{1/(m-1)}}, \qquad i = 1, 2, \ldots, c \;\; j = 1, 2, \ldots, n \tag{C.12}$$

The problem then is to find fuzzy clusters in the input patterns and determine the cluster centers v_i; accordingly. Applying a gradient-descent method to the objective function in Eq. (C.10) yields:

$$\Delta v_{ih} = 2\eta(u_{ij})^m [x_{jh} - v_{ih}] - \eta m(u_{ij})^{m-1} \frac{\partial u_{ij}}{\partial v_{ih}} \| \boldsymbol{x}_j - \boldsymbol{v}_i \|^2 \tag{C.13}$$

where $\boldsymbol{v}_i = [v_{i1}, \ldots, v_{ih}, \ldots, v_{iq}]^T$ and $\boldsymbol{x}_j = [x_{j1}, \ldots, x_{jh}, \ldots, x_{jq}]^T$. By differentiating Eq. (C.12) with respect to v_{ih} and using the chain rule, we obtain:

$$\frac{\partial u_{ij}}{\partial v_{ih}} = 2u_{ij}(1 - u_{ij})(m-1)^{-1} \frac{x_{ih} - v_{ih}}{\| \boldsymbol{x}_j - \boldsymbol{v}_i \|^2} \tag{C.14}$$

The fuzzy competitive learning rule, which moves the weight vectors toward their respective fuzzy cluster centers, is to use Eq. (C.13) with Eq. (C.14):

$$\Delta v_{ih} = \eta \gamma_m [x_{jh} - v_{ih}] \tag{C.15}$$

where,

$$\gamma_m \triangleq (u_{ij})^m [1 - m(m-1)^{-1}(1 - u_{ij})] \tag{C.16}$$

with $m \in (1, \infty)$. Thus, the objective function is minimized by the above update rule. If the update rule in Eq. (C.15) in batch mode (i.e., accumulate the changes Δv_{ih} for each pattern $\boldsymbol{x}_j$ before the weights are actually updated) is used, then the learning algorithm will correspond to the fuzzy c-means algorithm.

For a specified weighting exponent m, the term γ_m, which is determined by the membership grade u_{ij}, controls the amount of the weight v_{ij}, to be updated. Fig. C.6 depicts the function $\gamma_m(u_{ik})$ given by Eq. (C.16) for several values of m. As seen in Fig. C.6, the boundary conditions $\gamma_m(0) = 0$ and $\gamma_m(1) = 1$ are satisfied. For large m values, $\gamma_m(u_{ij})$ goes smoothly from 0 to 1 as u_{ij} goes from 0 to 1.

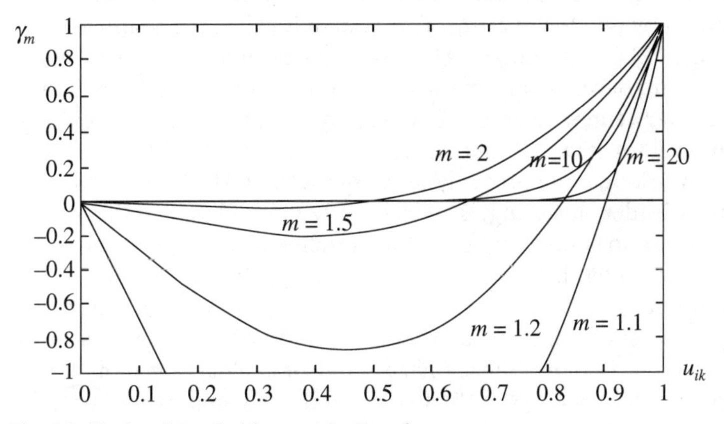

Fig. C.6. The function $\gamma_m(u_{ik})$ for several values of m

At small m values, it makes a rather sudden rise from near $-\infty$ to 1 over a narrow range of u_{ij} near 1. In the limit $m \to \infty$, $\gamma_m(u_{ij})$ just reduces to a step function rising at 1. In the other limit $m \to 1^+$, $\gamma_m(u_{ij})$ is negative infinite, except at the two end points. It should be also noted that $\gamma_m(u_{ij}) > 0$ if $u_{ij} > 1/m$ and $\gamma_m(u_{ij}) < 0$ if $0 < u_{ij} < 1/m$. This indicates that if membership grade u_{ij} is relatively large, then its corresponding weight vector is moved toward the current input vector; otherwise, the weight vector is moved away from the input vector. This is analogous to *Kohonen's learning vector quantization* (LVQ), in which the weight vector is moved away from the input vector when the current cluster of the winner is incorrect.

C.2.2 Adaptive Fuzzy Leader Clustering

In addition to the competitive learning rules, the ART network is another major clustering technique in neural networks. In this section, an ART like fuzzy neural classifier called the adaptive fuzzy leader classifier, will be introduced. There are many similarities between ART and the classical leader clustering algorithm, which works as follows:

Step 1: The leader clustering system begins with a cluster centered at the location of the first input pattern $\boldsymbol{x}_1$. This cluster is called the leader.

Step 2: For each of the remaining patterns, $\boldsymbol{x}_k$, $k = 1, 2, \ldots, N$, the following is performed.

1) Find the cluster closest to $\boldsymbol{x}_k$ (usually using Euclidean distance measures).
2) If the ratio of the distance between the cluster center and $\boldsymbol{x}_k$ is less than a prespecified tolerance value, the cluster center is updated to include $\boldsymbol{x}_k$ in its cluster by averaging the location $\boldsymbol{x}_k$ with the cluster center.
3) If the closest cluster ratio is too large, then this cluster is eliminated from the set of candidate clusters and the control returns to step 2a.
4) If the clusters are exhausted, add a new cluster with $\boldsymbol{x}_k$ as the cluster center.

Based on these steps, a fuzzy neural realization of the leader cluster algorithm has been proposed by Newton et al. [6]. This system is called the adaptive fuzzy leader clustering (AFLC) system. The AFLC system has an unsupervised neural network architecture developed from the concept of ART1 and includes a relocation of the cluster centers from FCM system equations for the centroid and the membership values. It can thus be seen as another fuzzy modification of an ART1 neural network or, more precisely, an integration of ART1 and FCM.

Adaptive fuzzy leader clustering, is a hybrid fuzzy-neural system that can be used to learn cluster structure, embedded in complex data sets in a self-organizing and stable manner. Let $\boldsymbol{x}_k = \{x_{k1}, \ldots, x_{k2}, \ldots, x_{kn}\}$ be the kth discrete or analog-valued input vector for $k = 1, 2, \ldots, N$, where N is the total number of samples. Figure C.7 shows the structure and operation characteristics of an AFLC system, where $\boldsymbol{x}_k$ is an n-dimensional discrete or analog-valued input vector to the system. The AFLC system is made up of the comparison layer, the recognition layer, and the surrounding control logic. As in ART1, the initializa-

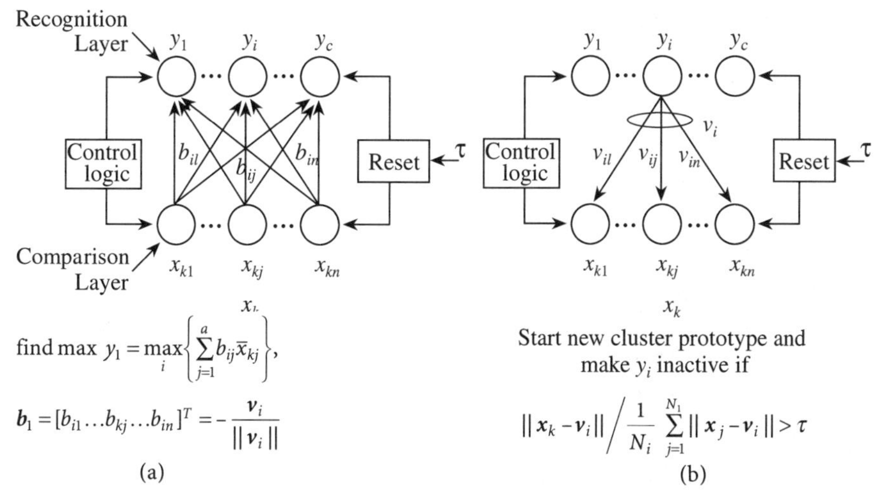

Fig. C.7. Operation characteristics of an AFLC architecture. (a) Initial stage of identifying a cluster prototype. (b) The comparison stage using the criterion of Euclidean distance ration $D > \tau$ to reject new data samples to the cluster prototype. The reset control implies deactivation of the original prototype and activation of a new cluster prototype

tion and update procedures in AFLC involve similarity measures between the bottom-up weights ($\boldsymbol{b}_i = [b_{k1}, \ldots, b_{k2}, \ldots, b_{kn}]^T, i = 1, \ldots, c$) and the input vector $\boldsymbol{x}_k$, and a verification of $\boldsymbol{x}_k$ belonging to the ith cluster by a matching of the top-down weights ($\boldsymbol{v}_i = [v_{k1}, \ldots, v_{k2}, \ldots, v_{kn}]^T, i = 1, \ldots, c$) with $\boldsymbol{x}_k$, where c is the (current) number of clusters and $\boldsymbol{b}_i = \boldsymbol{v}_i/\|\boldsymbol{v}_i\|$. The AFLC algorithm initially starts with the number of clusters c set to zero. The system is initialised with the input of the first input vector $\boldsymbol{x}_1$. As in leader clustering, this first input is said to be the prototype for the first cluster. The normalized input vector $\bar{\boldsymbol{x}}_k = \boldsymbol{x}_k/\|\boldsymbol{x}_k\| = [\bar{x}_{k1}\bar{x}_{k2}, \ldots, \bar{x}_{kn}]^T$ is then applied to the bottom-top weights by dot product. The node that receives the largest input activation is chosen as the expected cluster:

$$y_{i^*} = \max_i \{y_i\} = \max_i \left\{ \sum_{j=1}^{n} b_{ij}\bar{x}_{kj} \right\}, \quad k = 1, 2, \ldots, N \tag{C.17}$$

and the top-bottom weight vector corresponding to the winning node $\boldsymbol{v}_{i*}$ is chosen as the expected prototype vector (winning vector).

The recognition layer serves to initially classify an input. This first stage classification, activates the prototype or top-bottom expectation $\boldsymbol{v}_i$; for a cluster, which is forwarded to the comparison layer. The comparison layer serves both as a fan-out site for the inputs and as a location for the comparison between the top-bottom expectation and the input. The control logic with an input-enable command allows the comparison layer to accept a new input, as long as a comparison operation is not currently being processed. The control logic with a comparison-imperative command disables the acceptance of new input and initiates a comparison between the cluster prototype associated with y_i (i.e., $\boldsymbol{v}_i$)

and the current input vector. This is the so-called vigilance test in ART. The reset signal is activated when a mismatch between the prototype and the input vectors occurs, according to the criterion of distance ratio threshold, expressed by:

$$D = \frac{\| \boldsymbol{x}_k - \boldsymbol{v}_i \|}{(1/N) \sum_{\ell=1}^{N_i} \| \boldsymbol{x}_\ell - \boldsymbol{v}_i \|} < \tau \tag{C.18}$$

where N_i; is the number of samples in class i. If the ratio D is less than a user-specified threshold (vigilance parameter) τ, then the input vector is found to belong to the winning cluster y_i and thus the index i is the final network output; otherwise, if the criterion in Eq. (C.18) is not met, then the current winner y_i in the recognition layer is disabled and a new winner is found by repeating the above process. If no y_i satisfies the distance ratio criterion (i.e., all the nodes in the recognition layer are disabled), then a new cluster is created, its prototype vector is made equal to $\boldsymbol{x}_k$, and the index of the new cluster is the final network output. Note that a low threshold value will result in the formation of more clusters, because it will be more difficult for an input to meet the clarification criterion. A high value of τ will result in fewer, less dense clusters.

When an input is classified as belonging to an existing cluster, it is necessary to update the expectation (prototype) and the bottom-top weights associated with that cluster. First, the degree of membership $\boldsymbol{x}_k$ in the winning cluster is calculated. This degree of membership gives an indication, based on the current state of the system, of how heavily $\boldsymbol{x}_k$ should be weighted in recalculation of the class expectation. The cluster prototype is then recalculated as a weighted average of all the elements within the cluster. The weights are updated according to the FCM algorithm in Eqs. (C.11) and (C.12) with index k replaced by ℓ, index j replaced by k, and n replaced by N_i, where N_i; is the number of samples in cluster i. The update rules for an AFLC are as follows: The membership value u_{ik} of the current input sample $\boldsymbol{x}_k$ in the winning class i is calculated using Eq. (C.12) and then the new cluster centroid for cluster i is generated using Eq. (C.11). As with the FCM, m is a parameter that defines the fuzziness of the results and is normally set between 1.5 and 30.

The AFLC algorithm can be summarized by the following steps:

Step 1: Start with no cluster; $c = 0$.

Step 2: Let $\boldsymbol{x}_k$ be the next input vector.

Step 3: Find the first-stage winner y_i as the cluster prototype with the maximum dot product.

Step 4: If no y_i satisfies the distance ratio criterion, create a new cluster and make its prototype vector equal to $\boldsymbol{x}_k$. Output the index of the new cluster.

Step 5: Otherwise, update the winner cluster prototype associated with y_i by calculating the new centroid and membership values using Eqs. (C.11) and (C.12). Output the index of y_i. Go to step 2.

C.3 Fuzzy-Neural Models for Image Processing

C.3.1 Fuzzy Self Supervised Multilayer Network for Object Extraction

A multi-layer feedforward network with back-propagation learning has been used for image segmentation with limited success (e.g., see [7,8]). Since the use of multi-layer feedback networks requires a set of labelled input output data, a set of known images for supervised learning is needed and then the trained network for processing other images is used. The problem with this approach, is that it is valid only when the images to be processed are of similar nature. Also, the training images may not always be available in real-life situations. To attack these problems, a self organizing (or self supervised) multi-layer neural network architecture suitable for image processing, has been proposed by Ghosh et al. [9]. This network can be used for segmentation of images, when only one image is available, and it does not require any a priori target output value for supervised learning. Instead, the network output is described as a fuzzy set, and a fuzziness measure of this fuzzy set is used as a measure of error in the system (instability of the network). This measure of error is then back propagated to adjust the weights in each layer. Thus, the measures of fuzziness play an important role in the self-supervised back-propagation network and the useful ones will be described below.

C.3.1.1 Fuzzy Measures

Index of fuzziness
The index of fuzziness of a fuzzy set A having n supporting points, $X = \{x_1, x_2, \ldots, x_n\}$, is defined as:

$$f(A) = \frac{2}{n^k} d(A,C) \tag{C.19}$$

where *d(A,C)* denotes a metric distance (e.g., Hammimg of Euclidean distance) of A from any of the nearest crisp set C, for which:

$$\mu_c(x) = \begin{cases} 0 & \text{if } \mu_A(x) \le 0.5 \\ 1 & \text{if } \mu_A(x) > 0.5 \end{cases} \tag{C.20}$$

The value of k in Eq. (C.19) depends on the type of distance used. For example, $k = 1$ is used for a Hamming distance and $k = 0.5$ for a Euclidean distance. The corresponding indices of fuzziness are called the linear index of fuzziness $f(A)$ and the quadratic index of fuzziness $f_q(A)$. Thus, we have:

$$f_\ell(A) = \frac{2}{n}\sum_{i=1}^{n} |\mu_A(x_i) - \mu_C(x_i)| = \frac{2}{n}\sum_{i=1}^{n} \min\{\mu_A(x_i), (1-\mu_A(x_i))\}, \tag{C.21}$$

$$f_q(A) = \frac{2}{\sqrt{n}}\left[\sum_{i=1}^{n}\{\mu_A(x_i) - \mu_C(x_i)\}^2\right]^{1/2} = \frac{2}{\sqrt{n}}\left[\sum_{i=1}^{n}\{\min[\mu_A(x_i), (1-\mu_A(x_i))]\}^2\right]^{1/2} \tag{C.22}$$

Entropy:
The entropy of a fuzzy set defined by De Luca and Termni [10] (the logarithmic entropy) is:

$$H_\ell(A) = \frac{1}{n\ln 2}\sum_{i=1}^{n}\{-\mu_A(x_i)\ln[\mu_A(x_i)] - [1-\mu_A(x_i)]\ln[1-\mu_A(x_i)]\} \tag{C.23}$$

Another definition of entropy, the exponential entropy, given by Pal and Pal [11] is:

$$H_e(A) = \frac{1}{n(\sqrt{e}-1)}\sum_{i=1}^{n}\{\mu_A(x_i)\, e^{1-\mu_A(x_i)} + [1-\mu_A(x_i)]\, e^{\mu_A(x_i)} - 1\} \tag{C.24}$$

These measures will be used to compute the error or measure of instability of a multi-layer self organizing neural network.

C.3.1.2 Computation of the Measure of Instability of a Multi-Layer Self Organizing Neural Network

Consider a self-supervised multi-layer neural network for image segmentation or object extraction. Before describing the architecture of the network, we need to define a neighbourhood system. For an $M \times N$ image lattice L, the dth order neighbour N_{ij}^d of an element (i, j) is defined as:

$$N_{ij}^d = \{(i,j) \in L\} \tag{C.25}$$

such that $(i, j) \in N_{ij}^d$, and if $(k, l) \in N_{ij}^d$, then $(i, j) \in N_{kl}^d$.

Different ordered neighborhood systems can be defined considering different sets of neighboring pixels of (i, j). $N^1 = \{N_{ij}^1\}$ can be obtained by taking the four nearest-neighbor pixels. Similarly, $N^2 = \{N_{ij}^2\}$ consists of the eight-pixel neighborhood (i, j), and so on, as shown in Fig. C.8.

The three-layer version of a self supervised multi-layer network, is shown in Fig. C.9. In each layer, there are $M \times N$ sigmoidal nodes (for an $M \times N$ image), and each node corresponds to a single pixel. Besides the input and output layers, there can be a number of hidden layers (more than zero). Nodes of the same

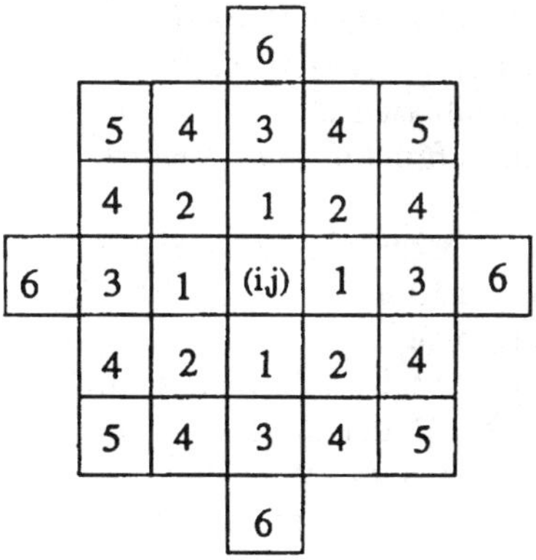

			6			
	5	4	3	4	5	
	4	2	1	2	4	
6	3	1	(i,j)	1	3	6
	4	2	1	2	4	
	5	4	3	4	5	
			6			

Fig. C.8. Neighborhood system N^d

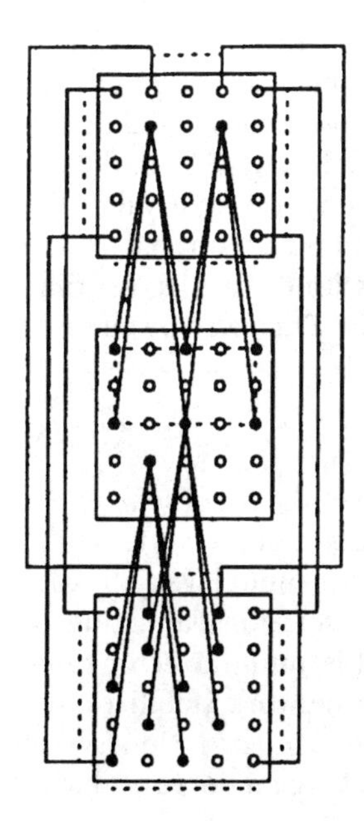

Fig. C.9. Schematic representation of self-supervised multiplayer neural network

layer do not have any connections among themselves. Such node in a layer, is connected to the corresponding node in the previous layer and to its neighbours (over N^d); thus each node in layer i $(i > 1)$ has $|N^d| + 1$ (where $|N^d|$ is the number of pixels in N^d) links to the $(i-1)$th layer. For N^1, a node has five links, whereas for N^2, nine links are associated with every node. However, for boundary nodes (pixels), the number of links may be less than $(N^d| + 1)$. Every node in the output layer is also connected to the corresponding node in the input layer.

The input to an input node is given as a real number in [0, 1] which is proportional to the gray value of the corresponding pixel. The input pattern is passed on to the output layer, and the corresponding outputs are calculated in the same way, as in a normal back-propagation network. In order to eliminate noise and extract spatially compact regions, all initial weights are set to 1. As a result, the total input (initially) to any node lies in $[0, n_\ell]$ (where n_ℓ is the number of links a neuron has); hence, the most unbiased choice for the threshold value θ for the unipolar sigmoidal activation function is $n_\ell/2$, which is the middle of most values of the total input range.

After the network output has been obtained for a given input pattern (image) through forward propagation, error back propagation for weight learning is performed. However, there is no target output to supervise the learning now. Here, the intention is to extract spatially compact regions through the process of self-organization, using only one noisy image. Under ideal conditions, the network is organized in such a way, that the output status of most of the nodes in the output layer are either 0 or 1, but owing to the effect of noise, the output is usually in [0, 1]; thus the status value represents the degree of brightness (darkness) of the corresponding pixel in the image. Therefore, the output status in the output layer may be viewed to represent a fuzzy set 'BRIGHT (DARK) pixels.' The number of supports of this fuzzy set is equal to the number of nodes in the output layer. The measure of fuzziness of this set, may be considered the *error* or *instability of the whole system,* as this reflects the deviation from the desired state of the network. Thus, without any a priori target output value, the fuzziness value is taken as a measure of system error and is back-propagated to adjust the weights so that the system error is reduced with the passage of time and in the limiting case becomes zero. The error measure E can also be taken as a suitable function of a fuzziness measure that is:

$$E = g(I) \tag{C.26}$$

where I is a measure of fuzziness Eqs. (C.21) and (C.24) of the fuzzy set.

After the weights have been adjusted properly, the output of the nodes in the output layer is fed back to the corresponding nodes in the input layer. The second pass is then continued with this as input. The iteration (updating of weights) is continued, until the network stabilizes, that is, until the error value (measure of fuzziness) becomes negligible. When the network stabilizes, the output status of the nodes in the output layer becomes either 0 or 1. Nodes with an output value of 0 constitute one group, and those having an output value of 1 constitute the other.

The weight update rules for different fuzziness measures will be derived next. According to the back-propagation algorithm, the general learning rule is:

$$\Delta w_{ij} = \eta\left(-\frac{\partial E}{\partial w_{ij}}\right) = \eta\left(-\frac{\partial E}{\partial y_i}\right)\left(\frac{\partial y_i}{\partial net_i}\right)\left(\frac{\partial net_i}{\partial w_{ij}}\right) = \eta\left(-\frac{\partial E}{\partial y_i}\right)a'(net_i)x_j, \tag{C.27}$$

where w_{ij} is the weight on the link from node j to node i, η is the learning constant, $a(.)$ is the activation function, x_j is the output of node j, net_i is the net input to node i (i.e., $net_i = \sum_j w_{ij}x_j$), and y_i is the output of node i [i.e., $y_i = a(net_i)$]. Hence, by deriving $\partial E/\partial y_i$ in Eq. (C.27) for node i, the learning rule of the corresponding weight can be obtained. Weight update rules can be derived using different fuzziness measures for the output layer, which will be described below.

C.3.1.3 Weight Update Learning Rules Using Fuzzy Measures

Learning rule for linear index of fuzziness
Let the error measure be:

$$E = g(f_\ell) = f_\ell \tag{C.28}$$

where from Eq. (C.21), the linear index of fuzziness is:

$$f_\ell = \frac{2}{n}\sum_{i=1}^{n} \min(y_j, 1-y_j), \tag{C.29}$$

where n is the number of output nodes. Then we have:

$$-\frac{\partial E}{\partial y_i} = \begin{cases} -2/n & 0 \le y_i \le 0.5 \\ 2/n & 0.5 < y_i \le 1\,. \end{cases} \tag{C.30}$$

Thus, from Eq. (C.27) the following learning rule is obtained:

$$\Delta w_{ij} = \begin{cases} \eta(-2/n)a'(net_i)x_j & 0 \le y_i \le 0.5 \\ \eta(2/n)a'(net_i)x_j & 0.5 < y_i \le 1\,. \end{cases} \tag{C.31}$$

Learning rule for quadratic index of fuzziness
Let the error measure be:

$$E = g(f_q) = f_q^2 \tag{C.32}$$

where from Eq. (C.22), the square of the quadratic index of fuzziness is:

$$f_q^2 = \frac{4}{n}\sum_{i=1}^{n} [\min(y_j, 1-y_j)]^2, \tag{C.33}$$

Then we have:

$$-\frac{\partial E}{\partial y_i} = \begin{cases} (4/n)(-2y_i) & 0 \le y_i \le 0.5 \\ (4/n)[2(1-y_i) & 0.5 < y_i \le 1\,. \end{cases} \tag{C.34}$$

Thus, from Eq. (C.27) the following learning rule is obtained:

$$\Delta w_{ij} = \begin{cases} -\eta(8y_i/n)a'(net_i)x_j & 0 \le y_i \le 0.5 \\ \eta(8/n)(1-y_i)a'(net_i)x_j & 0.5 < y_i \le 1\,. \end{cases} \tag{C.35}$$

Learning rule for logarithmic entropy
Consider the error measure be:

$$E = g(H_\ell) = H_\ell \tag{C.36}$$

where from Eq.(C.23), the entropy of fuzzy set is:

$$H_\ell = \frac{1}{n \ln 2}\sum_{i=1}^{n} \{y_i \ln y_i + (1-y_i)\ln(1-y_i)\} \tag{C.37}$$

Then we have:

$$-\frac{\partial E}{\partial y_i} = \frac{1}{n\ln 2}\ln\frac{y_i}{1-y_i}. \tag{C.38}$$

To expedite the learning, it is desirable to make a large weight correction when the network is most unstable (i.e., when all the output values are 0.5). For a neuron, the weight correction for its links should be maximum, when its output status is very close to 0.5, and minimum when its output status is close to 0 or 1. This can be achieved by taking (refer to Eq. (C.27)).

$$\Delta w_{ij} = \eta\left(-\frac{\partial E/\partial y_i}{|\partial E/\partial y_i|^q}\right)a'(net_i)x_j, \quad q>1 \tag{C.39}$$

where $|\partial E/\partial y_i|$ represents the magnitude of the gradient. According to Eq. (C.39), when $q = 2$:

$$\Delta w_{ij} = -\eta\frac{1}{\partial E/\partial y_i}a'(net_i)x_j = \eta(n\ln 2)\frac{1}{\ln[y_i/(1-y_i)]}a'(net_i)x_j \tag{C.40}$$

Learning rule for exponential entropy
Consider the error measure to be:

$$E = g(H_\ell) = H_\ell \tag{C.41}$$

where from Eq. (C.24), the exponential entropy of fuzzy set is:

$$H_\ell = \frac{1}{n(\sqrt{e}-1)}\sum_{i=1}^{n}\{y_i e^{1-y_i} + (1-y_i)e^{y_i} - 1\} \tag{C.42}$$

Then:

$$-\frac{\partial E}{\partial y_i} = -\frac{1}{n\sqrt{e}-1}\{(1-y_i)e^{1-y_i} - y_i e^{y_i}\}. \tag{C.43}$$

According to an argument similar to that for logarithmic entropy (refer to Eq. (C.39), the following learning rule for exponential entropy is obtained:

$$\Delta w_{ij} = -\eta\frac{1}{\partial E/\partial y_i}a'(net_i)x_j = -\eta n(\sqrt{e}-1)\frac{1}{\{(1-y_i)e^{1-y_i} - y_i e^{y_i}\}}a'(net_i)x_j \tag{C.44}$$

From the above four learning rules in Eqs. (C.31), (C.35), (C.40) and (C.44), it is observed that in each case, the expression for Δw_{ij} has a factor $\eta\alpha'(net_i)x_j$, which can be ignored in comparing different learning rates. The remaining part of the expression for Δw_{ij} will be referred to as the learning rate, because only that factor is different, for different measures of fuzziness. It can be easily verified that in each of the four cases the learning rate is negative for $y_i \le 0.5$ and positive for $y_i \ge 0.5$.

C.4 Fuzzy Neural Networks for Speech Recognition

C.4.1 Introduction

Since the back-propagation learning algorithm was developed, many neural network applications for speech recognition have been proposed [12]. Through the use of a back-propagation training method, a neural network can delineate an arbitrarily complicated pattern space that conventional methods of pattern classification may fail to partition. However, there are occasions when the ability to construct substantial complexity in the decision surface, is a disadvantage. In the case where the patterns to be classified occupy overlapping regions in the pattern space; for example, speech feature vectors are typically these kinds of patterns. It is possible to build a neural network that constructs a decision surface that complete "threads" through the *implied* boundary represented by the training patterns. However, this "tightly tuned" fitting to the training patterns may not be an optimal choice for later pattern classification, because some training patterns may represent outliers of the pattern ensemble. In this case, the robustness (i.e., the generalization capability) of neural networks is not as adequate as expected. This problem is essentially an over learning of the training set, which causes a drastic performance reduction when a slight difference arises in the testing set (e.g., speaking rate differences).

C.4.2 Problem Definition

Another problem arises when combining neural networks with a language model in which top-*N* candidate performance is required. This problem derives from simply using discrete phoneme class information as a target value in the conventional method. Thus, the neural network is trained to produce the top phoneme candidate, but not the top-*N* candidates. However, the top-*N* phoneme candidate information is very important when combined with a language model in continuous speech recognition. Once the lack of phoneme candidate information occurs, it may lead to a fatal error in recognizing continuous speech.

C.4.3 Fuzzy-Neural Approach

One way to overcome these problems is to take the fuzzy neural approach, which considers the collection of patterns, where each pattern identifies itself with a continuous membership value. With these membership values, it is possible to modify the back-propagation algorithm so that training patterns with large membership values play a more crucial role than those with small membership values, in modifying the weights of the network. As a result, the network is unlikely to be degraded by a few possible outliers in the training set. Thus, the fuzzy neural training method and the conventional neural training method for speech recognition, are both based on the back-propagation algorithm. However, they differ in how they give the target values to the neural network. In the conventional method, target values are given as discrete phoneme class information, that is, 1 for the belonging class and 0s for the others. In the fuzzy neural training method, the target values are given as fuzzy phoneme class information, whose values are given in between 0 and 1. This fuzzy phoneme class informs the neural network about the likelihood of the input sample to each phoneme class, in other words, the possibility of belonging to a phoneme class.

Fuzzy phoneme class information can be modelled by considering the distance, for instance, the Euclidean distance measure between the input sample and the nearest sample of each phoneme class in the training set. The assumption that when the distance of two samples is small, these two samples are considered to be alike, indicates that each sample has a possibility of belonging to the class of the other sample. However, when the distance of two samples is great, these two samples are considered to be very different, which indicates that each sample has no possibility of belonging to the class of the other sample. To model this fuzzy phoneme class information using the distance d, a likelihood transformation function $f(d)$, which can be considered a membership function, is adopted. By using a monotonous decreasing function such as:

$$f(d) = e^{-ad^2}, \tag{C.45}$$

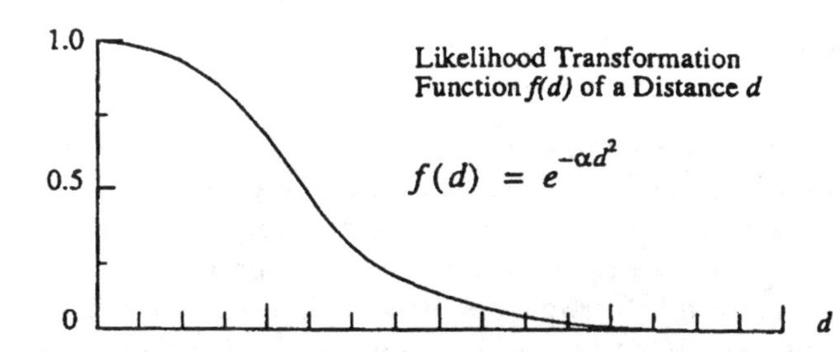

Fig. C.10. Likelihood transformation function

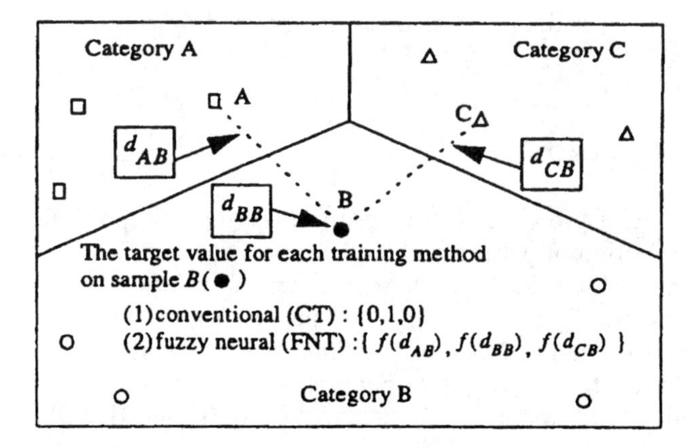

Fig. C.11. Basic concept of the fuzzy neural training method

where $a \geq 0$, as shown in Fig. C.10, it can easily model the idea that the larger the distance, the lower the likelihood, and the smaller the distance, the larger the likelihood. Using this function, fuzzy phoneme class information can be computed according to the distance between the input sample and the nearest sample of each phoneme class in the training set.

The basic concept of the above-mentioned fuzzy neural training (FNT) method as well as the conventional training (CT) method, is depicted in Fig. C.11. The target values in the CT method are given as discrete phoneme class information; that is, the target value of sample $B(\bullet)$ is given as $\{0,1,0\}$. The target values in the FNT method are given as fuzzy phoneme class information; that is, the target value of sample $B(\bullet)$ is given as $\{f(d_{AB}), f(d_{BB}), f(d_{CB})\}$, where $f(d)$ *is* a likelihood transformation function of distance d.

C.5 Fuzzy-Neural Hybrid Systems for System Diagnosis

C.5.1 Introduction

In the previous sections of Appendix C, the focus was either on bringing neural network techniques into fuzzy systems to form *neural fuzzy systems,* or on introducing fuzzy logic into neural networks to form *fuzzy neural networks.* In fact, a more advanced, straightforward approach is to put them together to form a fuzzy logic and neural network incorporated system, called a *fuzzy-neural hybrid system.* In such systems, fuzzy logic techniques and neural networks can be viewed as two individual subsystems. They do their own jobs by serving different purposes in a system. By making use of their individual strengths, they incorporate and complement each other to accomplish a desired task. A typical architecture of a fuzzy-neural hybrid system, is shown in Fig. C.12, where the neural network is used for input signal processing and the fuzzy logic subsystem is used for output action decisions. This system makes use of the strength of a neural network in its processing speed and the strength of fuzzy logic in its flexible reasoning capability for decision making and control. Of course, this is not the only structure for fuzzy-neural hybrid systems; fuzzy inference for input

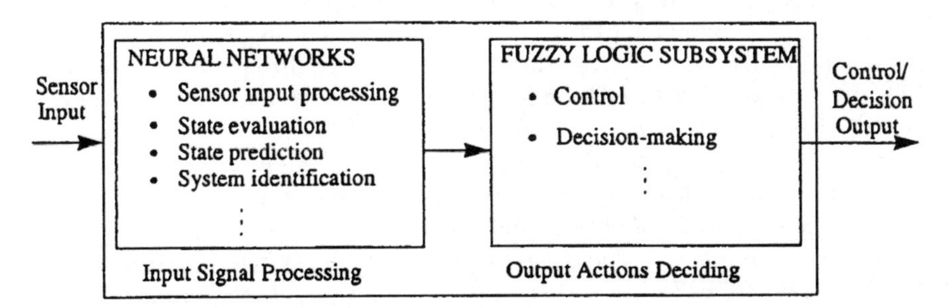

Fig. C.12. A typical architecture of fuzzy-neural hybrid systems

state evaluation and neural networks for control can also be used. Moreover, fuzzy neural networks or neural fuzzy systems in a fuzzy-neural hybrid system may also be included. In this section, a fuzzy-neural hybrid system for system diagnosis, as a case study, will be introduced.

C.5.2 Hybrid Systems

Real-time diagnosability of a system behavior has a significant importance in industry. Some applications have had a mathematical model available, and the use of explicitly rule-based algorithms has been a dominant force. However, hard logical decisions and exhaustive constraints often cause inflexible implementation of such systems in addition to the fast processing time requirement of real-time applications. In this regard, neural networks and fuzzy logic have been proven a viable alternative. A combination of these two paradigms can provide high speed, flexibility, and humanlike soft, logical decisions.

In this section, the fuzzy-neural hybrid system proposed by Choi et al. [13] will be introduced for real-time nonlinear system diagnosis. This system performs real-time processing, prediction, and data fusion for general real-world system diagnosis, which is usually complex, and a final decision should be based on multiple subsystem (channel) diagnosis. The architecture of the fuzzy-neural diagnosis system is shown in (channels) that could be contaminated with environmental noise. Each network is trained to predict the future behavior of one time series. The prediction error and its rate of change from each channel are computed and sent to a fuzzy logic decision output stage, which contains $(n + 1)$ modules. The $(n + 1)$st final-output module performs data fusion by combining n individual fuzzy decisions that are tuned to match the domain expert's needs. Thus, the basic idea is that the neural network's output predicts the normal system response and this output and the actual system response can be compared to evaluate the presence of the noise. In the case of a noticeable difference between the two responses, we can conclude "unstable" with the use of fuzzy logic.

The neural networks are trained in such a way, that each one models a channel transfer function. In case of a system abnormality, the trained neural network is expected to indicate the abnormality in advance. The predicted network's output is compared to actual system output according to the comparison results, humanlike decisions can then be implemented through fuzzy inference. The error signal (i.e. the deviation from the normal state) and the history of this error signal (i.e. the change in the error) can be part of the fuzzy input components. In a diagnostic task, one might also make use of the previous diagnostic results, to arrive at the current diagnosis. For example, if the previous diagnostic result was "unsafe," and both the error and the change in the error are increasing, then one sees that the performance of the system is getting worse. The individual fuzzy decisions $D_1(t), D_2(t), \ldots, D_n(t)$, resulting from each sensor signal, are then gathered in the fuzzy logic output unit for the final decision making. A feedback of the previous decision $D(t-1)$ can also be applied with the same philosophy as explained before. The functions of the neural network and fuzzy logic units, will be described in more detail in the figure below.

Predicting the behaviour of a complex system, can be viewed as a nonlinear mapping from past states to future states. In the fuzzy-neural diagnosis system shown in Fig. C.13, this is done by using neural networks. In other words, the job here is to predict the time series p time steps in the future from k samples of previous inputs,

$$\bar{x}(t+p) = f_N(x(t), x(t-1), x(t-2), \ldots, x(t-k-1)) \tag{C.46}$$

where f_N represents the map learned by the neural network. Once the network extracts the underlying nonlinear map, it is then able to predict normal behavior of the system equation. Several neural networks can serve this purpose, such as back-propagation networks, radial basis function networks, and so on.

Next, consider the fuzzy logic module found in each processing channel and the output stage in Fig. C.13. Each of them has the same basic architecture but may have different membership functions and fuzzy logic rules, thus giving each processing channel its own unique identity and behavior. Each accepts an error and a change-in-error signal (i.e. an e and $\dot{e}$ signal per channel) as well as

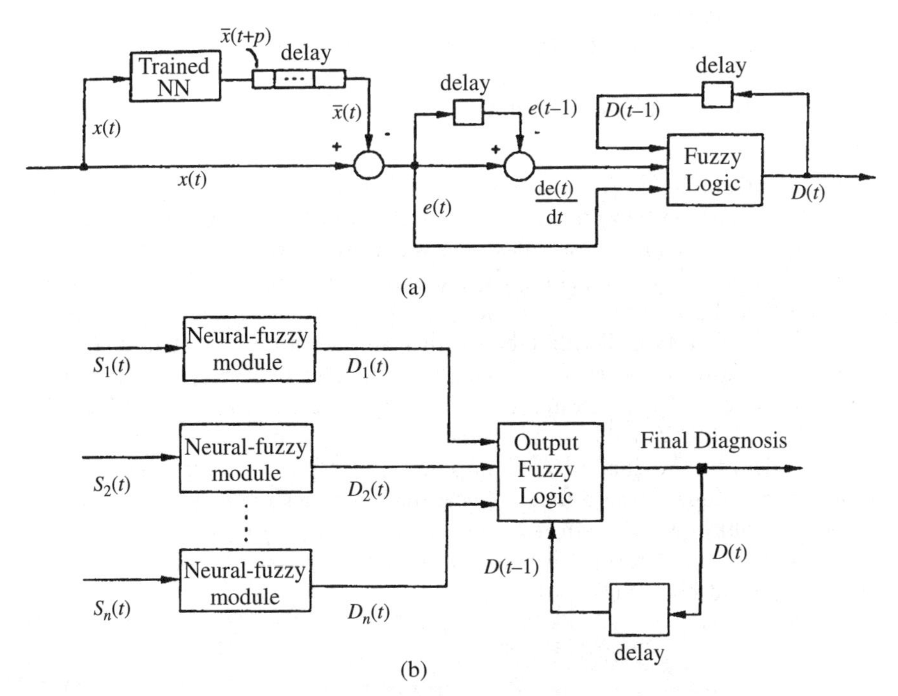

Fig. C.13. Architecture of the fuzzy-neural diagnosis system (a) Single module. (b) Multiple modules illustrating data fusion

Table C.1. Ezable of a fuzzy Control Rule Table for the Fuzzy Logic Modules in the Fuzzy-Neural Diagnosis System

(a)		$\dot{e}$			
		Low	Med	High	
e	Low	RS	RU	U	$D_i(t-1) = \text{RU}$
	Med	RU	U	U	
	High	U	U	U	

(b)		$D_1(t)$				
		S	RS	RU	U	
$D_2(t)$	S	S	RS	RS	RU	$D(t-1) = \text{RU}$
	RS	RS	RS	RU	RU	
	RU	RS	RU	RU	U	
	U	RU	RU	U	U	

a feedback signal $D_i(t-1)$ (for the ith channel) and then outputs $D_i(t)$ through fuzzy reasoning based on fuzzy logic rules. Some example rules are:

IF (e is Low) AND ($\dot{e}$ is NOT High) AND (D_i $(t-1)$ is Stable)
THEN D_i (t) = Stable;
ELSE IF ($\dot{e}$ is High) AND ((D_i $(t-1)$ is Stable) OR ((D_i $(t-1)$ is RELATIVELY Stable))
THEN (D_i (t)= RELATIVELY Unstable;
ELSE IF ...

A typical rule table for the fuzzy logic module in each channel is given in Table C.1 (a), where three fuzzy terms, "Low," "Med", and "High", are used for the error signals e and $\dot{e}$ and these rules have a common precondition, "$D_i(t-1)$ is Relatively Unstable (RU)". Similarly, there are fuzzy rules in the output fuzzy logic module that combines n individual fuzzy decisions for the final diagnosis. Table C.1 (b) is a typical rule table in this module for $n = 2$ (two channels), where four input fuzzy terms S (Stable), RS (Relatively Stable), RU (Relatively Unstable), and U (Unstable) are used for fuzzy inference of $D_i(t)$ and $D_i(t-1)$. The rules in Table C.1 (b) have one common precondition: $D_i(t-1) = \text{RU}$, where $D_i(t-1)$ denotes the final inferred output of the output fuzzy logic module. The exact fuzzy rules used in a fuzzy-neural diagnosis system can be provided by experts or obtained through learning processes. Also, the membership functions used can be determined heuristically or tuned by a learning procedure from trial e and $\dot{e}$ [or $D_i(t)$, $i = 1, 2, \ldots, n$].

C.6 Neuro-Fuzzy Adaptation of Learning Parameters – An Application in Chromatography

C.6.1 Introduction

Neural networks have the ability to classify, store, recall and associate information [14]. The use of fuzzy systems, on the other hand, has the advantage of bringing already available expert knowledge, directly into a system. It is evident that incorporation of fuzzy logic in neural networks, would improve their performance and there are many examples of fuzzification of neural networks. It is possible to fuzzify the weight matrix, input or output data, activation functions, almost every part of neural network. The new resultant system is more robust, has greater representation power and higher training speed [14].

There are three categories of fuzzy-neural networks that will be described in short, *fuzzy neurons*, *fuzzified neural networks* and *neural networks with fuzzy training*.

Fuzzy neurons are the equivalent of traditional neurons and can be divided in several categories. There are fuzzy neurons that have crisp input and output and they are used to evaluate fuzzy weights. The weighting process is replaced by the aggregation of the membership value of each input to a fuzzy set, resulting in an output in the interval [0 1] that could represent confidence level. Another type of fuzzy neuron uses the same philosophy as the above, but the inputs and output are not crisp, but fuzzy values. Finally, it is possible to use fuzzy rules instead of aggregation operations resulting in yet another kind of fuzzy neurons.

Direct fuzzification can be performed in a number of neural network types. One of the traditional problems addressed by neural networks is the problem of classification. The approach taken, is that during training the output node corresponding to the class of the input vector is set to 1, while all the others are zero. Similarly, a winner-take-all mechanism causes each input to be classified as belonging to one class. The introduction of fuzzy logic to this problem allows an input to belong to more than one class with a varying degree of belongingness. This new feature will not worsen the network's training, since the backpropagation algorithm will ensure that vectors that are more typical of their class (membership values near one) will influence the training process more, than vectors with many small membership values, since the error that backpropagates in the latter situation will be smaller.

Another kind of neural network that can be enhanced with the use of fuzzy systems is the Kohonen clustering network. This type of network suffers from some problems. They are heuristic, that is termination cannot be based on optimizing a model. The final weights depend on the input sequence. Initial conditions influence the final results. Training parameters must be varied from one set to another [14]. The introduction of fuzzy c-means clustering is one way to address the problems of the Kohonen networks and improving on the efficiency of the fuzzy c-means algorithm, using a linear weight update rule. One approach to combine the two systems, is to insert an extra layer to the output of the Kohonen network, so that the weights are not updated by a winner-take-all mechanism, but rather using appropriate membership values.

An additional method to enhance a neural network, is fuzzy training. There are a lot of ways to use fuzzy logic to train a neural network. One way is to use fuzzy teaching, since there are a lot of cases where the only training data available or at least easily obtainable, consist of fuzzy rules. These rules can be thought of as training vectors. Another way, which is going to be used here, is the fuzzy adaptation of the training parameters.

C.6.2 Fuzzy Training of Neural Networks

Almost all learning algorithms use one or more constants that define the speed and extend in weight matrix corrections. In most common algorithms, this constant is called learning rate. Setting a high learning rate tends to bring instability and the system is difficult to converge even to a near optimum solution. On the other hand, a low value will improve stability, but will slow down convergence.

Adaptive learning rates have been used for some time now. In most applications the learning rate is a simple function of time e.g. L.R. = $^1/_{1+t}$. These functions have the advantage of having high values during the first epochs, making large corrections to the weight matrix and smaller values later, when the corrections need to be more precise. Of course there are a lot of tests to be made, in order to determine functions that work well for each system.

Using a fuzzy controller to adaptively tune the learning rate, has the added advantage of bringing all expert knowledge in use. If it was possible to manually adapt the learning rate in every epoch, we would surely follow rules of the kind listed below:

1) If the change in error is small, then increase the learning rate.
2) If there are a lot of sign changes in error, then largely decrease the learning rate.
3) If the change in error is small and the speed of error change is small, then make a large increase in the learning rate.

We can consider the training process of a neural network as an observable and controllable system. We also have available knowledge from which to obtain rules and we are not really interested in an optimal solution, but in one that will perform well enough. This means that there are no performance or other requirements that prohibit the use of a fuzzy controller [15]. Such a controller would work for every system, without the need of fine-tuning learning rate constants or functions.

C.6.2.1 Description of the application

We are going to apply fuzzy training on a network that is used to approximate chromatography relative retention times of certain molecules from various

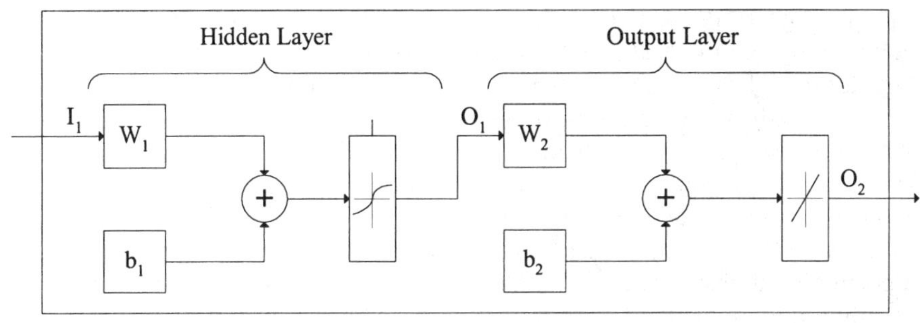

$O_1 = \text{tansig}(W_1 * I_1 + b_1)$
(Hyperbolic Tangent Sigmoidal transfer function)

$O_2 = \text{purelin}(W_2 * O_1 + b_2)$
(Linear transfer function)

W_1 is the hidden layer weight matrix
W_2 is the output layer weight matrix
b_1 the hidden layer bias vector
b_2 the output layer bias vector
I_1 is the input to the neural network
O_1 is the output of the hidden layer
O_2 is the output of the output layer

Fig. C.14. Neural networks implementation

chemical and physical properties – descriptors. Using principal component analysis, the input vector is reduced to size 15.

The network in consideration uses a *backpropagation with momentum* learning algorithm, has 15 input nodes (the principal components of the solute descriptors), 10 nodes in one hidden layer and one output node (Fig. C.14).

A fuzzy training algorithm was developed for Matlab called TRAINGDFPA (Gradient-Descent with Fuzzy Parameter Adaptation), using as a model the TRAINGDM (Gradient-Descent with Momentum) algorithm. The gradient descent with momentum training is used to increase the speed of the traditional gradient descent training. The change of each variable (weight matrix and bias vectors), is calculated using the following formula [16]:

$$dX = mc \cdot dX_{prev} + lr \cdot (1 - mc) \cdot \frac{dperf}{dX}$$

mc:	momentum constant
lr:	learning rate
dX_{prev}:	previous change
$\frac{dperf}{dX}$:	derivative of the performance function

Three new options are used in the TRAINGDFPA:

1) fis: Refers to the fuzzy inference system.
2) adaptFreq: Refers to every how many epochs the parameters are adapted.
3) maxFailError: Number of tries to reduce the learning rate, in order to avoid a large increase in error.

The TRAINGDFPA works similarly to TRAINGM with an added parameter adaptation function that is activated every time a prespecified time of epochs has elapsed. The output of the fuzzy controller is used to appropriately increase or decrease the learning rate and momentum. After the adaptation, the performance of the system is checked. If an increase in error above a pre specified threshold is encountered, then small reductions in both the learning rate and momentum are made until the error drops below the threshold, or a maximum of trials is reached. This is an added precaution used in many adaptive training algorithms, to ensure that the current learning rate will always remain sufficiently small, so that the algorithm will perform well.

If (time to adapt)

$$\text{Calculate} \frac{\Delta E}{\Delta E_{old}}, \frac{d\Delta E}{d\Delta E_{old}}, \frac{number\ of\ sign\ changes\ during\ v\ last\ epochs}{v}$$

Input these values to the fuzzy controller and get the output (cLR and cMC)

$\text{LR} = \text{cLR} \cdot \text{LR}$
$\text{MC} = \text{cMC} \cdot \text{MC}$

Ensure that MC is not greater than 1.
Ensure that LR is not less than $0{,}1 \cdot \text{LR}_{\text{initial}}$

End if

$$\text{While} \left(\frac{MSE_{new}}{MSE_{old}} \right) \geq 1.05 \text{ and number of tries} < \text{max number of tries}$$

$\text{LR} = 0.9 \cdot \text{LR}$
$\text{MC} = 0.95 \cdot \text{MC}$

End While

The parameter adaptation method uses a fuzzy controller that was developed to adaptively tune both the learning rate and momentum constants. The controller has 3 inputs, 2 outputs and a fuzzy rulebase of 11 rules. The rulebase was created after some experimenting on the general guidelines mentioned earlier. A detailed description of the controller follows.

C.6.2.2 Inputs

1) Change of error change (I1): Refers to the percentage of increase or decrease in the change in error. This input variable is fuzzified using five membership functions.
2) Change of speed in error change (I2): Refers to the percentage of increase or decrease in the speed of error change. This input variable is fuzzified using five membership functions similar to the above.
3) Number of sign changes to epochs (I3): Reflects the number of sign changes in error to the epochs since the last fuzzy adaptation. This input variable is fuzzified using only two membership functions.

C.6.2.3 Outputs

1) Learning rate change (O1): A percentage with which to increase or decrease the learning rate. This output variable is defuzzified from five membership functions and takes values between 0.7 and 1.1.
2) Momentum change (O2): A percentage with which to increase or decrease the momentum. This output variable is defuzzified from five membership functions and takes values between 0.7 and 1.1.

C.6.2.4 Rules

1) If (I1) has a large decrease AND (I2) has a large decrease AND (I3) is small then make a large increase in (O1) and a small increase in (O2)
2) If (I1) has a small decrease AND (I2) has a large decrease AND (I3) is small then make a small increase in (O1) and a small increase in (O2)
3) If (I1) has a large decrease AND (I2) has a small decrease AND (I3) is small then make a small increase in (O1) and a small increase in (O2)
4) If (I1) is steady AND (I2) is steady AND (I3) is small then make a small increase in (O1) and no change in (O2)
5) If (I1) is steady AND (I2) has a small decrease AND (I3) is small then no change in (O1) and a small increase in (O2)
6) If (I1) has a small decrease AND (I2) has a small increase AND (I3) is small then no change in (O1) and no change in (O2)
7) If (I3) is large then make a large decrease in (O1) and a small decrease in (O2)
8) If (I1) has a small increase AND (I3) is small then no change in (O1) and no change in (O2)
9) If (I1) has a large increase AND (I3) is small then no change in (O1) and no change in (O2)
10) If (I2) has a small increase AND (I3) is small then no change in (O1) and no change in (O2)
11) If (I2) has a large increase AND (I3) is small then no change in (O1) and small decrease in (O2)

C.6.2.5 Results

The neural-fuzzy system was tested several times in comparison to the simple neural network. Because the initialization of the weights plays an important role in training, both networks had the same initial weights during each test. After ten tests it was evident that the neural-fuzzy system performed better than its counterpart. There was, on average, a 73,0% decrease in epochs before the learning stopped, a 90,4% decrease in the training error and a 20,9% decrease in the testing error. It should be noted that the neural-fuzzy reached the error goal twice while its counterpart, never.

The parameters used for the training, are summarized in the following table. Wherever possible, the parameters have the same value for both training algorithms.

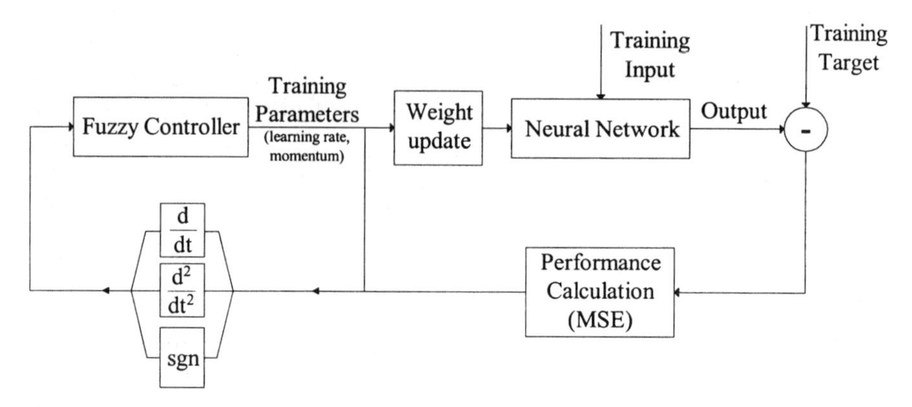

Fig. C.15. Training of a neural network using fuzzy training parameter adaptation

Table C.2. Performance of the neural-fuzzy and neural algorithms during 10 tests

Test No	Neural-Fuzzy				Neural			
	Epochs	Data MSE	Test MSE	Stopped	Epochs	Data MSE	Test MSE	Stopped MSE
1	1411	1,036E-04	0,018	Validation	5780	1,798E-03	0,027	Validation
2	2568	9,993E-06	0,017	Goal Reach	10000	2,720E-04	0,021	Max Epochs
3	1064	2,808E-04	0,042	Validation	4759	1,594E-03	0,059	Validation
4	1359	1,246E-05	0,008	Validation	10000	5,614E-05	0,009	Max Epochs
5	1922	2,237E-05	0,026	Validation	10000	5,582E-04	0,029	Max Epochs
6	968	2,561E-04	0,002	Validation	3584	1,890E-03	0,002	Validation
7	982	1,546E-04	0,023	Validation	2619	3,533E-03	0,029	Validation
8	930	6,868E-04	0,038	Validation	2006	3,363E-03	0,041	Validation
9	1096	4,487E-05	0,031	Validation	3543	3,676E-03	0,046	Validation
10	2324	9,996E-06	0,017	Goal Reach	10000	3,118E-04	0,025	Max Epochs

Table C.3. Training parameters used during testing of the neural and neural-fuzzy algorithms

	Neural	Neural-Fuzzy
Performance function	Mean Square Error	
Learning rate	0.007	0.007 (initial)
Momentum	0.8	0.8 (initial)
Maximum epochs	10000	
Goal	10^{-5}	
Minimum gradient	10^{-10}	
Adaptation frequency	–	1 per 5 epochs
maxFailError	–	50

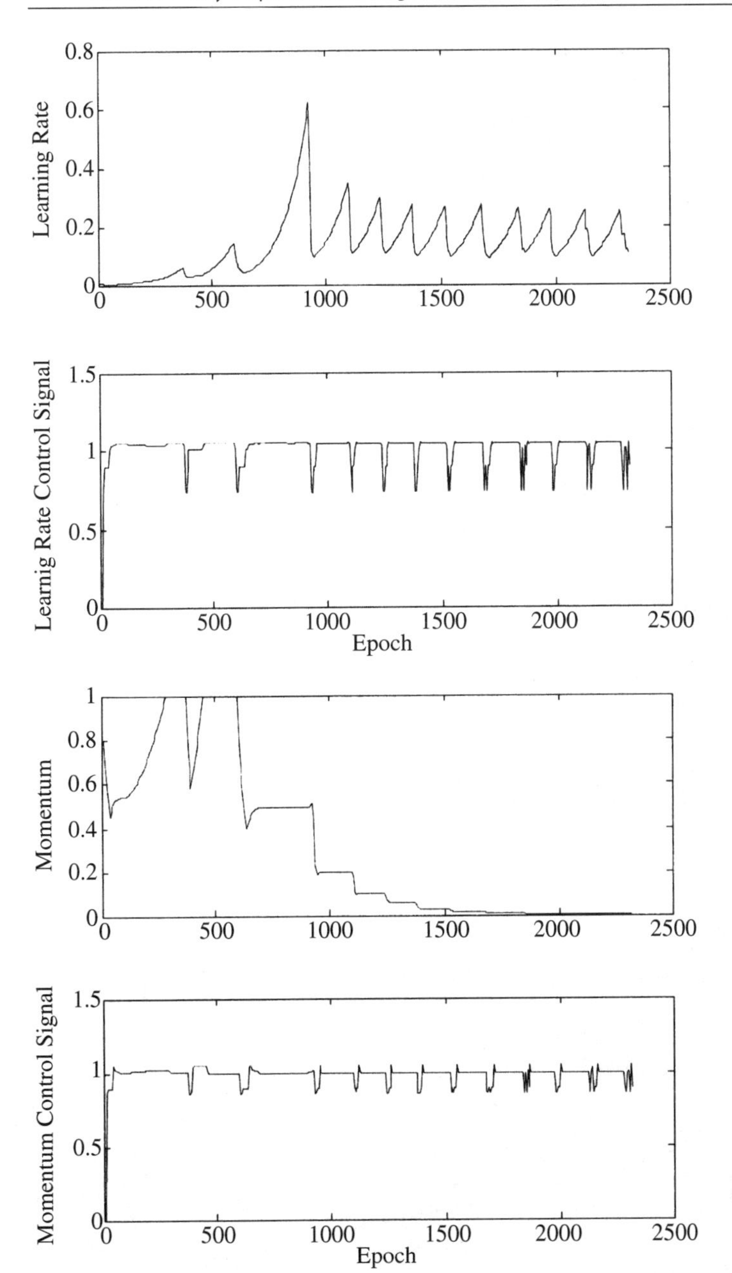

Fig. C.16. (a) Learning rate variations during one of the tests. (b) The fuzzy controller signal that corresponds to the learning rate. (c) Change of the momentum constant as a function of epochs. (d) The corresponding controller signal during one of the tests

These figures show the learning rate and momentum variations and the corresponding fuzzy control signal. It is possible to see that as the algorithm converges, the signal of the fuzzy controller becomes more and more erratic. It is also clear that the initial setting for the learning rate (0,007) was rather conservative, since that in the following epochs it never falls below 0,1. On the other hand a higher learning rate brought convergence problems to the simple neural training.

C.6.3 Conclusions

The use of a fuzzy adaptive training algorithm not only improves the training speed but may also increase the accuracy of the neural network. It is evident that achieving training time four times less than that of the traditional backpropagation algorithm is relatively easy. In time crucial applications better fine tuning of the membership functions and rules could result in even greater increase in training speed.

The change in execution time per epoch remained virtually constant since the added time for the evaluation of the fuzzy controller inputs and outputs is small and divided among more than one epochs. The improved speed may even allow to add more elements to input vectors or alternatively to set more strict

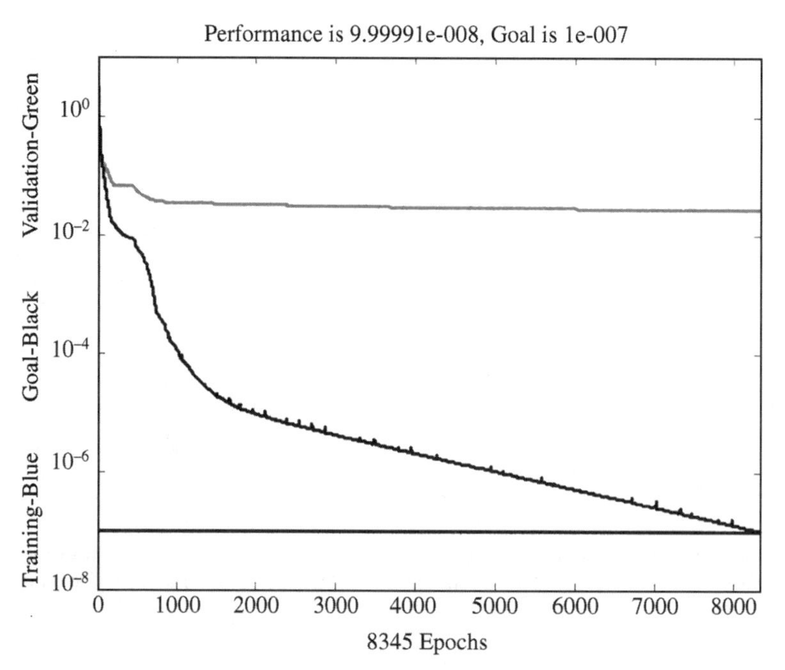

Fig. C.17. The evolution of the training and validation errors during one of the tests. The neural network using the fuzzy adaptive learning algorithm was able to reach the goal error in about 8300 epochs

error goals. During tests the network was able to reach a target of 10^{-7} error before the limit of 10000 epochs was reached.

Another advantage of the algorithm developed is that it can be incorporated into any neural network that uses backpropagation for training with no need for adjustments, since even when the fuzzy controller is not fine-tuned it can perform well enough.

References

[1] Duda R, Hart P (1973) Pattern Classification and Scene Analysis, New York, Wiley Interscience.
[2] Tou J T, Gonzalez R C (1974) Pattern Recognition Principles, Reading, MA: Addison-Wesley.
[3] Archer N P, Wang S (1991) Fuzzy set representation of neural network classification boundaries, IEEE Trans. Syst. Man Cybern., 21(4):735–742
[4] Simmpson P K (1992) Fuzzy min-max neurla networks Part I: Classification, IEEE Trans. Neural Networks, 3 (5): 776–786
[5] Jou C C (1992) Fuzzy clustering using fuzzy competitive learning networks, Proc. Int. Joint Conf. Neural Networks, vol. II, 714–719, Baltimore, MD
[6] Newton S C, Pemmaraju S, Mitra S (1992) Adaptive fuzzy leader clustering of complex data sets in pattern recognition, IEEE Trans. Neural Networks, 3 (5): 794–800.
[7] Baragughi N, Yamada K, Kise K, Tezuku Y (1990) Connectionist model binarization, Proc. 10th Int. Conf. Pattern Recognition, 51–56, Atlantic City, NJ.
[8] Blanz W E, Gish S L (1990) A connectionist classifier architecture applied to image segmentation, Proc. 10th Int. Conf. Pattern Recognition, 272–277, Atlantic City, NJ.
[9] Ghosh A N R, Pal S K (1993) Self-organization for object extraction using a multi-layer neural network and fuzziness measures, IEEE Trans. Fuzzy Syst., 1 (1) 54–68.
[10] DeLuca A, Termini S (1972) A definition of a nonprobabilistic entropy in the setting of fuzzy set theory, Inf. Control, 20: 301–312
[11] Pal S K (1989) Object background segmentation using new definition of entropy, Proc. Inst. Elec. Eng. 284–295
[12] Rabiner, Juang L B (1993) Fundamentals of speech recognition, Englewood Cliffs, NJ: Prentice Hall.
[13] Choi J J, O'Keefe H, Baruah P K (1992) Non-linear system diagnosis using neural networks and fuzzy logic, Proc. Int. Conf. Fuzzy Syst., 813–820, San Diego.
[14] Lin C T, Lee C S G (1996) Neural Fuzzy Systems, A neuro-fuzzy synergism to intelligent systems, Prentice Hall.
[15] Ross T J (1995) Fuzzy logic with engineering applications, McGraw-Hill.
[16] Demuth H, Beale M (2001) Neural Network Toolbox for use with MATLAB, User's guide of version 4, Mathworks Inc.

Subject Index